Eli Bowen

Coal and Coal Oil

Or, The Geology of the Earth

Eli Bowen

Coal and Coal Oil
Or, The Geology of the Earth

ISBN/EAN: 9783743326798

Manufactured in Europe, USA, Canada, Australia, Japa

Cover: Foto ©ninafisch / pixelio.de

Manufactured and distributed by brebook publishing software (www.brebook.com)

Eli Bowen

Coal and Coal Oil

OR,

THE GEOLOGY OF THE EARTH.

BEING A POPULAR DESCRIPTION OF

MINERALS AND MINERAL COMBUSTIBLES.

BY ELI BOWEN.
PROFESSOR OF GEOLOGY.

PHILADELPHIA:
T. B. PETERSON & BROTHERS,
306 CHESTNUT STREET.

PREFACE TO THE SECOND EDITION.

ABOUT seven years ago, after having devoted a long time to the investigation of the paleontology of the coal measures, and to mineral or fossil combustibles generally, both in their economic and philosophical aspect, I wrote most of the matter of the present work, under the title of the "Physical History of the Earth." As the most important feature of the book, I discussed at length the peculiar conclusions at which I had arrived, as the result of my long-continued investigations, in reference to the origin and manner of deposition of these valuable substances. The first edition of the book was rapidly exhausted; but for a year or more past its publication has been suppressed, to enable me to avail myself of a convenient time to revise it, and to make some important changes and additions. The book had created a very favorable impression on those who read it, and was warmly commended by the press; and since the present interest in coal and coal oil manifested itself, an unmistakable demand for it has been evinced all over the country. I had intended, at one time, to extract from it that portion treating of the phenomena of mineral combustibles, and offer it to the public in a separate volume; but on reflection, I deemed it more expedient to enlarge its scope and incorporate whatever additional remarks I had to offer in the present form—only venturing a prefix to the title which will express the most important feature of the work, and commend it more directly to those who now appear foremost to obtain it.

Many persons will doubtless feel some disappointment that, in speaking of coal and coal oil, I have in no instance referred to particular districts or localities. The phenomena which I discuss are, however, so widely diffused over the earth, that I could hardly have singled out any particular region, in the present excited and speculative feeling of our people, without betraying a partiality which would have been unjust to my character as a writer, and to friends and readers having rival interests elsewhere.

I have dealt with the subject only as a natural phenomenon. I myself own no stock in any company, and have no special sympathy, at this writing, with any particular oil or coal district.

Very respectfully,

ELI BOWEN.

Valencia, Schuylkill County, Pa.
JANUARY 20TH, 1865.

BREVIARY OF THE ARGUMENT.

(For a Complete Alphabetical Index of Contents, see page 489.)

———•———

THE FIRST DAY.—Description of the first day by Moses; the magnitude of the subject considered; impossibility of conceiving a beginning or an end; vanity of attempts to explore antemundane phenomena, but our duty to investigate those which immediately concern and surround us; man the servant of God, to whom he has confided the custody of his creatures; the antiquity of the earth; the meaning of the word *Day*—it contemplates circles of time, or lengthened cosmogonal eras; origin of calendar time—impossibility of the existence of solar or calendar days during the earliest stages of creation; differences of calendar days among different nations; the word day merely symbolically employed by Moses—is frequently used to express lengthened periods; harmony among the original promulgators of the divine word;—discord and misinterpretations among modern Christians the result of a neglect of Nature's laws; the Bible not addressed to one age, but to the people of all time—its truths manifested with the advance of our knowledge of Nature; the perversions of Infidelity, and the negligence of the Church; Religion should be based on Nature; God the creator of Nature and of Law; the Church errs externally, not internally; the spirit right, but its policy wrong—pride the basis of its inefficiency—cause for alarm for its ultimate safety; want of parallelism or conformable order in stratification proof of the antiquity of the earth; correspondence of the Mosaic days with cosmogonal eras; proof of the prophetic intelligence of Moses; the laws of paleontology—progressive creation, but not of development; simultaneous creation disproved; conciseness of the Mosaic narrative; the antiquity of the earth established by Astronomy; distance of the planets from the sun; the immensity of space, and of the universe; inability of the mind to comprehend distance, time, or magnitude; vision and mind alike circumscribed; diffusion of light through space; the Milky Way; vast extent of time required for the transmission of stellar light; Telescopic observations of Herschell; economical value of astronomical discoveries—calculations of eclipses, comets, and sidereal phenomena—Halley's comet; the seasons, precession of the equinoxes, the almanac; value of astronomy to marine navigation; the discovery of Neptune by Leverrier—wonders of mathematical demonstration; the origin of worlds; the Nebular hypothesis explained; resolution of nebula into stellar bodies no proof of weakness in the theory; Nebulæ abounding as independent bodies in space; worlds off-shoots from the sun; the planets all move in one direction only; their globular form; law controls all their movements; speciality of forms; the law of intermediate distance among the planets; the law of progressive density; the law of time in their revolutions around the sun; the former relation of the sun to other systems; original unity of worlds, embodied in God; his volition occasioned diffusion; the propensity to return into unity defined as gravitation; Moses describing cosmogonal phenomena in tableaux; worlds within worlds; life inherent in matter; worlds the mere compounds of aggregate atoms, united by affinity; Seers of ancient Greece; the Mosaic pictures consistent with cosmogonal facts; the Nebular hypothesis foreshadowed in Revelation; light inherent in Nebulæ; the earth in embryo; the interior heat; conclusion, and quotation from Milton's "Paradise Lost."

THE SECOND DAY.—The firmament, how produced; the aqueous origin of the earth; universality of water in earth, air, and animals; origin of rocks from interior calorific sublimation; chemical dissolution of water; evaporation of hydrogen, and its expansion into the dome-like firmament; the Mosaic contrasted with the Grecian Cosmogony; the concavity of the globe; the earth-animal of Kepler; original definition of the word firmament; Moses in advance of all succeeding ages; modern tendency of science to the confirmation of his Cosmogony; the birth of animated Nature ; terrific volcanic eruptions: crystalline and Plutonic rocks the solid base of the earth ; emergence of volcanic reefs and islands ; distribution of the primitive rocks in Asia, Europe, Africa, and America ; mineral characters and varieties of the volcanic rocks ; concluding remarks ; extract from Milton's "Paradise Lost."

THE THIRD DAY.—Appearance of dry land ; commencement of vegetable life ; the Metamorphic rocks, their character, position, and origin ; first appearance of limestone ; the Cambrian, Silurian, Devonian, and Carboniferous systems ; classification of Geologists ;—priority of animal over vegetable life denied ; the igneous or volcanic rocks referred to ; the Metamorphic rocks the basis of the Paleozoic formation ; propensity to introduce new words in geological nomenclature deprecated —a case in point ; the Geological Report of Pennsylvania ; rocks of the Metamorphic group described ; mineral veins ; copper and iron of Lake Superior ; iron of Pilot Knob, Cornwall, and Lake Champlain ; fossiliferous nature of the Metamorphic rocks ; the Silurian rocks, fossils and distribution of ; Silurian sandstones of the west ; Niagara Falls, description of, age, retrogression, discovery of ; the St. Clair flats ; emergence of the bottom of Lakes ; formation of conglomerate rocks ; the Devonian rocks described—old Red Sandstone—Hugh Miller ; distribution of old Red Sandstone ; America properly the *old* world ; the Andes, Rocky mountains, and volcanic peaks ; coal basins of Texas, of the Missouri, the Illinois, Michigan, the Alleghany, Rhode Island, British Provinces ; formation of conglomerate and sandstone ; character of the Devonian coal Lakes ; limestone generally absent ; depth of the northwest lakes ; gulf of the ancient sea in the south-west ; the drainage of the land in that direction ; vegetation of the Metamorphic era ; the Potsdam Sandstone ; identity of origin of various combustibles ; cosmopolitan distribution of coal ; no organs of fructification in the ancient grasses—their spontaneous development ; geological features of the coal basin of Rhode Island ; extraordinary disturbance, coal graduating into plumbago, graphite ; British lustre ; Metamorphic coal ; anthracite of France and Scandinavia ; number of fossil species of vegetation ; botanical systems of classification described ; description of the fossils of the coal formation—Ferns—Lycopodiaceæ—Lepidodendria—Sigillaria—Lycopodites—Ulodendron—Volkmania—Carpolites—Flabellaria—Noggerathia—Asterophyllites—Annularia—Stigmaria—Coniferæ—coal plants in general; resinous character of the vegetation—contemplation of Paleontology; the formation of coal—the peat-bog theory; estuary or drift theory; difficulties suggested; fossil trees in an erect position; theory of various Geologists—Hawkshaw, Buckland, Lyell, Rogers—objections suggested to all; fossils in *coal?* an error corrected; bark of fossil trees *not* coal—chemical nature of anthracite; composition of vegetation; no fruits, grains, nuts, or esculents in the ancient earth; coniferous trees furnished the bulk of the coal; description of existing coniferous trees; the wild pine—the Corsican pine—the Cluster pine—the Siberian pine—the Norway pine—American pines—red, pitch, white pines—the spruce—the silver fir—various other pines and firs—the larch—the cedar of Lebanon—the yew—the cypress—the juniper—the tallow tree, etc., transmission of the ancient seeds of vegetation—specific vegetation of geological eras—diversity of vegetation accounted for—distilling tar and turpentine, process described; deposition of the coal described; geographical relations of the northern lakes; character of the coal forests; extraordinary development of vegetation; nature of the ancient climate; large content of carbonic acid; fogs, mists, and radiated heat; the forests and earth converted into vast tar pits—the ground impregnated with liquid resin—accumulation of vegetable resin in the lakes—formation of seams of coal—overflow of the lakes—rubbish of the forests borne away—parallel overflows of the Mississippi—deposition of sediment over the coal—faults in coal veins explained—invasion of the sea—its ancient position to the coal basins—deposits of asphalt, chapapote, bitumen, etc., described—pitch lake of Trinidad—chapapote of Cuba—asphalt of New Brunswick—petroleum—oil springs—gas springs—a river on fire—intimate relation between coal and pitch—identity of origin—chemical transformations of vegetation into coal—mistaken inferences of geologists—a difficulty and false assumption explained—professional fallacy—the world controlled too much by mere ostensible learning—nature of Lignite or brown coal; amber, its origin; injecting railway cross-ties; conversion of bituminous into anthracite coal—a mistake corrected—theory of Rogers and Lyell; microscopic investigations—vegetable structure of coal denied—fire-damp explosions in mines—the Davy lamp—antiquity of the earth illustrated—geographical distribution of coal—concluding remarks—Milton's "Third Day."

THE FOURTH DAY.—The description of Moses; the light of the sun, moon, and stars appears for the first time, in full effulgence, the sun appointed to *rule* over the day, and with the light of the planets, to be for signs and seasons, days and years; misapprehension of the language of Moses; confirmation of the peculiar atmosphere of the coal era; the earth previously enveloped in mists and fogs; effect of such climate on vegetation and animals; elevation of mountain chains; the Alleghanies; diversification and refrigeration of the climate of the earth; the Mosaic cosmogony in advance of the Grecian philosophers; the Ptolemaic astronomy—the earth the supposed centre of the planetary system; discoveries and mathematical demonstrations of Copernicus; his laws described; false views of Cosmogony entertained by the Church; Tycho Brahe; laws of Kepler described; discovery of the Telescope; Galileo, and his discoveries; expansion of the knowledge of the universe; Sir Isaac Newton; discovery of the law of universal gravitation; description of gravitation; effect of this discovery; discoveries of the seventeenth century, the wonders of the air and ocean; the Gulf Stream of the Atlantic and its effect on climate; the atmosphere a great steam-engine; the quantity of salt held in solution by the ocean; currents of the air and ocean, how produced; weight of the atmosphere and the power it exerts; the description of Moses proved correct by the onward advance of science, and the increased knowledge of Nature; position of the Apostles to scientific dogmas; eloquent extracts from Paul, the Apostle; difference between spiritual and temporal contemplation of Nature; position of the Bible as to systems of Science; the Pleiades—accuracy of revelation; the cosmogony of the book of Job; opinion of Baron Humboldt; extract from the Psalms; the proof that Moses, in his *days*, contemplated lengthened periods; concluding remarks; Milton's Fourth Day.

THE FIFTH DAY.—The waters bring forth abundantly the moving creature that hath life, and fowl of the air; great whales created; the seas become filled with animal life; the four great divisions of animals; the Radiata, Mollusca, and the Articulata described; extent of animal life previous to the Tertiary era; the Spongiaria, Polypifera, and Infusoria; coral reefs; wonders of the microscope; fossil animalculæ; shells of Foramenifera; Echinodermata; descriptions of Molluscan animals; the Eucephalous and Acephalous classes; table exhibiting the distribution of fossil molluscs in each formation, and also the species of existing molluscs; the Articulata division; Annelida, Cirrhipoda, Crustacea, Arachnida, Myriopoda, and Insecta; table exhibiting the classes and characteristics of all insects; descriptions of various insects; the silk-worm and silk manufactures; butterflies, bees, wasps, and ants; governmental discipline; grass-hoppers and locusts; fallacy of the theory that animal preceded vegetable life; animalculæ in the juices of vegetation; vegetation of the metamorphoric era; vegetation the distinguishing feature of the entire Paleozoic formation; multiplicity of geological eras; description of the Secondary formation or fifth day; the new red sandstone; salt springs, building stones; fossil foot-prints of extinct animals; the Cheirotherium, the Labyrinthodon, the Microlestes, the Thecodontosaurus, the Apateon, the Archegosaurus, the Sauropos primævis; diverse views of Geologists; Lyell, Owen, King, Lea, Hitchcock, Humboldt, Rogers; sun-cracks and rain-drops; contradictions of the Bible; the doubters met on their own ground; the truth of revelation vindicated; the existence of *air breathing animals* prior to the new red sand stone disproved; sun-cracks, ripple-marks, and rain-drops accounted for; the non-luminous atmosphere of the coal period maintained; absurdity of geological dogmas; an original discovery; the Sauropus *moderni* or what-is-it? professional jealousy; economy of the Pennsylvania geological survey; discovery of the Cobham stone by Mr. Pickwick—the celebrated Pickwick controversy; explanation of the origin of rain-drops and sun-cracks; showers of sulphuric and carbonic acid; extraordinary meteorological phenomenon; credulity of the world; its readiness to receive the most absurd theories under the *name* of science; animals of the new red sandstone; foot-prints of birds; the reason of their early introduction upon the earth; Prof. Hitchcock's classification of the extinct bird-tracks; an inconvenience suggested; classes of birds; of fishes; of reptiles described; bones of different animals; the lias group of rocks; alternations of animal life; changes not development; the Oolitic rock; the Vestiges of Creation; changes in the ancient types of fish; a difficulty reconciled; absence of whales in the Oolite accounted for; supposed marsupial fossils; the Wealdon strata—its fossils; the Cretaceous rocks described; their origin and distribution; Cretaceous and Greensand of the United States; review of the secondary formation—the era of marine life; distribution of its strata; The Mosaic description confirmed; the fifth day of Milton's "Paradise Lost."

THE SIXTH DAY.—Land animals created; first appearance of man—he is clothed with dominion over the earth and all its inhabitants; man created in the image of God, and destined to increase and replenish the earth; the sixth day of creation, as described by Moses; the Tertiary formation described; its fossils and geographical distribution; its gradual passage into the Alluvium, or present geological era; Nummulite rocks; the London and Paris basins; character of Tertiary rocks; vegetation of the Tertiary; beds of brown coal and lignite; insects, polyps, animalcules, and fishes; the ancient and modern fishes; progressive development and degradation of species disproved; Hugh Miller; fossils of whales; their immense size; etc; peculiar significance of the language of Moses; the Bible requires no apologies in its behalf; the Tertiary pre-eminently the age of *land* animals; classification of Mammalia; the Diadelphian order; pouched animals; the Monadelphian order; the Testacea; the Ruminantia; the Pachydermata; the Edentata; the Rodentia; the Cheiroptera; the Amphibia; the Carnivora; general description of the animal kingdom, terminating with the Quadrumana, and the Bimana; the object of creation attained in man; the marriage of Adam and Eve; the celebration of the work of creation, and the dawn of the Sabbath.

THE SEVENTH DAY, OR SABBATH.—Description of Moses; the heavens and the earth finished; the seventh day sanctified; no rain in the earlier epochs; the basis of Christianity; the mystery of Nature; faith; the trinity described; the origin of worlds; the rebellion of Satan; the garden of Eden; the fall of Adam; and his punishment; man's position in relation to the Creator; the immortality of sin; the immaculate concep tion; the divinity of Christ; his character, motives, and doctrines; the triumph of man over sin; history of the Adamite race; the universality of the Noachian flood; Christian doubters; Noah's ark—its dimensions and capacity compared with the Great Eastern; speculations of Hugh Miller; estimates of Sir Walter Raleigh; number of species of animals on the earth; antediluvian giants; young, not adult animals in the ark; the arrangement of the animals in the ark; effects of the flood—how produced; the fountains of the deep broken up; subterraneous reservoirs of water; mountain glaciers; the Alps; movement of avalanches; icebergs off Newfoundland; icebergs in mid-ocean; character of icebergs; erratic boulders and moraines; Dr. Kane in the Arctic regions; the great Humboldt glacier: primitive rocks, and fossil animals; volcanic action in the polar regions; the flood produced by the breaking up of glaciers; rain generated; universal volcanic eruptions; river freshets; universality of the flood proved; submergence and elevation of continents and mountains; folly of contradicting the express language of the Bible; the universality of the flood denied, but yet established by geologists; submergence of continents explained; the Bible vindicated; fossils of the diluvium; the Niagara River; antiquity of the Mastodon; the original seat of the antediluvian race; the saltness of the sea; objections of Hugh Miller to the universality of the flood; all natural laws are miracles; insects in the ark; peculiarities of insect fecundation and generation; experiments of Messrs. Weeks and Crosse explained—spontaneous generation denied; fallacious positions of Hugh Miller—the tendency of his Reconciliation; supposed impossibility of accommodating all the animals in the ark; characteristic fauna of continents; animals of South America different from the older continents; analogy between the extinct and living species; fallacious conclusions and inferences; the extinct species destroyed by the flood and the existing species preserved in the ark, including many now extinct; the Mastodon, the Megatherium, the Glyptodon, etc., all modern; evidences of the flood considered; traditions of the Indians; opinions of Mr. Jefferson; the mounds of the West described; works of art, inscriptions, and skeletons; the golden anaglyphs of Chiriqui; the Indians of Central America; remarks of Columbus; the ruins of Yucatan, their Egyptian aspect;

opinions of Stephens; discovery of North America by the Northmen; colonization of Massachusetts and New York; America inhabited by white men in the ninth century; remarks of Humboldt; Behring's Straits; the Aleutian islands; probable Asiatic origin of the Aborigines of America; how they may have reached this continent; the ancient Pelasgi of the Mediterranean; the Egyptian origin of American Aborigines; the arts, religion, and civilization of the Egyptians; their veneration for animals; their religion affords the *key* to the introduction of the animals of the old continents into America; the dispersion of the family of Noah; superiority of the ancient races over those of the modern; the development of arts and civilization of description of the works of the Egyptians; the pyramids; the lake of Mœris, canals, palaces, etc.; their probable geographical knowledge and commercial enterprise; their colonization of America; the unity of races; opinion of Humboldt; the antiquity of American vegetation; the animals of America; succession of races of animals and of man; the workings of Providence; the seventh day or Sabbath; an institution for the ease of creation; the Sabbath of Nature; extract from Leviticus; the Sabbath founded in natural law; chemistry of vegetable growth; alternation of crops; necessity of the Sabbath to man; the Mosaic idea realized; the harmony of mankind and creation; man's mind god-like; disquisition on the order of Nature; the law of Nature the guide of human reason; man the concentrated essence of creation; God the God of Nature!

COAL AND COAL OIL;

OR,

THE GEOLOGY OF THE EARTH.

THE FIRST DAY—ASTRONOMICAL.

1 In the beginning God created the heaven and the earth. 2 And the earth was without form, and void; and darkness was upon the face of the deep. And the spirit of God moved upon the face of the waters. 3 And God said, Let there be light; and there was light. 4 And God saw the light, that it was good; and God divided the light from the darkness. 5 And God called the light Day, and the darkness he called Night. And the evening and the morning were the first day.

It is needless to remark, at the outset, that the subject which I propose to consider is the most interesting, comprehensive, and universal in the whole range of intellectual grasp. The mysteries of creation, and the external objects surrounding and involving us, having engrossed the attention of mankind, more or less, for several thousand years, it may appear rash and vain in an individual like myself, to enter the field of investigation; yet individuals like myself have an equal and co-ordinate interest in the subject with the greatest of their fellow creatures; and no effort, founded in good intentions, to throw new light on an old theme, can be regarded as altogether vain. While we have the benefit of the accumulated learning of past ages, it is perhaps questionable whether the aggregate amount of actual knowledge on this particular subject has been materially increased; or, if increased, whether it has been collated into a proper focus. We see with the same eyes—we are compelled to use the same bodily organs that were used by the "fathers and awful rulers of mankind." The Telescope and the Microscope have, indeed, revealed new creations and millions of new worlds; but instead of lessening the absolute mystery of God's work, they have only extended the boundaries of the field. Had we depended on our own resources—had we been cut off from the experience and fallacious deductions of former ages, physical investigations might, perhaps, have been directed into unexplored channels, and thus have opened up new wonders in the regions of creation, and thrown new light upon those now obscure. It may be said that, just now, all physical phenomena are surveyed through the glasses of the past; and though the rays of light are reflected and refracted by the most ingenious instruments, the varied colors we obtain, like those which span the vault of heaven, only remind us, after all, of the distance and the unfathomable depths, complications, and difficulties that encompass us on every hand.

"In the beginning God created the heaven and the earth." This implies that there *was* a time when the heaven and the earth did not exist—a fact which reason can at least entertain; for we *know* there was a time when *we* did not exist—when the rocks under our feet, and the mountains towering over our heads, did not exist. These things we *know*, and others we can believe. But was there a time when God himself did *not* exist? Who clothed *Him* with power to create worlds? Whence was *derived* that perfection of wisdom and of action that enables him to diffuse throughout illimitable space the unceasing harmony of Nature? The inspired writer says nothing on this point, for he well knew that human reason could not grasp it. It is one of those awful mysteries, the existence of which Knowledge, with her dim lantern, may descry, but cannot approach. Every attempt to go *beyond* the line of divine revelation, or to explore phenomena strictly ante-mundane, can only result in the aberration of reason or in the dreamy vagaries and sophistries of philosophy. From the lofty glacial peaks of Mont Blanc or the Himalaya—mountains that, like Castor and Pollux, belong equally to heaven and to earth—the eye may gloat on the scenic splendors that spread out before it, until their dim outlines fade into the blue azure of the horizon. It is precisely thus with the *mind*. Thought and vision are alike confined within certain limits, beyond which it is impossible to go. Vision may be extended by elevating the surveying position; and the Telescope can introduce us to new worlds in the regions of space. So, too, Thought may be rendered more comprehensive by mathematical formulæ and philosophical deduction; but after all, in either case, we only attain a dizzy precipice amid the dark, mysterious gloom. When, therefore, we speculate upon the *original Beginning* or the final *End*, we simply enter the precincts of those dark clouds that intervene between Death and the awakening resurrection—between the material world of which we *constitute a part*, and that unknown universe that lies beneath the dome of God!

> "The lamb thy riot dooms to bleed to-day—
> Had he thy *Reason*, would he skip and play?
> Pleased to the last, he crops the flow'ry food,
> And licks the hand just raised to shed his blood!
> Oh, blindness to the future! kindly given,
> That each may fill the circle marked by Heaven!"

As to the practical cosmogony of worlds, and especially that which we inhabit, human intellect has no restrictions imposed upon it beyond its own inherent weakness. God has deigned to enlighten us upon all points involving our immediate and future happiness; and it is our right and duty to extend our knowledge of his works by all the aids we can command, and which his benevolence has furnished. Living upon a planet, more or less influenced by all the others, we have a right to inquire into the relations they severally hold to each other and to us, but more especially into the varied phenomena immediately around us. Our interest in every thing God has created is direct and vital; and the investigation of Nature and its laws is, therefore, among the most rational and elevating duties which the Creator has allotted to us. It is, in fact, a special duty and pleasure, from which the inferior animals are all exempt; and hence we have been provided with organs of speech, of reason, reflection, observation, and mechanical power, by which we can render the earth we inhabit tributary to our gratification; by which we are able to cultivate and embellish it; and, while taking care of the creatures whom God has intrusted to our custody, also render them subservient to our wishes and pleasures.

The earth which we inhabit, we have every reason to believe, is infinitely older than the popular mind has been led to suppose. Instead of six (or, as some claim, seven or eight) thousand years, an examination of its rocky crust, and the laws controlling its primary structure, proves it to be of vast and utterly incalculable antiquity. In this respect, Geology occupies the same ground, in relation to *time*, that Astronomy does in relation to *distance* and *space*. The one can only compute the age of particular formations or eras by *millions of years;* while the other measures the distance which separates the revolving planets from the central sun by *millions of miles.*

It is true the Bible tells us God made the earth in *six days.* To doubt its holy authenticity would betray an irreverence unworthy the enlightened sentiment of the age—unworthy the granitic basis of Christian civilization. But may we not doubt whether we understand clearly what was intended to be conveyed? Are we perfectly sure that we have not misapprehended the true meaning? The word *day*, as at present understood, is generally restricted to express the diurnal revolution of the earth on its axis; or more properly, to that time during which one half of its surface is presented to the sun. This revolution occupies twenty-four hours, divided into darkness and light; but as only *one half* the earth is illuminated at a time, it follows that the portion of its orbit which is elevated above the rays of the sun must be involved in the darkness of night. This darkness at the poles continues without intermission for *six months*, until the earth, having attained the equinox, night again gives place to six months of uninterrupted day. Thus, while *our* days include but twenty-four hours, those of the polar regions have 8,760; or, while it requires three hundred and sixty-five of *our* days to make a year, it requires but one in the polar circles. We have no reason to infer that Moses was unacquainted with this phenomenon, since it will hereafter be seen that he describes many others altogether unknown to the age in which he lived; but it leads us naturally to the inquiry whether the word *day*, as used by him, does not more probably contemplate great geological or cosmogonal eras, or lengthened periods of time, than the twenty-four hours of our calendar. The Hebrews used the word to express *circles of events* and *periods of time*, without regard to duration, as contradistinguished from other circles and periods of time; and there is no doubt that its subsequent absorption into specific chronology, sidereal phenomena, and the diurnal revolutions of the earth, is mainly due to this cause. It does not follow, because these phenomena existed previously, that the same word was invariably used to designate them; and if it was, there is no reason to infer that its meaning was *confined* to the expression of twenty-four hours; for as the diurnal revolution of the earth *actually describes a circle*, or nearly a true circle, there would be sufficient and equal propriety in applying the word to *all other circles of time*, whether great or small. During and long after the time of Moses, much diversity existed among the different nations in regard to calendar time. Rómulus, 733 years before Christ, divided the year into *ten* months; but this was not a true year, because it requires, instead of the 304 days which he allowed, 365 days, 5 hours, 48 minutes and 51 seconds for the earth to make her annual revolution around the sun. Numa Pompilius, his successor, shortly afterward added two additional months; but the fractional parts were still wanting to make the year correct. Julius Cæsar, therefore, only forty-five years before the birth of Christ, made the year to include 365 days and 6 hours, and established every fourth year as a leap-year, so as to make up for the accumulating fractions of time. But the year thus became too long, and it was reserved for Pope Gregory XIII., in the year A. D. 1582, to ordain that the ten days between March 11 and March 21, should be omitted; so that, in that particular year, March 21 came directly after March 11. And to prevent future irregularity, it was provided that the first year of a century should *not* be a leap-year, with the exception of the first year of every fourth century. Thus, 1700 and 1800 were not leap-years, nor will 1900 be; but 2000 will be such. In this way true calendar time is obtained, or obtained as nearly as it is possible under the circumstances.

But, as contradistinguished from night, it appears that day originally expressed heat and warmth, or that which generates heat, warmth, or desire; and as applied in the Mosaic record to lengthened cosmogonal eras, during which the work of creation was proceeding, nothing could have been more appropriate; for as the earth, after its expulsion from the central nucleus of the universe, or the seat of the creative volition, could not have come directly under the *periodic* influence of the sun, in its embryonic condition, it is certain that no such days as our own could *then have existed*, and were therefore not contemplated by Moses, except so far as they implied warmth, heat, or life-infusing periods. It was only *after* the appearance of the sun, moon, and stars, on the fourth day of creation, that *our* days, properly so understood, could have existed. And even with us, the word has had various significations among different nations, and at different periods. The ancient Babylonians began their day at the rising of the sun, while the Jews reckoned theirs from its setting. The Egyptians began their day at midnight; and such is now the custom of the Spanish, French, English and American people.

"The evening and the morning were the first day."— Such is the expression of Moses throughout. *Evening*, with us, expresses the decline of day, while morning is, strictly, its *beginning*. But what of the intermediate time—the *day* proper, as distinguished from night? That such *days* did *not* exist during the *earlier stages of creation*, is sufficiently apparent by the express language and obvious meaning of the inspired writer. The morning, in the absence of day, stretched to the evening, and the evening to the morning. A *circle of time* is thus formed, significant of the first epoch of creation. But, in the face of the discrepancies among nations themselves as to the meaning of a simple word, and that word one of the oldest in all languages, where is the force of assuming inconsistency in the divine record? The mistake, and the corruption of words, is with ourselves; but truth is unchanging, and in proportion as we consult nature and physical law, our ability to see and comprehend it is increased. But while Moses uses the word *symbolically* and as a convenient *measure of time,* he also uses it in a more direct sense to express *lengthened periods.* This is perfectly conclusive in the second chapter of Genesis, where he says: "These are the generations of the heavens and of the earth when they were created, in the *day* that the Lord God made the earth and the heavens, and every plant of the field before it was in the earth, and every herb of the field before it grew: for the Lord God had not caused it to rain upon the earth, and there was not a man to till the ground. But there went up a mist from the earth, and watered the whole face of the ground." The word *day* is not only used here to express lengthened periods, but the *whole sentence expressly implies* lengthened periods. "There was no rain;" "there was not a man to till the ground;" "there went up a mist and watered the whole face of the earth"—these facts contemplate *long periods.* No one could suppose for a moment, that the ground could have been *cultivated,* since the land itself had only emerged from the dominion of the sea on the *third day.* And where would have been the necessity for *rain,* if the land had only emerged from the water one or two days previously? All this shows very conclusively that Moses used the word *day* to express *lengthened periods;* and it is impossible to contemplate his narrative in any other light. The word is often used in other portions of the holy book, as the "*day* of salvation" in second Corinthians; the "*day* of Christ," in the eighth chapter of St. John; the "*day* of retribution," in St. Paul, and elsewhere, both in the Old and the New Testament. The word is often used nowadays in a precisely similar sense. But St. Peter, as if to dispel all doubt on this point, declares that "one day is with the Lord as a thousand years, and a thousand years as one day."

Among the original promulgators of the revealed law

themselves, there never appears to have been the least misunderstanding or disagreement whatever. They were a unit in all the details and operations of their religious system. But it is with unaffected sadness that we now behold the Christian world very much and often bitterly divided in mere forms, as well as doctrinal opinions. When we come to a calm consideration of the peculiar circumstances under which the Book was translated at various times, and handed down from one generation and nation to another, invariably more or less affected by their varied ideas, habits, and nomenclature, I think we have a right to assume, not that there is or was error in the revelation itself, but that, if any really exist, it originated in our own translations and misapprehension, and is still, perhaps, sustained by the ignorance and questionable zeal of infatuated Theologasters—the dealers and venders of a spurious salvation.

However this may be, the antiquity of the earth is established by the physical witnesses everywhere around us; by the unimpeachable testimony of the great Creator himself, written on the faithful old rocks of the valley and the mountain, in language sufficiently plain and universal to be understood by the people of every clime, and tongue, and condition. And it may properly be assumed of the Holy Word itself, that it was not addressed to *one* generation alone, but to the people of *all time*, within the boundless eternity; and that, whatever appears obscure now, in the unceasing progress of human events—by the increase of illuminating force, instead of interposing imaginary contradictions, which all experience has shown to be transient and ephemeral, its solemn truths will be rendered more and more overwhelming to our improved nature and understanding. True Science and true Religion, if not really co-operative in their worldly missions, are at least never hostile or inconsistent with each other;

for while the one speaks in the poetry of inspiration of the beneficence of the Creator, and of his fatherly provision for the offspring of his hands, the other as gracefully unfolds the phenomena of his sublime laws, and with eloquence beyond human speech, points out the enduring monuments of his greatness, wisdom, and power.

But it may be remarked, with no disposition to find fault, that while the immediate votaries of Science and Religion are themselves generally content with the *presumed* harmony and compatibility of the leading dogmas of each, they have thus far failed to submit a scheme of Reconciliation, satisfactory in all its details to both parties, and more especially to the great mass of mankind, who, trembling in the feeble and unstable faith that is in them, naturally look to the researches of the learned for confidence and support. And if this observation may be continued a little further, it is at least questionable whether the mere *annunciation of individual belief* in an irregular and *undefined* Reconciliation, and which is exposed to all the involutions and obscurities of conflicting Theological, Geological, and Astronomical theories, be not, after all, more detrimental than beneficial to the progress of sound Religion, as well as to Science and Truth. Religion, indeed, can stand on no firmer or broader foundation than the recognized phenomena of Nature. The earth, the mountains, the rocks, the sea, the stars, and the overarching firmament, excite the wonder and the admiration of man. They speak to him in his hours of solitude. They are ever present in his walks. They do not bear the impress of human art. They bespeak a power infinitely higher and nobler than man. The air we breathe whispers of an all-pervading God! And these are his *works!* Should there be any fear of them—any mystery or concealment touching them? Should Religion hesitate to **explain them** from the pulpit, or to call willing Science to her aid? Would it not be wiser to disseminate tracts embodying the simple truths of Nature, rather than moral and effeminate fictions? There is no merit in the concealment of truth from the popular mind; nor can strength be gained by undue forbearance, or unmanly retreat from the field of investigation. The sermons which the Creator has written on stones, are more potent for good than all the cant flummery of pulpit Mawworms, or all the dignified mummery of scarlet-robed cardinals and pontiffs. If Religion be a serious reality, it must be exemplified in Nature, since it is only through the works of God that we are enabled to comprehend and approach him. The world is governed too much by mere men—too little by unerring Nature. All the wisdom of mortals is the veriest nonsense, if not derived from her teachings and counsels; and if this be true, how can we make an exception for *Religion?*

The simple truth is, that, long since perceiving the secret strength which Theology could derive by a closer and more familiar alliance with the leading dogmas of Geology, Astronomy, and the co-ordinate branches of Natural Science, and the freezing indifference with which these powerful allies have always been regarded, Infidelity has stepped forward, and suggesting plausible interpretations of the *material* scattered over the earth by the God of Nature himself, has thrown up thin partitions between them and Revelation, and boldly elaborated ingenious, inferential, and pseudo-philosophical hypotheses to destroy the harmony that should exist between them. The vail of *doubt*, thus thrown over Revelation, impairs the essence of true religion, subverts the moral sentiments, and gives free license to the human passions. While the world *seems* to be advancing in the scale of learning, and in all the arts and blandishments of civilized life, the *pro rata* of crime, wickedness, and folly, appear to be as great and universal now as it was several centuries ago. Immorality, indeed, may be more refined and polished to accord with the standard of a higher intellectuality;—but there seems to be no diminution of its universality or of its vital force, as compared with previous ages. Inherent in the blood and flesh of man, vice seems to grow with cultivation, like the seeds scattered by the farmer. Growing thus in the extended domains of *Christianity*, it presents a humiliating comment on its policy and boasted virtues. And yet, so long as the spirit of worldly pride is fostered *in* the church, what other result could be anticipated?

But to return to the antiquity of the earth: Had we no other evidence to destroy the theory of its simultaneous creation, with the first efforts of the divine volition, the mere want of *parallelism*, or conformable order, between the proximate layers of the several great formations, would alone suffice. For while, in the character of their fossil remains, and the circumstances of their deposition and elevation, they all preserve a peculiar individuality, they yet often occur in the utmost confusion to each other; thus pointing unmistakably to periods of alternating activity and repose—or more properly, lengthened periods of *night and day*. But not only is there a positive agreement, in the number of cosmical eras, with the six *creative* days of Moses, but there is also a regular correspondence, from first to last, in the prominent features of each. Geology, as a science, was wholly unknown to the era of Moses—so, too, was Astronomy, so far as true mathematical and telescopic investigations are concerned. And yet Moses describes phenomena which the brightest intellects of the world have been several thousand years in discovering. He indicates a progressive movement which the author of the *Vestiges of Creation* might have studied with profit, and which we may define with Pictet, thus: that the species of one geological epoch, as a general thing, lived neither before nor after that epoch;—that the differences between extinct faunas and living animals, are greater in proportion to their antiquity;—that the comparison of faunas of different eras shows that the temperature of the earth has been greatly varied;—that the species which lived in the present eras had a more extended geographical distribution than the species which exist now;—and that the faunas of the ancient strata are composed of animals of a more simple organization, and their degree of perfection increases in proportion as we approach eras more recent.

These phenomena are all indicated in regular order and succession, in the Mosaic account of the creation; and we thus perceive an onward, progressive movement—every era improving upon the last, and all contradicting the idea of *simultaneous action*, or what amounts to nearly the same thing, creative action of six day's duration. But I must here add a word of caution touching the character of this progress. While we discover a *progressive development* in the geological eras, as compared with each leading formation in ascending order, it is *not* a continuous and uninterrupted *chain* of development of species and type, from a low to a higher order, as claimed by the author of the *Vestiges of Creation*. It is, in fact, quite the contrary—inasmuch as every era was furnished with a new and *special* creation, thereby isolating and disconnecting it from those of the preceding or succeeding eras. This fact is clear and overwhelming to the geological investigator; and it is unmistakably embodied in the concise narrative of Moses. Had he introduced man in any other than the modern geological era; or had he given precedence to animal over vegetable life, or spoken of marine animals in the nascent seas of the earth—all this would have utterly destroyed the authenticity and integrity of his history. His narrative is therefore not only true in all its details and inferences, but it comprises one of the most remarkable, concise, and perfect descriptions of natural phenomena ever written; and it is not in the power of any living man to infuse into words any thing like the same amount of meaning.

Geology, however, does not stand alone in maintaining the vast antiquity of our globe. All the co-ordinate branches of science confirm it, and Astronomy establishes it on the fixed basis of mathematical demonstration. Mercury is the nearest planet to the sun—the intervening distance which separates them being thirty-seven millions of miles. Then comes Venus, revolving at a distance of sixty-eight millions; and then the Earth which we inhabit, ninety-five millions of miles distant. After the earth comes Mars, one hundred and forty-four millions from the sun; and then we have the eight asteroids discovered by Herschell, and which perhaps average two hundred and fifty millions of miles each. Next to these asteroids is Jupiter, whirling through space at a distance of four hundred and ninety millions of miles from the sun; then Saturn, nine hundred millions; then Uranus, nineteen hundred millions; and finally Neptune, twenty-eight hundred millions of miles from the centre of the solar system. As the lighthouse guides the benighted mariner in his trackless path along the tempestuous coasts of the ocean, so even these distant luminaries seem to guide us along the sloping shores of Time—for they remain forever "for signs, and for seasons, and for days and years." But as we turn from star to star, and from cluster to cluster, with all the aids that art and successive ages have placed in our grasp, we must cease further pursuit, to fall down and adore! The ocean of infinite Space expands before the view, and its azure hue separates our "mortal coil" from the unexplored Eternity!

The earth is but ninety-five millions of miles from the

sun around which it revolves—a distance which, compared with that of Neptune or Uranus, is comparatively insignificant. Yet, in an age of steam and magnetic telegraphs, no human being can realize such an extent of space. Ninety-five millions—*of miles!* It has been remarked that a cannon-ball, urged at the greatest velocity which such a projectile ever attained, would consume more than twenty years in penetrating such a space. Our nearest nocturnal neighbor, "the inconstant moon," is but two hundred and thirty-seven thousand miles from us—a distance so trifling that astronomers have actually mapped its geognostic configuration, and bestowed names upon its principal mountains, craters, deserts, and green plains. Some persons, indeed, indulging a taste for geology, have speculated upon its physical composition, which has popularly been referred to green cheese!—(curdled, very likely, from the *Milky Whey!*)

It has been ascertained, by various ingenious experiments, that light penetrates space with a speed something like 167,000 miles per second. This, multiplied by 60, would give a fraction over ten millions of miles per minute, or 600,000,000 per hour, and 122,400,000,000 per day. Strange as it may appear, we know that there are planets and clusters of stars and nebulæ so far beyond the reach of our telescopes, that, even at the extraordinary speed with which light diffuses itself, it must have required, in some cases, several millions of years for their light to reach the earth. Now, the Milky Way forms one of the grandest features of the firmament. It completely encircles the whole fabric of the skies, and sends its light down upon us, according to the best observations, from no less than eighteen millions of suns. These are planted at various distances, too remote to be more than feebly understood; but their light, the medium of measurement, requires, for its transmission to our earth, periods ranging from ten to many thousands of years. Such is the sum of the great truths revealed to us by the two Herschells, who, with a zeal which no obstacle could daunt, have explored every part of the prodigious circle. Sir William Herschell, after accomplishing his famous section, believed that he had gauged the Milky Way to its lowest depth, affirming that he could follow a cluster of stars with his telescope, constructed expressly for the investigation, as far back as would require 330,000 years for the transmission of its light. I am well aware that, to the minds of some readers, such statements will appear wild, if not perfectly incredible. But they will perhaps receive some degree of credence when it is remembered that the same scientific acumen and research which enabled Kepler, Newton, Leverrier, or Herschell to explore and resolve astronomical phenomena, also enables more humble workers in the vast regions of space to foretell, with unerring exactness, an eclipse of the sun or moon, the eccentric movements of comets, or any other sidereal phenomena. The arrival of the comet which figured so conspicuously in our horizon in the autumn of 1858, was correctly calculated several years in advance by nearly all the American and European astronomers. After the discovery of the law of universal gravitation had provided the key to unlock the door of astronomical mystery, Halley conceived the idea that all comets moved in obedience to such a law; and in speculating upon that of 1682, he not only explained phenomena previously noted by Kepler in 1607, and by Apian in 1651, but ventured to predict the return of the same body in 1758—a prediction fully verified. Its reappearance in 1835 had been predicted several years in advance, within a period of *six days!* Considering the eccentric movements, and the wide intervals of time between the arrival and departure of comets, the accuracy of these astronomical calculations is truly wonderful. But there is nothing in nature more regular than the precession of the equinoxes, the constantly changing seasons, or the daily rising and setting of the sun. The everyday operations of agriculture are very much influenced by the changes in the temperature of the earth, caused by the solar system of which it is a member; and the astronomical calculations usually embodied in the Almanac are consequently of great interest and importance to husbandry. But Astronomy subserves a still more important purpose in ocean navigation, where the commerce of the world and the lives of thousands of individuals are constantly exposed to the dangers of the sea. " That a man," says Sir John Herschell, " by merely measuring the moon's apparent distance from a star, with a little portable instrument held in his hand, and applied to his eye, even with so unstable a footing as the deck of a ship, shall say positively, within five miles, where he is on a boundless ocean, cannot but appear, to persons ignorant of physical astronomy, an approach to the miraculous."

But one of the most beautiful discoveries of the present age, was that of the planet Neptune, some twelve years ago. For many years after the discovery of Uranus, astronomers were perplexed to account for certain variations in the orbit of this planet—the calculations of old observations not agreeing with the new ones, and both alike inconsistent with known facts in reference to its orbital motions. Some accounted for the discrepancy by suggesting the influence of comets, or denying the universality of the law of gravitation; but Leverrier and Bouvier of France, and Adams of England, insisting on *the law* of the mutual dependence of planets upon each other, suggested the existence of *another planet* beyond Uranus. Acting on this theory, these gentlemen entered into the most complex and laborious calculations, to find out the position which this unseen disturber of Uranus occupied in the firmament. Leverrier finished his calculations, which exhibited a longitude for the planet of 326° 32', a mass two and one-half times that of Uranus, a distance from the Sun 36,154 times that of the earth, and a periodic revolution in 217,387 years. The position, magnitude, and general character of the unknown planet thus defined, Leverrier transmitted the result to *M. Galle*, of the Observatory at Berlin. On the very evening that the letter came to hand, Galle turned his great telescope in the direction indicated, and startling as the fact may appear, the planet was detected among the innumerable throng of the glittering stars, *within fifty-two seconds of the exact spot pointed out by Leverrier.* Such triumphs as these elevate man above the level of his nature. They exalt him above the fabled gods of heathen mythology. That we should thus explore the regions of space, measure the planets, and resolve their motions, is a fact which the Almighty author himself cannot but contemplate with benignant satisfaction!

The long-continued observations of astronomers, resolved into great mathematical and geometrical laws, not only establish the antiquity of the earth beyond all doubt or cavil, but also furnish the data by means of which we can form a satisfactory estimate as to the origin of the world itself. The most probable theory to account for the primary formation of worlds, is that originally suggested by Herschell, and afterward more fully elaborated by Laplace. It has been called the Nebular hypothesis. Shortly after Lord Rosse erected his gigantic telescope, and resolved the nebulæ in the constellation of Orion into clusters of stars, it was thought by some that the theory had been much impaired; for if the nebulæ could be resolved into stars in *one* instance, there would seem to be requisite only sufficient telescopic power to resolve *all* such aggregations in space into similar stellar bodies. This, indeed, might be true, and yet the force of the theory would not be destroyed; for it is only the *primary condition* of these bodies, *as* nebulæ, which the theory contemplates. But although the nebulæ of Orion has been resolved, there are many isolated and independent aggregations in space, which the telescope has *not* changed; and these can be accounted for in no other way than on the supposition of their having been thrown off from the primary planets, if not directly from the sun himself. These aggregations or swarms of floating vapor abound everywhere amid the universal space. Their elements, whatever they comprise, must be different from the surrounding ether, or else unite peculiar chemical compounds of the upper air. They assume every imaginable form, and appear in different degrees of density—passing from a mere thin film into fleecy atmospheres encircling central stars, like those of comets, and often emerging into the luminosity of actual stellar groups, as in the case of Orion. They have been thus traced in every form of condensation, from their primary embryonic condition into vast suns and systems like our own. It was owing to the *gradations* thus observed, in connection with other considerations, that the idea of the formation of worlds was suggested to the mind of Herschell. Regarding the sun as the primary centre of the planetary system, it is inferred that he threw off, at successive periods, while undergoing a process of condensation, as the inevitable result of a rotary motion, all the worlds that now revolve around him, including our own, as well as all the moons and satellites appertaining to them. The sun, therefore, must be regarded as the parent of all the stellar bodies comprising our solar system. This idea derives additional support from various collateral phenomena, among which may be mentioned, *First:* The singular and significant fact, that all the planets revolve in *one direction only*, viz., from west to east. As this is the direction which the sun himself observes, it is plain that it is the result of *law*—that the planets, as they were thrown off from his outer rim, were hurled through space in the direction of the parent body, and, while they evinced a constant tendency to *return home*, were as constantly *repelled by his superior motion*. *Second:* The planets all describe circles, with but slight elliptic deviations, and this form is due to their primary nebulosity exposed to a rotary motion. A drop of rain or a piece of molten lead, as it falls through space, assumes, a rounded or globular shape; and so these incipient worlds, as they were cast off from the rim of the parent sun, in the form of attenuated vapor, took the form which characterizes the earth, whereupon they were forever consigned to the positions they respectively occupy, and, once brought under the regular and undeviating laws of attraction and repulsion, underwent the changes, internal and external, which have resolved them into the physical properties essential to the universal harmony. Had the incipient worlds thus cast off *not* assumed the globular form, but spread out as mere irregular agglomerations or sheets of vapor, it is probable that they might have been again absorbed by the sun; but the *motion* which they all inherited from the parent body was sufficient to resolve them into globes, just as sheets of water, thrown off from a wheel in rapid motion, will part into thousands of independent but minute globes. The form which water thus assumes, is as much due to a *natural law* as that of gravitation itself. And there are few things in nature more interesting than this same law of forms, now termed *Morphology*. Every thing preserves an *individuality* of form and structure, from a grain of sand on the sea-shore, to a crystal of quartz, or of galena, or pyrites, in the solid rocks; from the flower or the fruit to the tree; from an insect to the soaring eagle; from the inferior animals up to man, and from mankind to worlds—all assume forms at once *specific* and peculiar to their kinds. *Third:* It will be observed, from the distances of the planets from the sun, as previously mentioned, that there is a regular or nearly regular increase, from one to the other, thus going to show that there was a *law* under which the expulsion of the planets from the primary solar nebulæ was regulated; that, in fact, they were not thrown off by mere chance or accident, but at regular intervals, and according to a fixed and previously arranged *plan*. This law (concerning the bearings of which we have still much to learn) was discovered by the celebrated *Bode*, and is expressed by saying "that the interval between the orbits of any two planets is about twice as great as the inferior interval, and only half the superior one." *Fourth:* Not only is there a fixed order in the *relative distances* of the planets from the sun, and from each other, but there is a similar order in their respective *times of revolution*. Thus, Mercury, the nearest planet to the sun, revolves around him in two months and twenty-eight days; Venus, in seven months and fifteen days; the Earth, in one year and six hours; Mars, in one year, ten months, and twenty-one days; Vesta, in three years, seven months, and twenty-one days; Juno, Ceres, and Pallas, in four years and some eight months each; while Jupiter occupies eleven years, ten months, and seventeen days; Saturn, twenty-nine years, five months, and twenty-four days; and Uranus eighty-four years and twenty-seven days! Here, it will be observed, the order of *time* is strictly in correspondence with that of *distance*, and we thus evolve a powerful inference in favor of the primary origin of all these bodies from the central sun around which they revolve—beginning with the most distant, as Uranus and Saturn, and terminating with the Earth, Venus, and Mercury. The weight of these orbs, as compared with an equal bulk of water, undergoes a change somewhat similar — diminishes outward from the sun. Mercury weighs more than seventeen times as much as water; Venus more than five times; the Earth more than four; Mars more than three; Jupiter more than one; while that of Saturn is barely half as great. Thus, an individual who weighs one hundred pounds on the Earth, if removed to Mercury, would weigh nearly four hundred pounds; while, if transferred to Saturn, he would dwindle down to *fifty pounds*. This law is accounted for in the superior density of all bodies toward their centre, and applies equally well to the Earth itself as to the planets in general. The nearer we approach to the centre of the Earth, the greater becomes the force of attraction; and it is precisely thus with the planets in their relation to the sun. The weight of the waters of the sea is such that, beyond a certain ascertained depth, the minute corals and mollusca that inhabit it, and fill the waters with their comminuted slime and chalk, cannot live. The radiation of light is a good illustration of the law which graduates the density of planets. To illustrate: let the letter A represent a light. Let B represent the surface upon which the light A is reflected, which we may suppose to be distant one hundred feet; at the distance of two hundred feet, the light of A would occupy at C four times the surface that it did at B, but would lose correspondingly in *density*. At the distance of three hundred feet, D, the light of A would cover nine times the extent of surface as at B, but with still diminished density; while at E, being four times the distance, the light would expand to sixteen times its original area; but in each case its density would decrease, like that of milk diluted with similar proportions of water. This law of variation according to the square of the distance, applies equally to all physical forces which are capable of radiation from a central nucleus, as gravitation, heat, electricity, magnetism, steam, sound, etc., etc.

While the human mind may *imagine* the origin of worlds, in the manner here suggested, and find much to support the hypothesis in the known and familiar laws of nature and matter, it is yet utterly incapable of proceeding further. It cannot comprehend—it cannot conceive, for a single moment, how, where, or when the great Creator obtained the seemingly varied materials with which he has constructed so many millions of radiant worlds. It cannot conceive any plausible explanation of *forces*, even. Things which we daily see and feel, as light, heat, air, electricity—or which we daily handle, as gold, silver, iron, sulphur, or any of the other sixty-two simple elements or substances with which we are acquainted—nobody can satisfactorily explain their primary origin, or trace them back to the laboratory whence they were evolved. We see them—observe their effects—and perceive that they all work agreeably to *law*. We thus detect a primary *cause*—a Creator who *acts* by a *plan*. We see him in all things—from the minute atom or particle to stupendous globes;—yet we cannot even imagine his vital or personal embodiment, much less can we imagine his origin. All these are points, the solution of which can only be determined by God himself;—they are not for man. It is enough for us to know that worlds exist within worlds; that whole systems and universes revolve in space, so far distant from our own that no conception can be formed of the intervening space; and yet, all that we can see or learn concerning them, shows that there is universal law, order, and harmony among them all.

Now, supposing the planets of our system (twenty-two in number, with perhaps an equal number of satellites or moons,) to have primarily belonged to the sun around which they revolve, and to have been successively thrown off in the manner indicated; we may assume that the sun previously stood in the relation to the other systems that these planets now occupy to him. In the beginning, therefore, the elements now comprising many globes, comprised but *one*. There was absolute *unity*, the embodiment of which was the Creator himself. In him all things centred, and upon the exercise of his volition, all things sprang or irradiated,—each particle of matter, or force, or principle, being clothed with powers for combination, affinity, and relation to other particles, in the union of which life was inclosed as in the seeds of fruit. After the planets had been thus irradiated, they were kept in place by an inherent propensity to *return back into unity* —a propensity which may be termed the law of gravitation. But, in consequence of the superior velocity of the primary sun around which they revolve, and the varied relations which they respectively owe to each other, there is no possibility of their fusion, and hence, the two opposing forces of attractive gravitation and diffusive repulsion, merely serve to keep them in constant motion around the primary source.

As to the earth itself: When we reflect how easily, by the mere change of temperature, water is converted into ice, or dissipated into steamy vapor—how copper, lead, or any other mineral or rock, may be made to bubble and boil in liquid fire at our feet, there is little real difficulty in the way of tracing the earth, according to the nebular hypothesis, throughout all the stages of a gaseous, a liquiform, and thence into that more dense, compact, and solid condition preceding the great purposes it was to subserve in the almighty design. The combination of oxygen with hydrogen, in certain equivalents, produces water; the compounds of carbon with both, form the great bulk of all vegetation and animal substances. These and other chemical compounds, comprise the solid body of the earth, and of the atmosphere around it; and hence, its origin from the nebulous elements abounding in universal space, is merely a chemical phenomena, so far as its aggregate particles have been formed and attracted to each other. If the embryo fœtus of animals be instinct with the vitality that is to bear them onward through the various stages of birth, of youthful development, and mature age; it may likewise be assumed that these nebulous bodies *inherit primarily* the light, or heat, or vital principle necessary for *their* subsequent development into planets, and for the discharge of all those functions in the future, which the unchanging laws of the universe impose. And the same law that originates and sustains organic life, and causes earthly and mineral atoms to cohere, has formed the world—for these are in fact but the great aggregate compounds of innumerable and independent atoms. A drop of water suspended from a trembling leaf, is as much, as completely a world to the animalculæ that inhabit it, as *our* world is to us;—and very likely the phenomena outside and around them, appear to their microscopic vision as stupendous, and wonderful as the other worlds that are suspended around us. The air, the ocean, and the earth, teem with minute worlds, inhabited by creatures so inestimably small that the highest powers of the microscope are often insufficient to reveal them, and yet our world is nothing but the combined aggregate of the whole, as the great universe itself is but the aggregate of millions upon millions of other worlds, many of them far exceeding the dimensions of our own.

As to the formation of the earth: if Moses, inspired by prophecy, could penetrate the narrow vistas of the future, there is every reason to infer that the past was not beyond the reach of his mental vision; but that, like

THE FIRST DAY—ASTRONOMICAL.

> "Caethus, the seer, *his* comprehensive view
> The Past, the Present, and the Future knew."

In ancient times the seers and soothsayers comprised a numerous, if not always a somewhat influential class; and though they did not invariably pretend to divine inspiration, they yet often directed the most important movements of armies and of states. The oracles of Greece were the most potent institutions of the confederation; and that of Delphi was endowed with the richest spoils of victory, and the most magnificent offerings of kings and states. As it would seem to be perfectly consistent with the whole theory and practice of prophecy, we have a right to suppose that the description he gives of the creation was revealed to Moses in a series of *Ideal Tableaux*, corresponding to the six calendar days or cosmogonal eras. These tableaux or pictures would, of course, delineate only the more prominent features of each day or formation, leaving the subordinate incidents or details in the background of Time.

When Moses, therefore, lifting the vail that revealed the unshaped embryo of the world, declares that "in the beginning God created the heaven and the earth, and that the earth was without form, and void, and darkness was on the face of the deep," he simply describes a phenomenon which the nebular theory reclaims from mysterious obscurity. "He *created* the heaven and the earth;" that is to say, he created the *substances* of which the heaven and the earth are composed: for if any thing *more* or beyond the embryo was created "in the beginning," there would obviously have been no occasion for the subsequent six days' creation, during which the earth was being gradually *developed* for its future purposes. The heaven was first in the order of time, and it was only the *earth* that was without form at the era which Moses describes—the heaven having evidently been complete. From this the substance of the earth was derived, and impregnated with the creative vitality—for "the spirit of God now moved upon the face of the waters;" or, in other words, God *moved the waters*, or caused motion to exist on the face of them, or throughout the entire body of the elemental mass. We must here bear in mind that words often lose their original meaning. While it is natural for us to suppose that the spirit of God moved like a spectre on the face or the surface of the waters, the truth is that the word *face* was formerly equivalent to our word throughout, and therefore implies that the spirit, moving as it did, communicated to the waters, or the half-liquid nebula, the force and strength and animation of his volition. In brief, the embryo earth thus became a living fœtus in the womb of the Universe, and had now entered the period of gestation.

"And God said, Let there be light; and there was light; and seeing that it was good, divided the light from the darkness." It will be borne in mind that light is not only reflected from nebulous bodies, but that, in consequence of its solar derivation, it forms an *inherent constituent* of their composition; hence, on attaining a certain degree of *density*, they are resolved into luminous globes. It was thus with the earth. Light, or rather heat, emerging from the elements thus chemically combined, was separated from the original gloom. It was not, perhaps, the flaming light which results from the combustion of gases, but rather the radiated *heat* of molten or liquid matter—such as is reflected from incandescent bodies, or such as now comprises the liquid atmosphere of the sun. The solar atmosphere which surrounds our earth had no existence upon it at that time—the heat having been confined to the centre while the opaque exterior was undergoing refrigeration. Consequently, there could have been no such *day* as ours, and for the sufficient reason that the world had not yet come under the direct and equable influence of the sun, but existed for an indefinite period as an incipient or embryo nebulous planet.

> "Let there be light," said God; and forthwith light
> Ethereal, first of things, quintessence pure,
> Sprang from the deep, and from her native East
> To journey through the airy gloom began,
> Sphered in a radiant cloud—for yet the sun
> Was not;—she in a cloudy tabernacle
> Sojourned the while. God saw the light was good,
> And light from darkness by the hemisphere
> Divided—light the day, and darkness night,
> He named. This was the first day, even and morn."
>
> *Milton.*

THE SECOND DAY—ASTRONOMICAL.

6 And God said, Let there be a firmament in the midst of the waters, and let it divide the waters from the waters. 7 And God made the firmament, and divided the waters which were under the firmament from the waters which were above the firmament; and it was so. 8 And God called the firmament heaven. And the evening and the morning were the second day.

THE earth which we inhabit occupies a position in that broad belt stretching over the firmament which, from its nebulous aspect, has been termed the Milky Way. Although this zone is studded with millions of worlds, many of them immensely larger than our own, its prevailing aspect, as indicated by the name, is that of nebulosity; and in connection with the supposed primary origin of our globe, this fact betrays a great deal of significance. From unshapen nebulosity, the earth passed into a fluid condition, having its inherent heat in the centre, as the axis. The earth is therefore described by Moses as a fluid—as a vast globe of water; and such, in fact, it was, and *is even now*. While water itself, in the form of seas and rivers, occupies more than three-fourths of the actual *surface* of the earth, its elements, in the form of hydrogen and oxygen, mainly fill the *air*. Its presence in organic life is equally as great as in inorganic substances. A man, for example, who weighs 154 pounds, is made up of 116 pounds of water—leaving but 38 pounds of dry matter; while in lower animals the quantity of water is still comparatively greater. Some aqueous animals, indeed, contain 99 per cent. of water in their solid composition. The original fluidity of the earth, therefore, as here described by Moses, is one of those singular and extraordinary facts which, although little suspected by the popular mind, is incontestibly established by the testimony of nature.

When the phenomena of the second day were revealing themselves to the mental vision of Moses, the young or embryo earth had already surrounded its interior heat by a wall of sub-aqueous granite—that is, while the exterior surface was a boiling sea, the interior heat was surrounded like the shell of an egg, by an earthy precipitate congealing into crystalline rock. This was the immediate result of calorific sublimation, as opposed to the external tendency to refrigeration. The pores of the rocky shell, however, were easily penetrated by the overlying water, while they also served as conduits of the interior heat. The carbon (of the heat) thus operated as a liberator of the oxygen of the water, and upon its partial decomposition, or rather its evaporation, the volatile gases were dispelled into the surrounding atmosphere. Water thus decomposed would leave behind its earthy precipitates, and evolve iron, sulphur, alumina, lime, or whatever solids remained in the rocks thus accumulating. In the mean time, the steamy evaporation compelled the gases to form new combinations with the ether (a substance believed to exist, but of unknown qualities), and nothing could have been more natural than for these (especially in view of the lightness of hydrogen) to arrange themselves into an arching vault over the vast plain of boiling water beneath —or rather, the vast *concave* surface of boiling oceans. Constantly and regularly widening, with the increased evaporation or liberation of the aqueous mass below, it was finally transformed into that magnificent ethereal dome, strung with myriads of glittering jewels, which we see suspended over us, and which, in fact, merely adapts itself, as Moses implies, to the convex surface of the earth. And here, again, we have another beautiful illustration of the divine origin of the Mosaic revelation. The ancients almost universally regarded the firmament as a *solid crystalline body*, to which the stars were attached as if by metallic rivets. At the same time the surface of the earth was very generally regarded as a *level plain*. Indeed, the Church of Rome, at one period, made it heresy to doubt this proposition, with others equally absurd. But long anterior to this, and in the face of all the academicians and philosophers of his own and subsequent ages, Moses enunciated a cosmogony utterly at variance with the old philosophy, and the truth of which seems to have been reserved for the Telescope, and other modern appliances, fully to establish. Now, in the original Greek, biased by the then prevailing speculations, the word firmament here used is translated to signify a *concave with a solid base;* but it appears that, on referring to the original Hebrew text, it is found strictly to imply a *vast space without limitation!* Even the great Kepler at one time regarded the earth as a stupendous animal, who, breathing in and spouting out the waters of the ocean like a whale, occasioned the ebbing and flowing of the tides! Yet, here we find Moses, several thousand years in advance, calmly and briefly detailing the cosmogony of the earth, and its varied creation, according to the concentrated and constantly accruing *facts* of six thousand years! He was, in fact, not only in advance of his own age, the age of Egypt in its highest power and glory; but he was in advance of Greece, by whom he was translated, in its highest power and glory; and he remains still in advance —far in advance of modern Science, and of an age unexampled in wisdom and intelligence! But, unlike preceding ages, the discoveries of modern times tend to the confirmation of his cosmogony; and every progressive step —every fresh incursion into the mysterious domain of Nature, only establishes the sublimity and depth of his penetration, and the absolute purity and extent of his inspiration.

But the process of evaporating or separating the waters was by no means a slow or tame one. It was accelerated by the most terrific volcanic action, the necessary prelude to the gigantic object to be attained. In anticipation of the birth of animated nature, with which the young earth was now pregnant, and in view of the great events to occur in the succeeding days, there crept beneath those vapor-enveloped seas a dull, complaining, and rumbling sound—faint and half-suppressed at first by the hissing waters, but rising anon in might and strength, like contending armies of defiant thunderbolts, whose glittering sabres throw off the vivid lightning's flash. Then followed fast, and then faster still, the stifling, choking volcano, every new explosion vomiting up lurid smoke, and scalding steam, and fantastic clouds of glaring fire! The earth shuddered! Its seas roll up in savage tempest-tossed billows; its submarine floors of granite break in deep and lengthened gulfs, while the elemental warfare still continues with redoubled strife within its cavernous bowels. Around—above—beneath—all was now one grand, seething, hissing, roaring, eruptive caldron! No animal—no living creature was on the earth. No fish was in the sea. No bird spread his soaring pinions in the air! No flowers bloomed. No trees waved their spreading foliage to the breeze;—but, amid the dusky atmosphere of poison—amid the detonations of hydrous oxygen, the stifling fumes of sulphur, the mephitic vapors of azote, and the scorching avalanches of porphyritic soot, the young Earth groaned in labor, and at every convulsive throe, fragments of molten rock, huge as high Olympus, fell in fiery tempests on her heaving breast!

Such were the extraordinary convulsions which characterized the second, and aided the operations of the third day. Toward the evening, when the waters began to withdraw like the ebbing of the tide, vast regions of submerged lands struggled with the shallow surfs to escape the dominion of the sea. The rocks thus emerging comprise the most extensive and wide-spread groups in the stratification of the earth, forming, as they now do, the central nucleus of the most elevated mountain chains that distinguish its surface, and the solid basis upon which all the others rest. They are the family of granites, and the immediate progenitors of the volcanic rocks that resulted from the disturbance of the primary shell surrounding the interior heat.

There is scarcely a district in Europe, Asia, or Africa, in which these rocks do not exist—though the more recent groups are often deposited over them in irregular patches. In America they constitute at least three-fourths of the surface, covering the greater portion of Canada and Russian America, and stretching in an uninterrupted belt along the Pacific Ocean, from South America to California, Oregon, and Behring's Straits. In the United States an irregular, though continuous belt of igneous rocks extends along the Atlantic slope, from Maine to Georgia; on the one side, running parallel with the Tertiary, which dips into the ocean, and on the other with a group of the New Red Sandstone, lying to the northwest. They also cover a large surface in Minnesota, Wisconsin, and the upper peninsula of Michigan; while they occupy nearly the whole extent of the Rocky Mountain country.

As these are the lowest rocks yet discovered in the earth, and the most widely distributed, there can be no question whatever as to their being also the oldest. Granite itself is composed of mica, quartz, and felspar—the latter predominating. Being crystalline and unstratified, it bears equal pressure in every position, and hence is largely employed in architecture.

The rocks comprising the inclosing circle of the interior heat, are usually termed *Plutonic*, to distinguish them from those immediately subsequent, which are *volcanic*. The two combined are termed *igneous* rocks, and are thus distinguished from the metamorphic. The Plutonic include granite, syenite, and eurite; the volcanic—which are the others differently amalgamated by fusion—are basalt, greenstone, claystone, porphyry, amygdaloid, and lava. Now, when the interior heat burst through the rind of granite, thousands of great streams of lava were discharged, in various quarters of the earth, from submarine volcanic craters. The fragments of rock were ground and shattered into pebbles and sand, in their upward passage, and their particles uniting with the liquid lava, spread over the bottom of the sea, the agitation and pressure of the water often giving them a wavy, irregular, and contorted structure. The ultimate effect of this universal and stupendous volcanic action (converting gases into liquids, and liquids into solids) was to bring from the watery waste a series of island reefs, united together by a reticulated chain of narrow peninsulas, stretching around expansive lakes, rivers, and basins, far into the ocean. These islands subsequently become continents.

Although basalt, greenstone, claystone, porphyry, amygdaloid, and lava, are the general names of the volcanic group, the various subsequent combinations of their particles by sublimation, have produced a large family of other rocks, bearing specific names. For instance, *clinkstone* belongs to the group, being a combination of felspa.

and quartz, and so named because of its ringing sound under the hammer. Trachyte and basalt often pass into clinkstone, by very gradual transitions. *Diallage* is synonymous with euphotide and gabbro, and is made up of euphotide and felspar, with occasional infusions of serpentine, mica, and quartz. *Dolerite* is merely another and a better name for greenstone (since it is not always recognized by color) and includes black augite, or hornblende and felspar, or augite, felspar and magnetic iron. Felspar forms an important ingredient in a great number of rocks. It is the base of felspar-porphyry, which contains crystals of felspar and quartz. *Hornblende* is composed mainly of black augite and iron, often mixed or imbedded in felspar. *Hornstone* is a species of flint, and only differs from compact felspar in its action before the blow-pipe. Obsidian is another name for lava. It is vitreous, and resembles the fracture of glass. *Ophite* is composed of hornblende and felspar, and becomes *serpentine* by the addition of *talc*. *Pearlstone* is somewhat similar to obsidian, and is thus named from its pearly lustre. *Pitchstone* is another name for volcanic lava, and is so called from its resinous lustre. *Porphyry* is a stone abounding in large isolated crystals of felspar. *Pumice* is a spongy trachyte. *Scoriæ* is volcanic cinder. *Serpentine* is a green rock, abounding in magnesia. *Noble serpentine* is a transparent mineral, sometimes affording gems. *Syenitic greenstone* is composed of felspar and hornblende. *Trachyte* is glassy felspar. *Amygdaloid* is so called from its cellular structure, the cavities of which are in the shape of almonds. In Lake Superior these cavities are often filled with chlorite, and sometimes with copper.

Besides the six or seven principal varieties of the volcanic rocks, it will thus be perceived that there are a great number of subordinate members of the family. These, however, only differ from the others in the proportions with which the original ingredients are united. And chemical analysis shows that the leading elements in all of them were silica, alumina, and magnesia—there being comparatively but a trace of lime, potash, and manganese. Many of them, however, had a large content of oxyd of iron, as, for instance, augite, chlorite, diallage, epidote, hornblende, hypersthene, mica, olivine, etc.

It may be suggested that because these rocks had not yet assumed the superficial characteristics of continents, they do not properly belong to the second day. But if they formed vast oceanic reefs, or incipient islands, it was sufficient, for Moses did not command the *dry* land to appear until the *third* day. In using the word *dry* land, he leaves us to infer the previous existence of rocky reefs and sub-marine continents; and because they did not rise up in tall cliffs, is no reason that they did not exist as rocks, or that they do not belong to this geological era. As well, indeed, might the alluvial silt of submerged river estuaries be ranked with the primitive rocks upon which it lays, as for those rocks to be included with more recent formations.

> Again God said: Let there be firmament
> Amid the waters, and let it divide
> The waters from the waters; and God made
> The firmament, expanse of liquid pure,
> Transparent, elemental air, diffused
> In circuit to the uttermost convex
> Of this great sound; partition firm and sure,
> The waters underneath from those above
> Dividing; for as earth, so he the world
> Built on circumfluous waters calm, in wide
> Crystalline ocean, and the loud misrule
> Of Chaos far removed, lest fierce extremes
> Contiguous might distemper the whole frame:
> And heaven he named the firmament: so even
> And morning chorus sung the second day.—*Milton.*

COAL AND MINERAL COMBUSTIBLES.

THE THIRD DAY—GEOLOGICAL.

9 And God said, Let the waters under the heaven be gathered together unto one place, and let the dry land appear: and it was so. 10 And God called the dry land Earth; and the gathering together of the waters called he Seas: and God saw that it was good. 11 And God said, Let the earth bring forth grass, the herb yielding seed, and the fruit-tree yielding fruit after his kind, whose seed is in itself, upon the earth: and it was so. 12 And the earth brought forth grass, and herb yielding seed after his kind, and the tree yielding fruit, whose seed was in itself, after his kind: and God saw that it was good. 13 And the evening and the morning were the third day.

WHEN Moses lifted the vail which revealed the picture of the Third Day, he saw the waters receding from the effects of the recent violent convulsions; and as they withdrew, the land gradually emerged, and appeared conspicuously in the scene. The command for the *dry land* to appear was based upon the capability its structure now possessed for drainage, and was only the necessary preliminary for that following, "for the earth or dry land to bring forth grass, and herb, and fruit."

After the land had thus emerged from the primitive seas, another group of strata was immediately commenced. This is called the *Metamorphic*, because the lithological nature of the rocks has been changed from an ordinary sedimentary to a crystalline, a compact granular, or fibrous structure. Beds of limestone were converted into white statuary marble, and particles of felspar and mica, or quartz and mica, united to form compact gneiss. The heated and agitated waters sweeping around, and permeating the long narrow islands and reefs, wore off their jagged edges and incoherent fragments, and carried away their comminuted debris into the adjacent seas, lakes, rivers, and basins, where they were spread out in thin laminæ, or rolled into contorted heaps by the motion of the water and the reciprocal vibration of the floor beneath. It was thus that the extensive layers of gneiss, hornblende, mica slate, talcose slate, and clay slate were deposited. Their position to the previous rocks is unconformable, but it is nevertheless easy to pass from them to the volcanic, and thence to the primitive rocks below. An illustration presents itself all along the zone of primitive rocks, previously referred to as extending parallel with the Atlantic coast, where most of them are adjacent and lead to the volcanic, and thence to those of the metamorphic varieties. This is owing to the fact that granite and the volcanic rocks formed the original reefs and islands, and hence now often appear on the *summits* of mountains, while the metamorphic occupy the *slopes* or the *valleys below*. These valleys, at the time of their deposition, were filled with water, and the debris from the islands gradually settled at the bottom. Now, although these rocks were elevated and *changed by heat*, it was not by means of direct *volcanic eruptions*, as in the previous case. It was by the more gradual *expansion of the interior heat*, the effects of which were felt at the *weakest parts*. This, owing to the pressure of the water, now that it had gradually collected into a great body, would be in the vicinity of the islands; and hence, while the heat was constantly baking and solidifying the soft mud in the lakes and basins, the islands and the basins themselves were gradually rising, higher and higher, from the water. In process of time, the water was entirely withdrawn; the vast regions of dry land were redeemed, and the earth was fully prepared, agreeably to the Divine command, to "bring forth grass, and herb, and fruit after his kind." And it may be added, that the silt thus exposed furnished a prolific soil for this purpose, containing, as it did, all the ingredients which the farmer, by the application of manures, now infuses into his soil.

After the deposition of the metamorphic rocks (formerly called the Transition), the earth seems to have generated or discharged a large amount of carbonic acid gas. It had most likely been generated in its bowels, and resulted from the spasmodic explosions of the interior heat. This is the more probable, from the fact that it has been diminishing in extent from the Paleozoic formation down to the present time. If it was the result of gaseous explosions during volcanic action, the fissures that disruptured the strata of the globe allowed it to escape in great abundance, and, combining with the air, it was again precipitated upon the earth in the form of vegetation, and to the seas in grass, moss, algæ, and half-vegetable corals and zoophytes. In fact, the *beginning* of organic life appears to have been simultaneous with the formation of beds of coal and limestone—the fossiliferous layers of which, alternating with slate, sand, and shale, distinct and in combination, comprise the main ingredients in the rocks known respectively as the Cambrian, the Silurian, the Devonian, and the Carboniferous—all of which are properly included in a common *Paleozoic era*, since it was begun and terminated by violent convulsions, and occupies a position *unconformable* alike to the preceding and subsequent rocks. This, I am well aware, is not the usual classification of geologists; for although it is the system of Conybeare and others, many assign a separate era to the old red sandstone and coal, and thus directly, as well as by inference, recognize the *priority* of the *marine animal life* of the Silurian group over the *vegetation of the land.* Such priority is in conflict with the *order* of nature and the authority of Moses, and I have never been able to discover sufficient grounds for its support.

In geological order, what are termed non-fossiliferous rocks occupy two distinct formations—the one termed Igneous, or Plutonic and volcanic (as previously mentioned), and the other Metamorphic, or stratified. The igneous rocks include granite and its varieties as *primaries;* and amygdaloid, porphyry, greenstone, and basalt as secondaries, or volcanic—all of which, bearing the traces of heat, exhibit no parallel lines of stratification. The metamorphic group comprises gneiss, hornblende, mica, talcose and clay slates; and although they are stratified, it is assumed that none of them contain fossil remains—a proposition, so far as *traces of vegetation* are concerned, which I propose hereafter to question.

The igneous rocks proper, it has already been remarked, form the inclosing rind of the earth. As they are, therefore, the immediate result of its primary nebulosity, undergoing the process of refrigeration, condensation, and chemical combination, they properly belong to the work of the preceding astronomical days. Although they had not yet *emerged* from the water, they nevertheless *existed* as rocks as fully as those now at the bottom of the ocean; and they therefore belong, *not* to the day that withdrew the water from them, but to the day during which they were *formed.* The metamorphic group is not only different in its origin, but different in structure and character. Being the *fœtus* which gave birth to organic life, it should be included with the Palæozoic formation, as the *basis* upon which to stand. There would, indeed, be no more propriety in dating the years, or geological eras of the animated earth, from the period of *gestation*, than there would be in the case of an animal or a human being. Having been elaborated from the abrasion of the previous rocks, they exhibit a striking transition from them into those of the Paleozoic, and really form the very bed upon which life was first introduced on the face of the earth.

The granite and volcanic rocks, therefore, belong to the evening of the second day; while those of the metamorphic introduce us to the morning of the third. This day I conceive to be represented by the paleozoic period; and I shall proceed to point out some of the varied phenomena which distinguished it.

The term *paleozoic* refers to the fossil remains of the ancient earth, and is usually applied by geologists to distinguish those great systems of fossiliferous rocks, known as the Cambrian, the Silurian, the Devonian, and the Carboniferous. I have enlarged its scope so far as to include those of the metamorphic, since these rocks form the basis of all the others, and very frequently are intruded between them. I should not have assumed this freedom (with the respect I entertain for the ancient landmarks), had I not good grounds to believe, in addition, that they were originally fossiliferous, and exhibit the metamorphosed remains of a primitive vegetation. Some geologists, it is perhaps proper to state, do not adopt the word, notwithstanding its peculiar significance as applied to a great paleontological formation. Among these is Sir Charles Lyell, always great in small things, and sometimes small in great ones, but whose extensive travels and amiable literary style entitle his works to all the respect they have received. Instead of the word Paleozoic, he adopts that of *Primary*—primary fossiliferous; but as this word is often applied to the Azoic, or non-fossiliferous strata, whose chronological priority none will question, its use is only calculated to mislead. Lyell, in fact, has invented more words, and made fewer *bona* *fide* discoveries in geological science, notwithstanding his extensive travels and experience, than any leading geologist in Europe or America. The truth is, that the accumulation of mere *words* is far in advance of geological progress; and such is the confusion and complication of nomenclature among geologists themselves, that it is almost impossible to describe or identify strata without resorting to technical tautology. The propensity for calling old things by new names, invariably betrays the weakness of those who indulge it.*

---

* I will here step out of my way to notice the most recent and the most absurd innovation of geological nomenclature ever attempted. The Geological Survey of Pennsylvania, commenced in 1836 and completed in 1859, is perhaps the most stupendous scientific trash ever collected together. The Surveyor, in his annual reports to the Legislature, arranged the leading geological formations of the State in *numerical order*. This, although a serious annoyance, and the sublimation of impracticable folly, was quietly submitted to, and our people in time learned to regard their native rocks as "No. 1," or "No. 2," etc. For my part, I never attempted to master this nomenclature, but adhered to the old family names—names which, like that of Weller, have long been read of in history; have, in fact, become classical in geological literature, and will stand with those of the sun and stars in cosmical philosophy.

After having, for many years, used these numerical terms in all his previous annual reports, the official surveyor again presents himself, after a long absence (which the geological interests of the State never appear to have realized), clothed from head to foot in a new suit of unmeaning words. Two thousand heavy, cumbrous pages teem with the words Primal, Auroral, Matinal, Levant, Surgent, Scalent, Pre-meridian, Meridian, Post-meridian, Cadent, Vergent, Ponent, Vespertine, Umbral, Scral, etc. These terms, I believe, are based upon the *idea of a Paleozoic day*, and may be supposed to represent the *hours*. According to his own explanation, obscurely inserted in the preface to the first volume, they respectively imply the Dawn, Daybreak, Morning, Sunrise, Mounting Day, Climbing Day, Forenoon, Noon, Afternoon, Declining Day, Descending Day, Sunset, Evening, Dusk, and Nightfall! Although these absurd terms are applied to a Paleozoic Day (a day which, by the way, he fails to describe in his report, except so far as pertains to the coal formation—for the mere incidental and obscure references to the characteristics of the other fossiliferous rocks cannot be regarded as a *geological*

This increasing disposition to embarrass and complicate the Natural Sciences should be checked. More time and trouble are often wasted in overcoming the barriers of nomenclature thus surrounding ideas and facts, than in comprehending the naked facts themselves. Words should elucidate, not perplex and confound. The old geological nomenclature of England is that of the English language everywhere. It originated in her mines and description), they could, with much more propriety, be applied to *geological time as a whole*—weeks, months, years; in which case granite would be *Primal*; basalt, amygdaloid, porphyry, and other volcanic rocks, would be *Auroral*, or day-breaking rocks; Talc, mica, and the clay slates would be *Matinal*; the Cambrian group would become *Levant*, or sun-rise rocks; the Silurians would be *Surgent*, or mounting-day strata; and the Devonian, *Scalents*, or climbing-day rocks! The distinction between sun-rising, mounting, and climbing, is not very perceptible as applied to *rocks*, but the idea is poetical! Should another stratum of rocks, *intermediate between these*, ever be discovered, their proper appellation would be *Snail*-ent, or *crawling*-day rocks. In ascending order, the conglomerates and old red sandstones of our mountains would belong to the *Premeridian*, or forenoon rocks; and the seams of coal lying upon these would become *Meridian*, or noon rocks,—furnishing the material, as they might, for cooking the noon meal! We now *descend* in the geological day, but still continue *upward* in geological order. The new red sandstone and magnesian limestone, according to this patent adjustible, double-acting, reciprocal, and self-revolving nomenclature, ought to be the *Post-meridian*, or afternoon group. Here we enter the great Secondary formation, the lower Triassic precincts of which would be *Cadent*, or declining-day rocks. Next we have what is termed in England the Oolitic group, comprising upper, middle, and lower strata, and these, with the Liasic, *may be Vergent*, or descending-day rocks. The difference between declining and descending day rocks, would appear to be that so long existing between tweedle-dum and tweedle-dee. Still going *downward* in the day, but *upward* in the geological scale, we arrive at the end of the Secondary formation, consisting of several layers of cretaceous and fresh-water strata. These are possibly *Ponent*, or sunset rocks. The Eocene beds of the Tertiary are *Vespertine*, or evening rocks; the Miocene, *Umbral*, or dusk; and the Pleiocene, *Seral*, or nightfall rocks. After this, we obtain a night's repose, and then awaken early in the morning to the modern or present geological day!

quarries, where geological science itself had birth, and where it must always flourish in native vigor. Geology, indeed, is nothing more nor less than the *accumulated experience* of the miner, aided by the observations of the traveler. The shafts that pierce the bowels of the earth two thousand feet beneath the surface—the connecting gangways, and galleries, and tunnels—what would have been known of the interior crust in the absence of these? They furnish the basis upon which the so-called Geologists weave their finely-spun theories and speculations; and they as often provide the data by which they may be blown away as bubbles of the air!

The paleozoic formation, by the arrangement which I propose, includes and begins with the strata of the metamorphic group. These, I have already mentioned, embrace gneiss, hornblende clay, mica slate, talcose schist, and clay slate; but included among them, and derived from them, are various other minerals, bearing individual names. Of these we may mention talcose gneiss, a stratified talcose granite; talcose schist, consisting of quartz and talc, or talc and felspar; quartz, consisting of silicious grains, stratified; crystalline limestone, or marble; mica schist, or slate, consisting of laminated mica and quartz; hornblende gneiss, comprising felspar, quartz, and hornblende; chlorite schist, abounding in scales of chlorine; chiastolite, similar to ordinary clay slate, but including slender rhomboidal crystals of chiastolite; actinolite, a foliated slate, containing more or less felspar, quartz, or mica. The composition of gneiss is precisely similar to that of granite; but being stratified and laminated, the different mineral ingredients, instead of being thoroughly intermixed, are disseminated in irregular parallel seams. The group, in fact, embraces very nearly the same material as those of the preceding plutonic and volcanic formation, the main difference being in stratification. Marble, or crystalline limestone, unknown to the previous groups, makes its first appearance among these strata, and distinguishes every subsequent geological formation, either as marble or ordinary limestones. I have already called attention to the content of oxyd of iron in the previous rocks. They nearly all contain more or less, while a few have as high as from twenty to forty per cent. It rarely occurs, however, in large deposits, or in a state of very great purity in those rocks; but their disintegration, during the metamorphic period, permitted the earthy material with which it was united, to escape and to be carried away by the water, leaving the iron behind to accumulate in large layers and deposits, which it could readily do from its high specific gravity. The great iron and copper regions of Lake Superior, running close together, and nearly parallel with each other for more than one hundred miles, are surrounded by primitive or igneous rocks; but the mineral lodes often traverse sedimentary rocks which have been altered by the trap dykes that ramify the entire formation. The copper often occurs in a state of absolute purity, while the iron is little inferior, yielding from sixty to seventy-four per cent. of metallic iron in the furnace. Both these minerals must have been sublimated in the Plutonic rocks that surround them. The copper, from its metallic purity, must have been injected in a liquid state, like that tapped from a smelting furnace, and thus forced between the strata and through the cavities and pores of the overlying sedimentary rocks, all of which became metamorphosed and very much disturbed and contorted by the operating heat. But before the copper was injected, it is probable that the iron of the adjacent region had been collected together in the bottom of a lake, having been precipitated as an oxyd upon the evaporation of the water of the seas during the second day, while the heat accompanying the volcanic dykes no doubt served to change the sedimentary character of the rocks themselves, and to sublimate both the minerals at nearly, if not the same time. Phenomena somewhat similar will also account satisfactorily for the origin of the great iron deposits of the Pilot Knob in Missouri; that of Cornwall in Pennsylvania, and of Lake Champlain in New York. Indeed, so far as *aqueous* action is concerned, all the deposits of hæmatite (hydrous oxyd of iron) found in metamorphic rocks, have undergone a process not much dissimilar, hence their purity and comparative freedom from extraneous substances.

As to the *fossiliferous character* of the metamorphic rocks, I shall have some remarks to offer hereafter, in connection with the coal formation. In the mean time, I may here observe that plumbago and anthracite occur to some extent in them, and I regard this fact as sufficient to establish the previous existence of vegetation, the remains of which have been nearly obliterated in consequence of the heat which metamorphosed them.

Immediately above the metamorphic group is that denominated the Cambrian, after a district in Wales, in which they were first noticed. There strata, however, are not unfrequently distinguished as the "lower Silurian"—a name derived from the ancient *Sylures* of Britain, the applicability of which to geology, like a great many other terms, is not very apparent. It is a marine formation, consisting of slates, lime, and sand, variously intermixed, but generally in separate layers. Some of these are fossiliferous, affording specimens of coral and molluscan brachiopoda. The upper, or Silurian proper, is not very dissimilar in lithological character, but is much more prolific in marine fossils, principally molluscous. Fish, so called, but partaking more of the nature of lizards, first appear in this era; while corals, serpula, and trilobites are abundant.

The aggregate thickness of these rocks varies from six to ten thousand feet, and they cover very extensive areas both in Europe and America. In the United States the Cambrian or lower Silurian rocks extend in a continuous and gradually expanding belt from the State of Alabama, northeast through Georgia, Tennessee, North Carolina, Virginia, Pennsylvania, and New York. They form the north shore of Lake Ontario, and thence pass along the St. Lawrence river, on both sides, to Montreal and Quebec, and finally disappear in the Gulf of the St. Lawrence. They also cover a vast region of country in Wisconsin and the British possessions north of the Rocky Mountains. Detached deposits occur on the Ohio river below Louisville, and near Nashville, Tennessee, and are generally prolific in their characteristic fossils. The *upper Silurian* strata are still more extensive. Arkansas, Missouri, Indiana, Ohio, and New York, have more or less of surface covered with them; while all the lakes, except that of Superior, are nearly surrounded by them. The beautiful white sandstone and fossiliferous limestone so extensively used for building purposes in Detroit, Chicago, Cincinnati, and St. Louis, are characteristic of this group of rocks. They are also very plentifully distributed in Russia and Siberia, in England and Wales, and more sparingly in isolated patches in France, Germany, Portugal, and Prussia. They occur to some extent in the East Indies, along the River Ganges, and the southeastern portion of Bengal. In Australia these rocks form long belts, rising parallel with and often including the gold region.

Rocks of the Silurian group are cut through and handsomely exposed at the celebrated Falls of Niagara. They not only form the perpendicular shores of that river and of the adjacent lakes, but extend all over the surrounding country, in some directions for hundreds of miles.

The valley of the Niagara river, some fourteen miles between the two lakes of Erie and Ontario, was originally much wider than it is now. The terraces of the ancient shores of the stream are plainly discernible; and they abound with shells of *Unio, Cyclas, Valvata, Planorbis,* and *Helix,* all of recent species. Beyond these terraces are others, still more ancient; and there is good ground to believe, from the sand, gravel, and mud distributed over the surface, that the adjacent lakes, at a period not very remote in geological time, extended all over the surrounding plains.

Attention was first directed to these falls by a French missionary, named Hennepin, in 1678. The French explored the whole country from the Gulf of the St. Lawrence to the head of Lake Superior, and held possession of it at the time that the English planted their Colonies in America. Hennepin wrote a short account of the cataract, and accompanied it with maps of the lakes, and a pictorial sketch of the Falls. According to this sketch, there were at that period *three distinct cataracts.* In addition to the "American" and the "Horse-shoe" Falls, there was then one which might be called the "Table-Rock" Falls, because it was evidently in that vicinity. This latter cataract was occasioned by an obstruction in the Rapids, which diverted the water around the Horse-shoe Falls, and hurled it over the precipice at the Table-Rock, which is directly opposite the American Falls. Table-Rock is traversed by fissures and cracks which permit the periodical detachment of fragments; and the shores of the river below are strewn with the masses thus worn off.

The Falls of Niagara are precipitated over a stratum of limestone eighty feet thick, and a stratum of argillaceous shale, under the limestone, also eighty feet thick. The perpendicular descent of the water is therefore about one hundred and forty feet. The limestone contains numerous cavities filled with sulphate of lime, sometimes called selenite, alabaster, or gypsum. These nodules of alabaster, varying in size from a walnut to a cocoanut, are extensively used by the lapidists at the Falls in the production of carved ornaments, which are sold to travelers to commemorate their visit to the place. The underlying shale contains a large amount of iron pyrites, the decomposition of which, on exposure to the air, hastens the disintegration of the rock. Now, the violent whirling and agitation of the water below, as it falls over the precipice, wears away the crumbling shale, and thus undermines the stratum of limestone. It is in consequence of this wearing away of the shale that visitors are enabled to pass *under the Falls*. But owing to the peculiar cellular structure of the bed of limestone, the undermining process cannot extend beyond a certain limit without producing fractures, which are materially extended by the weight and velocity of the water. Large masses of rock are therefore detached, from time to time, and the gradual retrogression of the cataract is thus rendered certain, and plain to our comprehension.

The extent of this erosion and retrogression was estimated by the celebrated Mr. Bollewell, in 1830, as equal to an average of one yard per annum. It is supposed that the original site of the falls was at Queenston, seven miles below; and if so, the erosion, at the annual average estimated, would have occupied about 12,000 years. Sir Charles Lyell, however, made a careful examination of the Falls and of the country around it, during his visit to America in 1841. He estimated the erosion at one foot per year, according to which it would have required 36,000 years for the Falls to change their original for their present location.

If the reader has ever passed over the waters of Lake

Huron or St. Clair, or over the great western railway of Canada, from Toronto to Detroit, he will remember the vast marshy flats that encompass those waters. As far as the eye can see, there is often nothing but these flats. They just emerge sufficiently above the level of the water to support a rank, long-bladed grass. Millions and millions of acres of land, flat as the surface of the lakes, are thus ramified and encircled by the water; but their rank vegetation gradually raises the surface higher and higher, and in time the land will be wholly redeemed from the dominion of the water. Now all the praries of the great West were once in the condition of these lake flats. They have been redeemed from primitive lakes and seas by vegetation—the mould and lime of which constitute their characteristic agricultural fertility.

That the surface of the ancient seas and lakes has been gradually reduced, and the water emptied into narrow river channels, will be very apparent to the geological observer in traveling through the South and West. The Valley of the Mississippi was once a series of great lakes, covering the adjacent Silurian country for hundreds of miles around; and these basins were finally drained in the manner just described. An illustration of the diminution of the water, as well as of the ceaseless economy of nature in the formation of rocks, may be observed on the shores of Lake Ontario. Standing by the water's edge you will see the larger pebbles coated with innumerable parasitic pebbles, varying from a pin's head to particles as large as chestnuts. These are joined together by a cement of lime held in solution by the water. As the small pebbles are rolled up on the shore, the limy concretion attaches them to the larger stones; and when the lake is ruffled by storms, the excited waves roll in larger pebbles, and they are thus intermixed with the sand and cement, and all united together. The result of this operation is the formation of beds of *conglomerate rock*; and on casting your eye up to the ancient terraces you perceive beds of such rocks, varying from ten to one hundred feet in thickness. You thus perceive not only the gradual withdrawal of the water from the shores of the lake, but also the whole mechanical process of the *origin* of its rocky strata. In time all these lakes will dwindle into mere rivers, and the adjacent flats will become teeming prairies like those of Illinois.

Next above the Silurians, we have another group, scarcely less extensive, either in bulk or geographical distribution. In England it is called the *Devonian* system, after a county in which it furnishes the prevailing rocks. It is, however, more generally and familiarly known to the public through two of its most important representatives—the *Old Red Sandstone* and the *Carboniferous* or *mountain limestone*. It is almost useless to mention special localities, for they are strewn all over the world in one form or another. They occur in great abundance on the western slope of the Ural Mountains, and far interior from the eastern shores of the Baltic Sea, in Russia. They encircle and underlie all the coal fields of England, Scotland, Wales, Germany, France, and indeed nearly every coal basin on continental Europe. The old red sandstone of Caithness and Cromarty, in Scotland, has been clothed with geological interest, from the explorations and poetical descriptions of the late Hugh Miller. In some respects he was a geologist in the fullest sense of the term; in others he was quite the reverse. He always spoke of the rocks, however, with the enthusiasm of an investigator, and the familiarity of one who loved them. The old red sandstone, wherever it occurs throughout the civilized world, will form an enduring monument to his memory. I shall probably have occasion, in the subsequent pages of this book, to combat some of his latest geologico-theological propositions, inasmuch as they are in conflict with divine revelation; but I may say here, in advance, that I entertain the highest regard for his memory, and whatever remarks I shall address to his writings, must be understood as applying to the school of geologists still living, of which he was an ornament and a literary expounder.

In the United States, the rocks of this group constitute the rim or boundaries of coal basins—hence the universal use of the term "basin" as applied to coal. The formation of these basins or lakes, did not differ materially from those of the metamorphic and Silurian eras, nor yet from those now existing. While, however, the others were all marine, and directly appertained to the primitive seas, those of the Devonian period were sometimes inland, (like those of lakes Superior, Huron, or Erie,) and sometimes marine, like the great basin of the Gulf of Mexico; and again, partaking of the alternate nature of both, like the estuaries formed by great rivers where their waters are emptied into the ocean. During the period now in question, by far the largest portions of Europe, Africa, and Asia were still under the dominion of the sea. In Europe, portions of England, France, Prussia, Austria, and Turkey, were still submerged; but the greater part of Russia proper had been redeemed by the Silurians, while the whole of Lapland, Finland, Norway, and Sweden, had appeared during the metamorphic era. In Africa, the whole country from the great desert of Sahara south to the Cape of Good Hope, had already emerged from the sea; but the great desert itself remained a desert of water until after the Tertiary formation. So in Asia, nearly the whole surface of Tartary, and the vast region in Siberia, east of the Ural Mountains, and bordering the Arctic Ocean, were unreclaimed until the dawn of the modern geological era.

Although America is popularly termed the *new world*, geology proves it to be of much greater antiquity than either of the other geographical divisions of the earth. By far the largest portion of it had appeared during the metamorphic era; and at that time it comprised at least *twice as* much surface as Europe, Asia, and Africa combined. All that portion now comprising the coast line of the Pacific, in South America—or more properly, all that belt of country, comprising the mountain system of the Andes, and traversing nearly the whole length of South America, parallel with, and from sixty to one hundred and thirty miles from the coast line,—had been elevated during the primary eras. From Terra del Fuego, north through Chili, Peru, Ecuador, New Grenada, and Venezuela, and thence over the narrow strait separating the two oceans, to the range of Rocky Mountains (a mere continuation of the Andes, their geological structure being similar;) thence northwest through California, Oregon, and Washington, to Russian America, where the formation greatly expands, and finally sinks into the ocean—all this vast region had been elevated during the primary epochs of geology. And it may be added that volcanic action is still occasionally aroused, not only in the Andes, where the loftiest volcanic peaks in the world are to be found, but all along the coast, and even amid the wide expanse of waters; and such action is still gradually but certainly making new acquisitions of territory. Indeed, a large extent of country, east of the Andes, passing through Patagonia, Buenos Ayres, Bolivia, and the western part of Brazil, has been reclaimed from the ocean since the Tertiary period. The country drained by the Amazon and its tributaries, is generally alluvial, and has been converted into dry land within the modern era of geology—the whole of that vast domain—larger than the States of New York, Pennsylvania, Virginia, and Ohio, combined—having previously been occupied by lakes emptying into the Atlantic Ocean. The largest portion of Brazil, however—that lying along the Atlantic, and for the most part drained by the river La Plata—is of primitive origin, as are also the larger portions of Guiana and Venezuela;—but, with these exceptions, nearly the whole remaining surface was submerged until within a period comparatively recent. The vast expanse of territory north of lake Superior, and indeed all around Hudson's Bay, is likewise anterior to the Devonian era; and with the exception of the prairie regions of the West, already mentioned, it may be safely assumed that most of the states comprising our confederate cluster, were still covered over by the sea, or with great interior lakes emptying into, and liable to be invaded by the sea, during the deposition of the Devonian rocks.

Now during the particular geological era we are considering, there was a small basin running from the Rio Grand river, in the southern part of Texas, northeast to the Red river. This basin at one time received the waters of both these streams, besides those of the upper Colorado and Brazos, which now pass through it to the Gulf of Mexico. Coal is found at both ends of it; and when it was finally elevated, its waters were discharged in the rivers now flowing through and around it. Another basin is traversed by the Arkansas river, further east, and lying in the state thus named. Another, larger and better defined than either, occurs in Iowa and Missouri, into which the Missouri river originally emptied, and through the western boundary of which it now passes. A fourth basin, still larger and better developed, is in the state of Illinois, the capital of the state being very nearly in the centre. This great basin received the waters of the Missouri, those of the upper Mississippi, of the river Illinois, of the Ohio, the Tennessee, and many other smaller ones. It was a great basin, and considerably larger than our existing lake Superior. Further north, bounded on the west by the entire length of lake Michigan, and on the northeast by lake Huron, is a fifth basin, not so large nor so well defined as the last mentioned; but indicating, from its proximity to the existing lakes, their former extension over the Devonian rocks that are now intermediate between them. The last, and by far the largest basin, is that comprising the Alleghany Mountains, beginning in the southwest on the head waters of the Mobile river, in the state of Alabama, and in the northwest on the Tennessee river (very nearly adjoining the Illinois basin,) and thence running northeast with the Alleghany Mountains, through Kentucky, Virginia, Maryland, and Pennsylvania, and finally terminating in a broad expanse in the state of New York. A sixth, but very small basin, occurs in Rhode Island, connected with the sea by a broad inlet; and another, or rather several of them, in the British Provinces of Nova Scotia, New Brunswick, and Prince Edward's Island. These latter basins literally sink into the ocean and reappear in the island mentioned. All the basins thus enumerated contain coal.

The same means by which the waters of lake Ontario are daily forming conglomerate rocks, and strewing them in layers on its shores, were in operation, though on a much larger scale, during this Devonian period. The great bulk of the rocks comprise conglomerates and sandstones, interstratified with layers of slate and shale, and concretionary and blue crystalline limestones. In many instances the limestones are wanting; or where they occur the conglomerates are omitted, or appear in thin layers. No limestone (or but a very small and impure seam) exists in the Devonian rocks underlying the anthracite coal basins; and very little is found in those of the bituminous, east of Pittsburg. This singular fact may be accounted for in this way: The Alleghany basin comprised many subordinate ones, communicating with each other, like those of Superior, Huron, Michigan, and Erie. Lying adjacent to the main chain of basins, were several outliers, like that of lake Michigan.. These outlying basins are those of the anthracite coal regions; another is the semi-anthracite of Broad Top, further west; and another is that of Cumberland, in Maryland. These are all on the eastern slope of the Alleghany, while the *main basins* now occupy the *summits* of that mountain or its western slope. The northeastern terminus of the Devonian rocks was originally surrounded by those of the igneous formation. Their debris was drained into these lakes, the waters of which held silex in suspension. The fragments of rock, as they were moved about by the water, became rounded and angular, and were finally deposited in layers, and cemented together by the silicious secretion of the water. The vast beds of conglomerate rock and coarse sandstone, underlying the anthracite coal, were formed in this manner; and the whole process is exactly similar to that now daily illustrated on the shores of lake Ontario. While the larger pebbles were thus converted into rocks, the smaller ones were hurried on from one basin to another by the current of the water; and the fine sand went still further. This gradation from a coarse to a finer conglomerate, and thence to ordinary sandstone, is beautifully illustrated in the coal measures. The thickness of the rocks also diminishes with the extent the debris was transported. The conglomerate of the anthracite basins are much thicker than the same strata elsewhere, while toward the southwestern termination of the Alleghany basin, they scarcely occur at all, or when they do, it is in the form of sandstone. The conglomerate having been mainly retained by the upper basins, and especially by those of the anthracites, where it originated, its deficiency in those below, was supplied by layers of limestone, which increase in thickness, southwestward from Pittsburg, with the decrease of the silicious rocks. There was thus a reciprocal movement between the two ends of the great basins—one supplying silicious, and the other calcareous matter to the water. The limestones embrace two kinds.; that called the concretionary was abstracted from the disintegration of the adjacent Silurian rocks, which formed the rims of all the basins below Pittsburg;—while the blue crystalline was derived partly from the influx of the sea into the lower basins, and partly from the calcareous belt washed down from those above. These limestones are remarkable for their cavernous structure, and their mineral ores. The celebrated Mammoth Cave of Kentucky, and (I believe) those of Weyer and Madison, in Virginia, are in rocks of this group. All of them abound in stalactites of the most beautiful and picturesque form. In the vicinity of Galena, in Illinois, these rocks contain lead; and in other places, besides that mineral, they afford copper, zinc, manganese, and iron.

It may be observed, that while the anthracite basins were at the head of the others, they were also the *deepest*, and hence required a great deal more earthy material to fill them up. Lake Superior is likewise the largest and deepest of all those below it—while St. Clair and Erie are the smallest and the shallowest. Superior is 1000, Huron 900, Ontario 600, and Erie but 20 feet deep. In time, Erie will be so far filled up that its present bottom will become like the St. Clair flats, and its main current will ultimately degenerate into a mere river, similar to that of the Detroit, the St. Mary's, or the St. Lawrence. It must be borne in mind, that these Devonian lakes drained the surrounding country, in the same manner as existing lakes and rivers, and therefore it is highly probable that, before the elevation of the Alleghany Mountains, and while the sea was still undulating over the country now comprising Alabama, Mississippi, Louisiana, and part of Tennessee, Arkansas, Texas, and nearly the whole of Nebraska, Kansas, and the Indian reservations, the whole system of *drainage was in that direction*, that is, west and southwestward, and all the adjacent country, including the coal basins of Illinois, Missouri, and the lower Alleghany, were, from time to time, invaded by the sea. In fact, the whole interior region between Canada on the north, the New England States on the east; Georgia and the Carolinas on the south, and the Rocky Mountains on the west, was a vast shallow marine gulf, scarcely inferior to that of Mexico, receiving the drainage of the metamorphic country, previously elevated in the north, east, and west. The Silurians had afterward converted much of this great basin into land; but I beg the reader to understand that, by far the largest portion of it, during the Devonian period we are now describing, *was still under water*, while the flat marshy land itself, was constantly liable to inundation from oceanic tides, storms, and crust oscillations. This fact sufficiently understood, the reader will readily comprehend the phenomena to which I will presently invite his attention—phenomena, the solution of which, constitute the most difficult problems in theoretical Geology.

After the Devonian basins began to fill up, and their broad margins had already been converted into marshy flats, precisely similar to those of St. Clair, vegetation flourished immediately in the most extraordinary luxuriance—the resinous juices of which, by a process of fermentation and combustion hereafter to be described, were subsequently converted into layers of mineral coal. But before I enter upon a description of the origin of coal, it is expedient here to attend to some other matters as a preliminary step. We shall therefore defer further remarks on this point, to consider briefly the paleontology of the rocks we have thus far encountered, or rather that portion comprising fossil botany; for although the Silurians and Devonian systems afford specimens of Zoophytes and Mollusca, and in some regions fish of a peculiar type, they may more properly be reviewed in the Fifth Day. The reptiles and land animals—foot-prints of which are supposed to have been found as low down as the Silurians—will also receive attention; but in the mean time, we may premise that vegetation constitutes the prevailing characteristic of the era under consideration, and that whatever animals existed were of a low and humble type, and confined to the sea altogether. The theory of land animals, of rain-drops, sun cracks, and other visions of the geologists, as referred to the Devonian rocks, we shall show to be in direct conflict with the Bible, and to have not the shadow of foundation in fact. It will then be demonstrated, what few geologists have yet conceded, that vegetation necessarily *preceded* animal life on the earth, agreeably to the Mosaic requirement.

Now, the vegetation of the metamorphic rocks must have consisted mainly of flowerless grasses, perhaps not dissimilar to that which flourishes spontaneously over the St. Clair flats, and along the marshy bottoms of rivers and oceans. The great heat and moisture of that period, must have added very materially to their growth; while the subsequent metamorphism of the inclosing rocks, converted the grass into a species of impure coal, and thence into earthy plumbago. It is because of the heat to which all these rocks were exposed, that all traces of vegetable structure have been obliterated; but the coal and plumbago themselves evince an unmistakable vegetable origin.

The oldest fossiliferous rock yet discovered in the United States (or in Europe), is supposed to be what is called, in New York, the Potsdam sandstone, because of its occurrence at a town of that name. It extends from that state to Michigan and Wisconsin; and though its lithological nature varies considerably, it is believed to be well identified. It belongs to the lower Silurian group. While it contains a few marine shells, it also affords specimens of fossil plants—(*Scolithus linearis*) and occasional fragments of anthracite. Finding the flora and the fauna of the ancient earth thus associated in rocks of the *same age*, we may safely assume, for the present, the anteriority of the one over the other. But it may be objected that the grasses, the remains or fossil impressions of which are thus found, are algæ or *aqueous plants*, and do not therefore constitute the "dry land" vegetation of Moses. It so happens that we can afford to dispense with all such marine plants, and present something more tangible and formidable.

In casting our eye over the innumerable basins of coal, lignite, asphalt, bitumen, pitch, and various other combustible substances imbedded in the rocks of the earth, nothing could more astonish us, than their seeming *identity and similarity of origin*, under circumstances of extreme lithological diversity. While the combustibles themselves all point to a common *vegetable* source, they yet exhibit the most singular and variable diversity in their geological positions, in their degrees of mineralization, their density, purity, and inflammable properties, as well as in the local circumstances attending their deposition, and their geographical distribution.

There is hardly a state or kingdom on the face of the earth that is not provided with these substances, in one form or another; and the reason of this universal distribution may be found in the fact, not generally recognized or considered, that coal, with its characteristic deviations of quality, has been produced in every *successive formation from the metamorphic rocks to those of the Tertiary*

—extending even to the present time, and is, without doubt, still undergoing the slow processes of conversion from an immature state to that reserved for it hereafter in the undeviating economy of Nature. It is not to be supposed that those who come after us are to be left without fuel! Their harvest is maturing, while ours is being *consumed.*

The "dry land" of the metamorphic and Silurian eras having appeared on the morning of the Third Day, the earth began at once to bring forward its vegetation. This consisted, as I before remarked, of flowerless plants, denominated *Agamiæ* (or concealed marriage). And it is a singular coincidence, which should not escape observation, that no means of fructification have been discovered in this species—hence the name. Now, while Moses speaks of the "*seeds* of the *herb* and the *fruit* as being within themselves in the earth," he says nothing of the *reproductive organs* of the *Agamiæ*, or grasses, but allows us to infer their *spontaneous* growth from the dry land itself. And such would appear to be the fact. The earth must have contained *within itself* the germ that gave vitality to these grasses, leaving the "herbs and fruits" to reproduce themselves by means of fructiferous seeds. Brongniart says of the *Agamiæ*, that it is a term which only expresses our ignorance—but that the class comprises the different families confounded under the names of *algæ, fungi,* and *lichens.* They may be described as forming cellular tissue, or interlacing tubular filaments, without vessels properly so called; they never present true leaves, and their organs of reproduction consist only of very fine seedlings, which appear to develop themselves without fecundation, and are immediately inclosed in membranous conceptacles, analogous to the filaments of that tissue which composes the whole of the plant. The only fossil plants of this class known, are some *con-*

fervœ (slender weeds), and several *algœ* (water-grass and sea-weed). These weeds or grasses are plentifully distributed in the shales of the old red sandstone. They are of rarer occurrence in the Silurian strata, and have never been found in those of the metamorphic. The extreme delicacy of their structure, however, will readily account for their omission in these rocks. It was only under circumstances of repose and quiet, such as generally marked the deposition of the soft shales and mud of the Devonian strata, and those underlying the coal seams, that *their preservation* in a fossil state could be secured. I have several specimens of both these grasses in my collection; and I have seen thousands of them, all tangled up, upon slabs of slate around the coal mines. Their general structure, I repeat, is very similar to the rank marshy grasses that flourish on the flats of St. Clair.

In Rhode Island and Massachusetts there is a coal basin of considerable extent, which occupies a geological position among the upper strata of the metamorphic group. The coal, no doubt originally very impure, has been completely destroyed in consequence of the heat to which the inclosing rocks were exposed, and the contortions and twisting which they suffered in their elevation.

" In the course of two miles of the cliffs of the east coast, the conglomerate beds are six times thrown up, and as often descend below the tide level. Then occur a numerous suite of twisted and contorted schists, of gray laminated slates, whose surfaces singularly resemble the grain of bird's-eye maple; and again, another series of green, talcose, contorted schists, crowded with crystals of iron pyrites; crossed in every direction by innumerable veins of white quartz, and succeeded by compacter beds, which almost possess the qualities of sandstone. Perpendicular upthrows and heaves, and again the reverse movements, divide the whole series into large and separate sections, rising above or sinking below water level. The inclination of the respective masses is continually changing. To the rocks we have enumerated succeed a *melange* of metamorphic slates, of gray fissile beds, of conglomerate, quartz veins, and black shales; of veins and filons of asbestos, and of talcose laminated strata; undulating, fractured, contorted, inverted—in short, disposed with such absence of order as to defy the pen and pencil of the geologist to delineate. * * There are many features here that have no parallel in our ordinary Secondary (Paleozoic?) coal fields. Among these are the *vast assemblage of talcose waving slates;* the *veins and seams of asbestos*, abundant even *among the coal shales*, and occasionally penetrating the *anthracite coal itself*; the quartz veins also in the coal; the unusual appearance of vegetable remains on these greenish-gray, schistose laminæ; the traversing veins of white crystalline quartz, and the *plumbaginous nature of nearly all the out-crops of coal.* * * There are three coal seams proved on the western side, occurring at a distance of ninety feet from each other, and dipping at an angle of 38° to the centre. Toward their out-crops, all the strata evince the effects of great pressure and squeezing; producing corresponding irregularities in the thickness of the coal beds."*

This formation extends into the neighboring state of Massachusetts. Its plumbaginous character throughout, is not devoid of interest.† At Wrentham, in Massachusetts, are several seams of highly plumbaginous coal. At Mansfield, Dr. Jackson mentions a bed of coal which "*was found to have been altered*, and was like graphite or plumbago." At some recent openings in Rhode Island,

---

* R. C. Taylor.—*Statistics of Coal.*
† Geological Survey of Massachusetts.

the coal has been observed to pass either wholly or partially into graphite. It is remarkably light, spongy, and cellular (showing the effects of heat), and forms an article of sale, under the name of *British Lustre*. Asbestos occurs abundantly, running through the slates which adjoin the coal or graphite bed.

Notwithstanding the overwhelming indications of the metamorphic origin of this coal, some geologists of distinction are disposed to rank it with those lying upon the Devonian rocks. There are, however, some exceptions; but how any one of ordinary practical geological acumen can assign it a position so high up, with all the rocks of the previous group in and around it, is a mystery to me. The whole formation is surrounded by granite, and the veins of coal themselves are traversed by asbestos, talcose schists, and other true metamorphic rocks; while the changed condition of the coal indicates the *heat* to which it was exposed.

Nor is this a solitary example of metamorphic coal. There are many such formations in different quarters of the earth; and were the coal which they afford of any commercial value—were it not converted, as it often is, into other substances, by the heat which contorted, twisted, and uptilted the strata, there is little doubt but that the necessities or the cupidity of man would long since have revealed and explored many other regions that now slumber in the original obscurity of their primitive basins.

Some of the anthracite basins of France, Ireland, and Sweden, are in rocks analogous in age and character, to those of Rhode Island. Throughout the greater part of Scandinavia, comprising Sweden, Norway, Lapland, and portions of Finland, the metamorphic and igneous rocks abound. In the midst of them occur basins of anthracite, and in some cases, the coal is found lying on the gneiss rock, with the characteristic metamorphic slates above.

The coal, of course, is often changed into graphite, like that of Rhode Island; while fragments are not unfrequently disseminated in the adjacent slates.

But while the coal is thus found in the metamorphic rocks, we must not overlook the important fact that these are always in their *true geological position*—that is, in immediate proximity to the granite and igneous rocks. If, indeed, they occurred in the regular Devonian or the Silurian group their metamorphic character would of course be seriously compromised;—but it so happens that the containing rocks of the coal rest on the preceding igneous and granite rocks, and there is thus established a regular geological *order*. This is particularly the case in Scandinavia, where it has before been remarked, that the igneous rocks largely predominate. It is scarcely less so in Rhode Island and Massachusetts—the adjacent state of New Hampshire, being celebrated as the "old Granite State." Indeed, the entire surface of New England, with the exception of a portion of Connecticut, is covered by the primitive rocks; and it is over these that the coal basin of Rhode Island occurs in regular superimposed order. The same primitive rocks extend parallel with the Atlantic coast, from Maine to Georgia; and both anthracite and black lead are found in them, at different places. The coal basin near Richmond, in Virginia, although resting on granite, is a recent deposit. A series of volcanic dikes, during the era of the New Red Sandstone, metamorphosed a small upper seam of the coal in this basin, and converted it into coke. The great bulk of the coal below, however, was unaffected, in consequence of the heat having been confined to the surface alone.

Inasmuch as the formation of coal from the vegetation of the Paleozoic period, or Third Day, constitutes its distinguishing feature, it may be considered advisable to dwell somewhat minutely on its interesting phenomena—especially in view of the fact, that it forms debatable ground in reconciling the accuracy of the Mosaic cosmogony, with existing geological facts and theories.

The whole number of fossil species of vegetation thus far found in the strata of the earth, is estimated at about two thousand, of which more than five hundred belong to the coal measures. The very extensive mining operations, in the coal basins of Europe and America, are making constant additions to the fossils previously known. In the anthracite measures of Pennsylvania, at least two hundred specimens, previously unknown, have been discovered within the last six or eight years. In a collection of seven hundred specimens—(but many of them duplicates, and referable to the same species,) I have some fifty or more which appear to be new, and undescribed. In time, by the additions thus being made, the number of distinct species appertaining to the coal strata, must be greatly increased; and instead of five hundred, there will probably be double that number.

The fossil plants, like those of living species, are variously classified by Botanists, and there is consequently a good deal of complication and confusion in dealing with them. This appears to be one of the necessary concomitants of all the branches of Natural Science, and is perhaps the only reason why they are so much neglected by the popular taste in favor of other and less practical studies. In the sixteenth century, the celebrated Conrad Gessner, of Germany, proposed a method of botanical classification, founded on the nature of the flower and fruit, and the relation which different species occupied to the same genera. In other words, he traced species into genera, and by this means was enabled to describe, with more intelligence than had ever been done before, all the plants known at that time. Nearly a century afterward, an eminent French botanist, named Tournefort, Professor of Botany in the Garden of Plants of Paris, wrote a work, in which he described over 10,000 species of plants, resolvable into 700 genera. Among these were several thousand species entirely unknown to Botany before, having been collected by Tournefort during extensive travels on the continent of Europe. In this work, a new and more precise system of classification was adopted, which gave to Botany the rank of a distinctive science. In the year 1734, the celebrated Linnæus appeared, and he relieved the infant science of much of the confusion of nomenclature that existed up to that time. The systems of Linnæus and Tournefort were both founded on the same, or a very similar basis—Tournefort adopting the corolla, and Linnæus the stamina, for separating the leading divisions of the classes of plants. By this arrangement, plants having one stamen were ranked under the class *Monandria;* plants having two stamens were classed with the *Diandria;* those with three stamens belonged to the third class, or *Triandria;* those with four stamens to the fourth class, or *Tetrandria*, and so on. The name of the class is thus generally expressive of the position the plants occupy in the scale—though there is a little obscurity in some of them. The arrangement was simple, and for that reason popular; but, with the increase of number, and the complication of structure of the plants themselves, it finally proved defective. It furnished little or no information regarding the plants thus classified, beyond the *name* of the class to which they belonged. To find out their peculiar structure, organization, and properties, other means had to be resorted to; so that, while the systems were both simple and beautiful, they were yet of little practical value in the identification of species.

Under these circumstances the system of Jussieu (embracing two or three distinguished botanists of that name), called the method of *natural varieties*, is that now most generally adopted by botanists. It differs altogether from those of Linnæus and Tournefort;—the divisions not being founded on a single organ, but on a combination of features characteristic of the plant or family. Agreeably to this arrangement, plants are separated into *two* great divisions, the first consisting (as previously mentioned) of such as are composed entirely of cellular tissue, are destitute of vessels, and whose embryo or germ has no cotyledons or seed leaves, whence they are termed acotyledonous. They are also named *cryptogamic*, from the obscurity of their fructification. The other division is not only more numerous, but comprises plants of a higher and more complicated structure. Being furnished with cellular tissue and tubular vessels, and the embryo having one or more cotyledons, or seed leaves, they are called *vascular* or *cotyledonous*, and are sub-divided into *dicotyledonous* or exogenous, and *monocotyledonous*, or endogenous classes. The first class of the *cryptogamia* comprise the families of *confervæ* and *algæ*, which have hitherto been referred to. The next class, called *cellular cryptogamia*, comprise the extensive family of mosses and liverworts. The third class, *vascular cryptogamiæ*, includes the families of *equisetaceæ*, or horse tails; the *ferns*, very numerous; the *marsailliaceæ*, or pepperworts; the *characeæ*, or charas, and the *lycopodiaceæ*, or club-mosses. A fourth class, called *phanerogamæ gymnosperms*, comprises the families of *cycadeæ* and *coniferæ*, or fir tribes. The fifth class constitutes the *monocotyledonous phanerogamiæ*, and includes the families of *naiades*, of *palms*, of *liliaceæ* or lilies, and of *canneæ*, or canes. The sixth and last class, denominated *dicotyledonous phanerogamiaæ*, embrace the families of *amentaceaæ*, or the birch tribe, the *juglandeæ*, or walnut, the *acerineæ*, or sycamore, and the *nympheaceæ* or the water-lily tribe. These families, it may be observed, afford an almost innumerable variety of individual types or species.

Of the fossil plants comprised under the class *cryptogamiæ*, and the tribe of *equisetaceæ*, perhaps the most numerous in the coal, are those of the *Calamites*. They vary in size from small reeds, not more than the eighth of an inch thick, to trunks two or three feet in diameter. The smaller impressions leave a delicately-polished surface on the slates, and are often strewn over each other like tangled ribbons. Their general appearance is somewhat similar to the stems of Indian corn, except that they are more conspicuously furrowed, but like corn they have regular joints where the leaves were attached, which vary only with the age or development of the tree. The leaves were also narrow and verticillate, somewhat in the manner of corn; but they are seldom attached to the stem of the fossil. From the fact that these beautiful fossils are nearly always surrounded by small seams of coal, and occur in great abundance in nearly every vein, there can be no doubt but that they contributed largely to its formation. I have found fine specimens of calamites in the solid sandstones, over and under the veins of coal. The woody structure is always converted into *shale*, but the outside is coated with a thin seam of pure coal. The calamites also flourished before the deposition of the Devonian coal, and became wholly extinct in the subsequent era of the new red sandstone.

Another fossil, very numerous in the anthracite regions, is that of the *equisetum*. There are several species, the most common of which is termed the *columnare*. They have a close resemblance to the larger calamites, but their columns are not so long and slender, and where they are interrupted by joints, they terminate in two-sided pyramids intersecting each other. The fossils indicate trees of considerable size, and they are often found surrounded by thin sheets of coal. They became extinct after the coal, but reappeared in a greatly diminished form during the Tertiary period, and at this time comprise the small species of plants called the horse-tails, which flourish in our ponds and river flats.

By far the most numerous, the most beautiful, and the best-preserved fossils in the coal measures, are those of the family of ferns. I have more than two hundred specimens in my collection, belonging to this extensive family. Most of the species, however, are only represented by their leaves, or by the slender stems to which they were attached. Notwithstanding their great abundance, I do not believe their vegetation contributed materially, if any thing, to the formation of the coal. I shall presently give my reasons for this opinion, since, from their abundance in the coal measures, nearly all geologists have inferred that they contributed the great bulk of the solid coal.

The fossil ferns comprise the following genera, all of which are determined by the character of their leaves or fronds : *pachypteris*, or thick fern ; *sphenopteris*, or wedge fern ; *cyclopteris*, or circular fern ; *glossopteris*, or tongue-shaped fern ; *neuropteris*, or nerve fern ; *odontopteris*, or tooth fern ; *anomopteris*, signifying secret fern ; *tœniopteris*, or wreath fern ; *pecopteris*, of unknown significance ; *longchopteris*, or spear-shaped ; *schizopteris*, or divided fern ; *otopteris*, resembling the ear ; and *caulopteris*, a stem-like fern.

Nearly all these varieties of the fern have representatives in the coal ; but such is the diversity of structure among them, that a description here would be tedious and unprofitable. Of the pecopteris, there are no less than sixty-two species, well identified, in the coal ; of neuropteris, some forty-three species ; caulopteris, five or six ; and cyclopteris, seven. Most of the others have from one to five species, while a few only are unrepresented. Such, however, appear in the subsequent eras of the new red sandstone, the oolite, and the chalk, and many, if not all of them, survive at the present time. But all the Ferns now existing are very small and dwarf-like, while it is supposed that those of the coal attained the proportions of ordinary young forest trees. This, however, in my opinion, is very much exaggerated.

The next family is that of the *Lycopodiaceæ*, or the Club-mosses. These are well represented in the coal, and comprise a series of very beautiful and interesting fossils. I have twenty-five or more specimens, representing as many different species. The most numerous are those of the Lepidodendrons, or scaly tree—so called from the imbricated structure of the bark, or their resemblance to the scales of fish. These trees may indeed be allied, as the botanists allege, to the club-mosses of the present day; but it seems to me that a closer relation exists between them and our existing *yellow pines*. Were the tender shoots of these pines, or those of from one to ten years old, buried in mud, and then subjected to heat and pressure, they would stamp impressions on the baked slate exactly similar to those of certain species of the fossil Lepidodendrons. In both cases, the imbricated scales and scars were produced by the detached leaves, or needles, which originally surrounded the stem in regular order. The interior of the conical scales or lines is sometimes very curious—the leaf-dots hanging like miniature chandeliers by means of little threads or chains. The scars assumed their rhomboidal form in consequence of the bulging out of the leaf-stalk, and the lines that would otherwise have been continuous, straight, and parallel, like the ribs of Sigillaria, are thus forced apart at intervals, and then again united. The variation in the structure of these rhomboidal cavities or scales, is due to the varying dimensions and age of the tree. In the young ones they are small and close together, and the lines describe regular angles; but with the increased thickness of the bark, and the expansion of the tree, they grew larger and larger, and always varied with the species.

The Lycopodites are smaller than the Lepidodendrons, and had their leaves attached to the stem in two opposite rows, leaving obscure parallel scars in the fossil, after it became flattened by pressure. The stem of the Ulodendron is covered with rhomboidal plates, broader than long, in the interior of which, or on the raised surface of the fossil, as the case may be (for there are always two casts to every fossil), are large scars, the whole being very similar to the cones of pine trees. The Lepidostrobus is an ovate or cylindrical cone, composed of imbricated scales, encircling a woody axis, the seeds of which are oblong and solitary. There is some dispute among botanists in regard to these cones. Some twenty specimens in my collection are exactly similar to those of certain species of pine; and, although the classification seems to forbid it, I cannot help believing that all these fossils appertain to the family of resinous pines *existing at the present time*, and that they formed *by far the largest portion of the material of which the coal is composed.* The whole family of *Lycopodiaceæ* disappeared after the deposition of the coal; but it is supposed that they are represented by the club-mosses of the present era—a proposition which, at least, admits of considerable doubt.

It is a singular fact, that the most abundant and the most important and conspicuous fossil in the coal measures is that of which the least is known. This is the family of Sigillaria. Fossil botanists claim forty different species in the coal. I have at least that number in my collection, among which are several new species not yet named or classified. There can be no doubt whatever but that these trees, with those just mentioned, furnished the great bulk of the coal. They grew to enormous dimensions, rivaling the venerable pines of our forests.

Their fossils sometimes occur in the shales over the coal for a continuous length of eighty or a hundred feet; and I have often seen slabs from three to six feet wide, entirely covered with their deep parallel furrows. *I have never yet met a specimen of Sigillaria that was not coated and surrounded with thin sheets of coal.* They were formerly assigned a doubtful position in botanical classification, or a position intermediate between the divisions of Cryptogams and Monocotyledons; but recent investigations seem to justify their association with the Dicotyledons. I can venture no opinion as to the rank to which they are entitled; but I am very certain that all of them *secreted resinous or oily juices*, and that they were so far Dicotyledons as to resemble existing species of the pine. The Sigillaria were conical trees, the bark of which was deeply furrowed and ribbed. These ribs are from an eighth of an inch to two inches apart, and run parallel with each other, lengthwise with the tree. Between the ribs, in the concave furrows, are scars at regular intervals. From the resemblance these scars bear to the stamp of a seal, in sealing-wax, they are called Sigillaria. The furrows vary in width, as well as the space between the scars, in proportion to the dimensions of the tree. The scars are the marks left by the leaves after they had become detached; and this is a *characteristic of all resinous pine trees*—the leaves of one year falling off when those of a new year appear. In this respect, the Lepidodendria, the Sigillaria, and perhaps the Stigmaria, all resemble the pines of the present era. The scars of the Sigillaria vary in different species. Sometimes they are round, or rhomboidal; sometimes there are two, closely attached, and forming a heart; sometimes they are long and slender, or consist of two little dots, in the same scar; sometimes the scars are half-covered by an arching roof, or are surrounded by circular indentations. Again, there are occasionally two circles, one on top of the other, and the leaf dot upon the last. Sometimes the scars are placed in arching cavities, like eyes; while in others, the ribs are omitted, and the dots are surrounded with delicate lines, radiating as from a central nucleus, or waving around them in graceful curves. The fossils are invariably highly sculptured, and constitute by far the most varied, important, and interesting class of the ancient flora. The Sigillaria made their first and last appearance during the coal period. And the fact is not without significance, that the three principal classes of the coal vegetation, viz., the Calamites, the Lepidodendria, and the Sigillaria, are all confined to that era—a mere trace only of the former appearing in the new red sandstone.

Besides the Sigillaria, there is another genus, termed *Volkmannia*, of doubtful affinity. They are leaves, with a striated stem, and articulated; and are usually found in large whorls. They are supposed to be leaves of the calamite, but I think this exceedingly doubtful. Under the name of *Carpolithes*, are included all the fruits of the ancient earth, for which no specific names are provided. These are scarce in the coal, as might be readily inferred from the character of the vegetation. I have two very fine specimens, however, which resemble the chestnuts of our forests. I have frequently noticed an obscure fossil, varying in size from a chestnut to a walnut, but much flattened, which may be a fruit. The surface is always very smooth and shining, and the only feature which distinguishes it from a leaf is its thickness. I have also specimens of a fruit, which is perfectly round, and the interior exhibits a dotted or cellular structure. These are an inch and a quarter in diameter, and resemble the large ink-balls of the oak more than any thing else that I can now think of.

Of plants of the true Monocotyledonous class, there are comparatively few in the coal. Of the palm family, the leaves of the *Flabellaria* occur sparingly. They are petiolate and fan-shaped, contracting and plaited at their base. The *Nœggerathia* are more numerous—five or six species having been found. They are also petiolate and pinnated; leaflets obovate and nearly cuneiform against the edges of the petiole, but toothed toward the apex, with fine diverging veins. *Zeugophyllites* is another genus, of which two species are known in the coal. They are described as petiolated, pinnated; leaflets opposite, oblong, or oval, entire, with a few strongly-marked ribs, confluent at the base and summit, and all of equal thickness. The *Sternbergia* is a slender, naked, and cylindrical stem, terminating in a cone, marked by transverse furrows, but with no articulations. There are three or four species in the coal; but they are of doubtful affinity in botanical arrangement. *Poacites* are all monocotyledonous leaves, with parallel veins, simple and of equal thickness, but not connected by transverse bars. There are several species in the coal. Of fruits, there are two species of *Trigonocarpum*, and two of *Musocarpum*. So far as known, these are the only species in the coal measures, properly belonging to the division of Monocotyledons; but the seplants were very numerously represented during the Tertiary period by the families of Palms, Zostera, and Naides, and by more than one thousand living species at the present time. They flourish best in tropical climates, and comprise the great bulk of the vegetation of those regions.

Intermediate between the Monocotyledon and the Dicotyledon divisions, are a few families of doubtful affinity, but of considerable abundance in the coal. The *Asterophyllites* have stems scarcely tumid at the articulations, but branching; leaves verticillate, linear, and acute, with a single midrib, quite distinct at the base; fruit, a one-seeded ovate, compressed, nucule, bordered by a membranous wing, and emarginate at the apex. Twelve species occur in the coal; and they are *not* confined, as Mr. Lesquereux says, in the Geological Report of Pennsylvania, "to the upper coals." They are found in nearly every vein where other fossils abound, without regard to "high or low coal." *The Annularia* is a family which includes some beautiful fossils, some of which resemble the Asterophyllites. The stem contains numerous clusters of leaves, which radiate around it in the form of a star. The stem is slender and fragile, articulated, and has opposite branches springing out from above the leaves. The leaves are verticillate, flat, usually obtuse, with a single midrib united at the base, of unequal length. There are six or eight species in the coal. The *Phyllothera* have a simple, straight, articulated stem, surrounded at regular distances by a sheath, having long linear leaves, which have no distinct midrib. There is but one species in the coal. *Bechina* have a branched, jointed and articulated stem, deeply and widely furrowed; the leaves are verticillate, very narrow, acute, and ribless. One species only in the coal. These plants may, perhaps, be ranked as Cycadæ,—and if so, they constitute the only representatives of that group in the coal. They diminished in the new red sandstone, and increased during the Oolitic era; they again decreased, and nearly became extinguished in the Cretaceous, but again expanded, greater than before, in the upper Tertiary. They now threw out several lateral branches, which includes the families of Poplars, Willows, Elms, Sycamores, Magnolias, Oaks, Birches, Maples, and numerous other trees of existing forests.

Among the other branches or families of the Dicotyledons, there are but two or three represented in the coal—but they constitute the most important sources of its vegetable material. Of these, the *Euphorbiaceæ* is represented by the *Stigmaria*. This is described by botanists as having a stem originally succulent, and marked externally by roundish tubercles or scars, surrounded by a hollow, and arranged in a direction more or less spiral, having internally a woody axis, which communicates with the tubercles by woody processes. The leaves arising from the tubercles are succulent, entire, and veinless, except in the centre, where there is often some trace of a midrib. There are five or six species in the coal. I may add to this botanical description, that the Stigmaria is almost invariably,—in fact, so far as I have been able to observe, I may say, without qualification, it *is* invariably found in the slates underlaying the veins of coal. Mr. Lesquereux, however, is of a different opinion, and says that he found them somewhat plentifully in the top slates of the mammoth vein at Minersville, and in the roof of the South Salem vein at Pottsville. I must beg leave to say to Mr. Lesquereux, that there *is no* mammoth vein at Minersville, or, rather, that it does not outcrop, and has never been worked there. The nearest white ash coals to Minersville are those of Wolf Creek, at which place the so-called mammoth vein is *divided* into two or three distinct veins. It therefore has no existence at or near Minersville. Although it could not, therefore, have been the *mammoth* vein to which he alludes, it does not follow that he is mistaken as to finding the Stigmaria in the top slates. But when he speaks of finding them in the South Salem vein *at* Pottsville, another doubt arises. The south veins of the red ash coals, in the Sharp Mountain, at Pottsville, are tilted over; and that which was originally the *bottom* slate is now the *top slate*. It is therefore very natural to find the Stigmaria in what *seems to be* the top slate of these veins, but, in reality, they are in the original bottom slate, where they grew; and it so happens that they there *occur in extraordinary abundance*

—they are, in fact, the prevailing fossils on the slope of the Sharp Mountain.*

Some geologists have suggested that the Stigmaria are the roots of the Sigillaria. It may be remarked, in support of this view, that the slate in which the Stigmaria are found, is materially *different* from that inclosing the Sigillaria. The bottom slate of all coal veins is baked mud and clay, and, on exposure to the air, it decomposes into *irregular* lumps. That of the top, and overlying the coal, is invariably *laminated*, and readily splits into *thin sheets and slabs*. Now, the tree may have had its roots in the bottom shale, and after the deposition of the coal, when it fell to the ground, the top slate, in the form of sediment and silt, may have buried the trunk and its branches—leaving the vein of coal between them. I have occasionally noticed, at the coal mines, what *appeared* to be the stumps and roots of the Stigmaria; but I never could discover any traces or tendency to *pass into Sigillaria*. The characters of the two families, considered as fossils, appear to be distinct. The Stigmaria, unlike many other fossils, is often surrounded by leaves, which branch out from the stem a distance of from four to ten inches. These leaves are spear-shaped, soft, and succulent. Some persons have styled them *rootlets;* but, if so, it seems preposterous that they should have served to support such gigantic trees as the Sigillaria. I have several specimens of Stigmaria, deprived of these leaves, which I cut out of *solid sandstone*. It is not possible that they could have *grown there*—or at least not likely. One or two specimens of the stem are *converted* into finely comminuted sandstone. I think it very likely from what I

---

* Mr. L. may, however, have reference to the old Salem Slopes near Pottsville, abandoned many years ago. If so, I have only to say that I have never detected the Stigmaria among the characteristic fossils of that vein.

have seen of it, that this plant was a species of *vine*, which flourished in the soft mud and marshes in immediate proximity to the coal basins, and was thus liable to be buried beneath the veins of coal, where we now find it. This view is supported by the observations of several persons in the mining regions of England, and is only contradicted by the mistaken inferences of theoretical speculators in Geology and Paleontology.

Of Coniferæ, there are several families in the coal. The *Pinites* have axes composed of pith wood in concentric circles, bark, and medullary rays, but with no vessels; walls of the woody fibre reticulated. There are three or four species in the coal, but the wood only is known. The *Auraucaria* have axes composed also of pith wood, in concentric circles, bark, and medullary rays. *Sphenophyllum* have branches deeply furrowed; leaves verticillate and wedge-shaped, with dichotomous veins. There are ten species in the coal. The Coniferous plants flourished to a very great extent in the coal, as well as in the new red sandstone and the oolite. They disappeared during the cretaceous era, but came forward in great abundance in the upper Tertiary. They are now represented by the extensive family of the pines, and flourish all over the world,' in cold as well as warm climates.

We have thus briefly glanced at the leading families of the vegetation which furnished the material of which coal is composed. There are a great number and many varieties of species which it would be tedious and useless to describe here, since they are nearly all comprehended in the classes already specified. Thousands of leaves, stems, fruits, and flowers occur in fragments and matted heaps in the slates that accompany the coal; and although the internal structure and woody fibre have been compressed and superseded, or changed into shale, the *pitch* or oil which permeated the *pores of the plant* has glued them to the slates, and thereby imprinted their external characters as distinctly as an engraving or lithograph on a sheet of paper. The comparative anatomist is often compelled to identify animal species by an isolated tooth or a fragmentary bone. In like manner, the fossil botanist has sometimes to rely upon a scattered leaf or an altered woody structure; yet, so regular and undeviating is Nature in all her works, that the sequel frequently establishes the correctness of scientific generalization, notwithstanding the obscure data on which, in such cases, he is compelled to rely. Nor can we, in view of their bearing upon the past history of the earth, regard these magnificent fossils as the result of accidental circumstances. There are hundreds of coal basins distributed over the surface of the earth, and these fossils invariably occur *in all of them*. They are therefore too wide-spread and universal to be regarded as accidental, either in the manner of their deposit or their preservation. All the works of Nature betray a *design*—an *intellectual plan;* and what we sometimes regard as deviations or apparent contradictions of harmony and uniformity, only expose the feebleness of our faculties of analogy and perception. The hieroglyphics of Egypt, Nineveh, and Babylon, inscribed on the marble panels of their ruined palaces, obelisks, and catacombs, are perhaps the most ancient memorials of the human race that have been transmitted to succeeding ages. And what is the moral lesson which they teach? Simply that the sculptured rocks *record faithfully* the vanity, folly, and ambition of nations and of individuals, and survive the evanescent glory of both! They tell us that the spoils of victory—the dignities of office—the gains of craft;—the buoyancy of youth, and the severity of age;—the fears of the wicked, the pains of the afflicted, the chains of the enslaved—that all the pride, and power, and majesty of long lines of kings, are wholly obliterated in the dust of their bones, leaving not a vestige behind but the polished stones upon which they carved their names—faithful but melancholy sentinels on the frontiers of the Past, to tell marching years the story of their passing away! And such, in some respects, are these wonderful fossils; but *they* bear no such miserable comment. Instead of adding to the posthumous glory of man, they reveal the majesty of the living God! They perpetuate no local or ordinary event; they do not speak of war, and captives, and blood;—but they exhibit pictures of the young earth, when the creative volition of the great Architect was first displayed. They are the picturescripts, the universal hieroglyphics, of Nature, and record the wise forethought and the unsolicited benevolence of the great Jehovah!

While we might naturally expect some diversity of opinion regarding the character of the vegetation the remains of which have been thus preserved in the crust of the earth, there is an equal if not a greater conflict of opinion as to how it was accumulated into separate seams, and thence transformed into mineral coal. The geological theories cannot all be correct; but it may be safely assumed that a certain amount of probability appertains to each, since they are *all* based on the vegetable origin of the coal itself. The theories may be divided into two leading divisions,—the first comprising the *Peat-bog Theory* and the other the *Estuary, or Drift Theory*. There are others intermediate between these, or partaking of some of the features of both, which we shall notice.

The Peat-bog theory contemplates an extensive level marsh, traversed by numerous springs of water, or its permeation by the waters of an adjacent river or ocean. During the coal period, it is inferred that such marshes supported a luxuriant vegetation of the character already mentioned; that successive crops of such vegetation fall down and were buried in the bogs; that after the accumulation of a great stratum of carbonaceous matter, consisting of arborescent plants, leaves, and grasses, the bog was overflowed by the adjacent sea, the waters of which deposited over it layers of mud and silt, which now contain the fossils, and then successive layers of sandstone, (or limestone,) and clay, and shale. The solid trunks and woody fibre thus buried between layers of sand, gravel and mud, which afterward hardened, caused fermentation to ensue in the vegetable material, thereby converting it into a species of lignite; and the continuance of pressure and fermentation, finally resolved it into bituminous coal. After the first seam of coal had been thus deposited, the waters of the sea were withdrawn, and *another* peat bog was again commenced, precisely as before. There being in some coal basins, as many as from sixty to one hundred distinct and separate seams of coal, it is necessary to suppose, on the basis of this theory, that the sea overwhelmed the successive layers of peat in the order in which they were accumulated. The idea of such regular and periodical invasions of the adjacent sea, is rather too stupendous to be seriously entertained. But granting the probability of the thing, is it reasonable to suppose that the vegetation, scattered about in irregular heaps of trees, and stumps, and leaves, should be reduced to a *uniform level* over the bog, and the mud *evenly deposited* over it, without *intermixing* and *ramifying* the pores and layers of the peat? If the peat, before the invasion of the sea, had already assumed the external form of a smooth and compact layer, then the sediment might not so readily have intermixed; but if such a seam had been previously formed, how are we to account for the *splitting of numerous coal veins*, and their *reunion* at irregular intervals? This is a phenomenon well known to practical miners, but seldom contemplated by theoretical geologists. A vein of coal which is at *one point* ten feet thick, at *another* may be but five feet. This may be explained thus; the seam of coal parts in the middle, by the intrusion of a sandstone or other rock, and thus forms two apparently *distinct seams* of coal. The space between them often expands from twenty to eighty yards, and may continue for many miles before they *again converge to each other*. The same division often occurs on sinking downward through a vein—expanding and contracting, as the case may be, in every direction in which it is pursued, and varying from a mere thin slaty wedge, to coarse, well-defined, sedimentary rocks. Now, all this is incompatible with the idea of a *periodical submergence* of the peat-bogs, because the veins thus separated are made to *undulate with the contraction and expansion of the wedges of rock between them*—whereas, it is necessary to premise, on the basis of this hypothesis, that the floor of the bog was always perfectly level. But, whether level or not, if the sea overflowed the bog, it would scatter its sediment equally, and not leave isolated deposits from twenty to one hundred yards high *at one place*, and no sediment whatever at the other places.

The Estuary, or Drift Theory, is mainly founded on the objections arising against the other—that is to say, the utter improbability of the repeated invasions of the sea and adjacent fresh waters to deposit the rocks that alternate with the coal seams. To account more satisfactorily, therefore, for the interposition of these rocks, it is supposed that the vegetation was transported by rivers from the beds where it grew, and deposited in the basins or estuaries formed at their junction with the sea. On this hypothesis, the alternation of marine and fresh-water deposits, between the veins of coal, is easily accounted for; but the liability of the loose and fragmentary material of the vegetation to become intermixed with the mud and sediment held in suspension by the water, and thus to destroy the purity of the coal, is an overwhelming and a fatal objection. The great rafts of logs and leaves, as they were hurried on to the sea, would become coated and surrounded by mud; and, as they gradually sunk to the bottom, they would be still further involved in it. But as trees and stumps have been found in *an erect* position, *under* and immediately *above* veins of coal, it is inferred that they never could have been transported in this manner; and geologists have therefore been gradually abandoning the theory. In Nova Scotia, Professors Darwin and Lyell have detected sixty planes of successive vegetation among the strata of that region; and in some of these planes they found fossil trees, in an erect position, and in others the stumps and roots of Stigmaria. These planes of vegetation conform to the stratification of the inclosing rocks, which lean at an angle of about $30°$ to the horizon. Mr. Hawkshaw, of England, in 1839, described five fossil trees discovered in a cutting on the Manchester and Bolton Railroad. These trees stood erect over a bed of coal eight inches thick. The largest measured five feet in diameter at the base, and was eleven feet high. He conceived it probable that the trees grew where they were found. In a subsequent paper, after having found another fossil tree, standing *over* the same coal seam, Mr. Hawkshaw observes: "If the coal be considered as the debris of a forest, it is difficult to account for not finding more trunks of trees than have been discovered in our coal basins; and it is only, perhaps, by allowing the original of our coal seams to have been a combination of vegetable matter, analogous to peat, that the difficulty can be solved." After Mr. Hawkshaw's first communication, Mr. Beaumont, in a paper read to the Geological Society of London, upon the subject of the same trees, states several objections to the *Drift Theory of Coal,* and conceives that the vegetation grew *where it is found.* He thinks that it must have flourished on swampy islands, and consisted principally of ferns, calamites, coniferous, and other trees, which operated, through their decay and regeneration, to form peat-bogs; and that the islands, by subsiding, were covered over with drifted sand, clay, and shells, till they again became dry land, and supported another vegetation; and this process, he supposes, was repeated as often as there are coal seams! Dr. Buckland, in commenting on this hypothesis, observes that "in denying altogether the presence of drifted plants, the opinion of the author seems erroneous; universal negative propositions are in all cases dangerous, and more especially so in Geology. That some of the trees which are found erect in the coal formation have not been drifted, is, he thinks, established on sufficient evidence; but there is equal evidence to show that other trees and leaves innumerable, which pervade the strata that alternate with the coal, have been removed by water to considerable distances from the spots on which they grew. Proofs are daily increasing in favor of both opinions, namely, that some of the vegetables which form our beds of coal grew on the identical banks of sand and silt and mud, which, being now indurated to stone and shale, form the strata that accompany the coal; whilst other portions of those plants have been drifted to various distances from the swamps, savannas, and forests that gave them birth; particularly those that are dispersed through the sandstones, or mixed with fishes in the shale beds." In these views of Dr. Buckland, Sir Charles Lyell would seem to concur, as, in quoting the above passage in his *Elements,* he says that "it can no longer be doubted that both these opinions are true, if we confine our attention to particular places." Another paper, on the subject of the same fossil trees found on the Manchester and Bolton Railway, was read cotemporaneously with the last communication of Mr. Hawkshaw. The author, Mr. Bowman, is of opinion "that the theory of the subsidence of the land during the carboniferous era, receives much support from the phenomena presented by these fossil trees." He does not deny that plants may have been carried into the water from neighboring lands; but he conceives it difficult to understand whence the vast masses of vegetables necessary to form thick seams of coal could have been derived, if drifted, and how they could have been sunk to the bottom without being intermixed with the earthy sediment which was slowly deposited upon them. Another difficulty of the Drift Theory, he says, "is the uniformity of the distribution of the vegetable matter throughout such great areas as those occupied by the seams of coal." Mr. Bowman believes that the coal has been formed from plants which grew on the areas now occupied by the seams; that each successive race of vegetation was gradually submerged beneath the level of the water, and covered up by sediment, which accumulated till it formed another dry surface for the growth of another series of trees and plants, and that the submergences and accumulations took place as many times as there are seams of coal. In reviewing the foregoing facts and opinions, Dr. Buckland conceives that a luxuriant growth of marsh plants, as *Calamites, Lepidodendra, Sigillaria,* etc., may have formed a superstratum of coal, resting on a superstratum of the same, composed exclusively of remains of *Stigmaria;* and, in accounting for the marine and fresh-water strata alternating with the coal beds, he appeals to the intermitting and alternate processes of subsidence, drift, and vegetable growth.

Prof. Rogers, in introducing his own hypothesis, says of the foregoing, "that they do not attempt to account for some of the most remarkable relationships among the strata, such as the extraordinary frequency, beneath the coal beds, of the Stigmaria clay, the very general occurrence of laminated slates immediately above the seams and the singular contrast which these underlying and overlying rocks present, in the variety and condition of the imbedded vegetable remains. Nor do they explain satisfactorily why the coal itself contains so few traces of the forest trees of the period, either in a prostrate or erect position; while thin broken stems are mingled with the fragmentary parts of the Stigmaria, in more or less abundance, in all the coarser rocks." This is very true. Any theory which contemplates *merely the coal itself*, must necessarily be unsatisfactory, incomplete, and defective; for it so happens that the coal is nearly always associated with, and frequently graduates into, the adjacent slates, which are also of equal and sometimes greater thickness. But besides the coal, we must also account for the numerous deposits of soft, unctuous, and shelly coal which occur so frequently in the vein, and which sometimes extend several hundred yards in length, entirely displacing and superseding the pure coal. What is this substance? That it is not *pure coal* is sufficiently plain; that it is not *slate* is equally plain; that it is of *vegetable* origin no one will deny. I shall not anticipate here the remarks which I propose to make hereafter; but I may merely suggest that this *unctuous shale is the true peat of the peat-bogs;* but that the *coal itself is quite another substance.*

Prof. H. D. Rogers, in his voluminous Report on the Geology of Pennsylvania, promulgates a theory which comprehends some of the leading features of drift, estuary, and peat bogs; and though portions of it are original with himself, and extremely curious, the whole may be regarded as a fair exposition of the views now generally held by geologists on this subject.

"Let us imagine," says Mr. Rogers, "the areas now

covered by the coal formation, to have possessed a physical geography, in which the principal feature was the existence of extensive flats, bordering a continent, and forming the shores of an ocean, or some vast bay, outside of which was a wide expanse of shallow but open sea. Let us now suppose that the whole period of the coal measures was characterized by a *general* slow subsidence of these coasts, on which we conceive that the vegetation of the coal grew;—that this vertical depression was, however, interrupted by pauses and gradual upward movements of less frequency and duration, and that these nearly statical conditions of the land alternated with great paroxysmal displacements of the coal, caused by those mighty pulsations of the crust which we will call earthquakes. Let us further conceive, that during the periods of gentle depression, or almost absolute rest, the low coast was fringed by great marshy tracts or peat bogs, derived from and supporting a luxuriant growth of *stigmariæ* (sigillaria and lepidodendria,) and that along the landward margin, and in the dryer places of these extensive sea morasses, grew the *coniferæ, tree-ferns, lycopodiæceæ*, and other arborescent plants, whose remains are so profusely scattered throughout the coarser strata, between the coal seams. In this condition of things, the constant decomposition and growth of the meadows of stigmariæ, would produce a very uniform extended stratum of pulpy, but minutely-laminated pure peat. This would receive occasional contributions from the droppings by the scattered trees of their leaves, fronds, and smaller portions, which, being driven by winds, or floated on the high tides, would lodge among the stigmariæ in the marshes, and slightly augment the deposit. These leaves and fronds, covered over more or less rapidly by the growing stigmariæ, or varying in their tendency to decay, according to the abundance or deficiency of their juices, would, when thus inclosed, pass at once either to the pulpy state, and ultimately form coal, or by the more rapid extrication of their volatile portions, remain as mineral charcoal, and preserve their vegetable fibrous structure. In both of these conditions of coal and charcoal, we often find the smaller parts of plants retaining their organized forms among the laminæ of the purest coal seams. Upon this view of a gradual accumulation from the *stigmariæ*, assisted by the deciduous parts of the trees, it is altogether unnecessary to suppose that any portions, even the upper layers of the coal beds, derived their vegetable matter from the stems of the trees themselves. Thus the absence of trunks and roots from the coal, is reconciled with the occasional occurrence of their fronds and lighter extremities. Upon no other hypothesis respecting the physical condition of the region which produced the coal vegetation than that here imagined, can I explain the singular infrequency of fossil trunks standing on or in the coal, or account for their occasional occurrence, as in the instances described by Hawkshaw and Bowman. No other supposition seems to furnish a cause for the absence or all traces in the coal itself of the larger parts of arborescent plants, and for their equally remarkable abundance in a broken and dispersed state in the overlying strata."*

Assuming such to have been the condition of the surface during the tranquil periods of accumulation of each coal bed, Mr. Rogers conceives the alternating strata to have been produced in the following extraordinary manner : " Let us suppose an earthquake, possessing the characteristic undulatory movement of the crust, in which I believe all earthquakes essentially to consist, suddenly to have disturbed the level of the side peat-morasses and adjoining flat tracts of forest on the one side, and the

---

* Geological Survey of Pennsylvania, vol. ii.

shallow sea on the other. The ocean, as usual in earthquakes, would drain off its waters for a moment from the great Stigmaria marsh, and from all the swampy forests which skirted it, and by its recession stir up the muddy soil, and drift away the fronds, twigs, and smaller plants, and spread these and the mud broadly over the surface of the bog. In this way may have been formed the laminated slates, so full of fragmentary leaves and twigs, which generally compose the immediate covering of the coal beds. Presently, however, the sea would roll in with impetuous force, and, reaching the forest land prostrate every thing before it. Almost the entire forest would be uprooted, and borne off on its tremendous surf. Spreading far inland, compared with its accustomed shore, it would wash up the soil, and abrade whatever fragmentary materials lay in its path, and, loaded with these, it would then rush out again, with irresistible violence, toward its deeper bed, strewing the products of the land in a coarse promiscuous stratum, imbedding the fragments of the broken and disordered trees. Alternately swelling and retiring with a suddenness and energy far surpassing that of any tide, and maintained probably in this state of tempestuous oscillation by fresh heavings of the crust, the waters would go on spreading a succession of coarser or finer strata, and entombing at each inundation a new portion of the floating forest. Upon the dying away of the earthquake undulations, the sea, once more restored to tranquillity, would hold in suspension at last only the most finely-subdivided sedimentary matter, and the most buoyant of the uptorn vegetation—that is to say, the argillaceous particles of the fire-clay—and the naturally floating stems of the plants. These would at last precipitate themselves together by a slow subsidence, and form a uniform deposit, exhibiting but few traces of any active horizontal currents, such as would arise from a drifting into

COAL VEINS—HOW DEPOSITED. 103

the sea from rivers. The chief portion of the coarser fire-clay would settle first, and then the more impalpable particles, in company with the stems and leaves of the uprooted vegetation. Thus we may account for the constant reproduction of the peculiar soil of the coal seams, and for the preservation, particularly in its upper clayey layers, of the Stigmaria (Sigillarià); the simple consequences of the final subsidence of those materials being the production of the necessary substratum of another coal marsh. The marine savannahs becoming again clothed with their matting of vegetation, and fringed, on the side toward the land, with wet forests of arborescent ferns and other trees, all the essential conditions and changes that constituted this wonderful cycle in the statical and dynamic processes belonging to each seam of coal, and the beds inclosing it, would be completed, and ready to be once more renewed."*

This, indeed, is a very extraordinary theory. An earthquake for every seam of coal! In those basins where there are from fifty to one hundred seams, large and small, there must have been an equal number of earthquakes! In Nova Scotia there were at least seventy such earthquakes —yet, singular as the fact may appear, there is little or no *disturbance of the strata!* In the Anthracite regions, according to Mr. Rogers' showing, there are some fifty coal seams, consequently there must have been *fifty or more successive earthquakes!* and these earthquakes were not confined to a particular region; they are not local in their operations; but extended alike all over the coal regions of Russia, France, England, Wales, Germany, and America; and by a singular coincidence, always occurred *immediately after the deposition of a vein of coal!* It does not appear to have occurred to Mr. Rogers, that while

---

* Geological Survey of Pennsylvania, vol. ii.

these earthquakes were arousing the ocean, and causing him to uproot forests and scatter their trees and stumps broadcast over the coal, the *coal seam itself* would have suffered material *damage!* The earthquake must have been tame and amiable indeed, if, with the undulations and the wide fissures they usually produce in the crust, they did not contort and ruffle the strata of coal, so that the subsequent layers above would have occurred in *unconformable order.* If earthquakes aroused the ocean in the terrible manner described, it is at least wonderful how most of the coal basins maintained their horizontality, and the regularity of the superposition of *one seam over another.* This is, at least, wonderful!

It will be observed that Mr. Rogers has omitted to account for the phenomenon of the fossil trees occurring *in situ* over small seams of coal. It may be doubted whether such a phenomenon could exist amid the savage depredations of the ocean and the wave-like flexures of his earthquakes. Yet, they *do* exist; and there is too much significance in them to be cavalierly passed over.

Now, we are told by all the Geologists whose opinions have been quoted, and by many others whom we have not thought it worth while to quote, that fossil trees have been found in the coal measures. They go further: they state that fossil trees, in an upright position, have been found below, above, and passing *directly through* small seams of coal. All this I can readily believe, because I have myself seen such fossil trees. The inference created by the annunciation of the fact of the existence of such trees *on*, *under*, and *in* the solid coal would naturally lead to several conclusions, which all the geologists themselves appear to have arrived at: *first*, that the coal was derived from such trees; *second*, that the trees grew where they are found; and *third*, that such trees flourished on the peat-bogs which they describe. All works on Geology teem with descriptions of these fossil trees, and wood cuts exhibiting the quarries and situations in which they were found. Since Mr. Hawkshaw's discovery in 1839, more than two hundred other trees and stumps have been found in England, France, Scotland, and Nova Scotia. Nearly all these occur in the coal measures, either immediately over, under, or in the coal vein. Those that I have seen in the anthracite coal regions, are directly *over* the vein of coal. Assuming, then, that they all *grew* on the peat marshes that produced the coal, is it not singular that none of the *trees themselves should have been converted into coal?* Is it not incredible that all the trees thus described with so much learning and scientific acumen, instead of being *coal*, are converted into *shale*, or *silex*, or *carbonate of lime!* Prof. Lyell mentions one solitary instance, among the numerous fossil trees and forests which he describes, of trees being converted *into* coal. These were found in a vein of coal in Wolverhampton, in England. There were no less than seventy-five trees, with their roots attached, occupying a space of one-fourth of an acre. The trunks, broken off close to the root, were lying prostrate in every direction, often crossing each other. One of them measured fifteen, another thirty feet in length, and others less. "They were," says the Professor, "*invariably flattened to the thickness of* one or two inches, and converted into coal." In the case of Mr. Hawkshaw's trees, they are described as having a thin "*coating of coal*, so friable that it crumbled to pieces on removing the shale." And the solid trunks of *these* trees, like all the others described by Mr. Lyell, were also converted into shale, or sand, or lime.

I have myself seen thousands and thousands of fragments of fossil trees, and occasionally their solid round trunks; but never, in a solitary instance, was the *woody structure* converted into coal. I have made inquiry, and wherever such cases were reported, have gone to some trouble to ascertain the facts. In every instance I found that the woody trunk, when flattened, was merely *surrounded by coal*, but the *interior* was converted into dark shale. The Scientific Association of Pottsville, in one of their publications, proclaimed that they had discovered fossils *in* the coal, and thus created the inference that such fossils were *themselves* coal. On examining the specimens, I found numerous stems of plants, and limbs of trees *in the coal*, and every where surrounded by it; but the stems and fragments of *trees were themselves converted into slate.* If a scientific body, operating in the midst of the most prolific coal field in the world, and surrounded by specimens of the coal vegetation from all quarters of the country, could thus inadvertently create a false impression by means of its authorized publications, I feel that it would be no discourtesy to the well-known judgment and critical scrutiny of Mr. Lyell to infer that, in this instance, *he* may also have fallen into an error. I have seen the flattened trunks of Sigillaria, extending for fifty feet over veins of coal, exactly in the manner described by him; but on close examination they were found to be *surrounded* by coal, while the interior *woody structure* of the trunk was *invariably* converted into *slate*. I have observed on more than a thousand occasions, the solid limbs of trees imbedded in the slates of the coal veins, converted into sandstone, iron pyrites, and slate; but *never* have I seen them carbonized. And whenever they occur in the coal itself, which is very rarely the case, they still maintain their slaty character. But while the interior of the fossils is *always* slate, sand, or clay, the *outside* is as invariably coal, especially where the wood has been flattened by pressure. Of the thousands of specimens of Sigillaria, Stigmaria, Lepidodendria, and Calamites, that I have found in the coal mines, I have never yet met any that did not exhibit traces of coal on the *outside*, while

AN IMPORTANT MISTAKE CORRECTED.

most of them were coated with a thin stratum of it. This thin coating of coal is supposed by nearly every geologist who has described it, to be the *bark of the tree.* But a moment's reflection and a closer scrutiny will effectually dispel this idea. The scars of the Sigillaria, and the exterior marks of the Lepidodendria and Stigmaria, were all produced by the *detached leaves.* They therefore occur on the *bark*, exactly as similar scars are produced on the outer covering of the recent shoots of pine trees. Now in all fossils the coating of coal is invariably and necessarily *over these scars*, and they can seldom be *seen at all* until the *coal is removed.* The coal exhibits no trace of woody structure; instead of being fibrous, it is decidedly *vitreous, resinous, and brittle,* and invariably increases in *thickness* with the line of pressure to which the tree had been exposed. This proves, *first,* that the coal had been in a soft and viscid condition; and *second,* that it was expelled from the interior woody cells of the tree in the form of turpentine, oil, bitumen, or a peculiar resinous tar. The proofs of this are overwhelming and undeviating. Sometimes, indeed, the juices may have been extracted, for the most part, *before* the tree became fossilized, in which case the amount of coal is small; but whenever the tree was exposed to pressure while its juices were retained, the coating of coal on the outside is *uniformly present* in greater or less abundance. I therefore lay it down as a broad axiom in my experience, that *pure anthracite coal is the chemically-changed resinous matter discharged by the coal vegetation;* and that it is impossible to detect in such pure coal any *traces of vegetable structure.*

The elementary substances which enter into the composition of all vegetables, are confined to a small number —as oxygen, hydrogen, nitrogen, carbon, lime, silex, alumina, magnesia, potash, soda, and iron. These constitute the greater portion of the list. Plants, however, are endowed with the powers of assimilating and combining these various substances into compounds, assuming various forms and properties. The chief of these vegetable compounds are gum, sugar, farina or starch, gluten, albumen, fibrina, extract, tannin, coloring matter, bitter principle, narcotic principle, alcohol, acids, oils, wax, resins, gum resins, balsams, camphor, caoutchouc, cork, lignin, or woody fibre, sap, proper juice; while the simple or uncombined products are carbon or charcoal, the mineral alkalies, earthly and metallic oxydes.

It is unnecessary here to remind the reader of the large family of trees existing at the present era, which discharge the various kinds of gums, glutens, starches, albumens, and acids, known in commerce, manufactures, medicine, and domestic economy. The variety is altogether innumerable, and the extent incalculable. But while the animal kingdom, including the human species, is now and always have been wholly supported by the vegetation of the earth, it is a singular fact that the coal vegetation resembles that now existing only in its ferns and coniferous trees—trees upon which animals cannot subsist. We may look in vain among these fossils for any thing resembling the grains, fruits, nuts, and roots, upon which animal life is now sustained. There is no trace among them of wheat, rice, corn, or the grasses known to agriculture. There are no remains of the date, the palm, or the potato; of carrots, turnips, radish, cabbage, beet, lettuce, or rhubarb; of peas, beans, cherries, strawberries, or gooseberries; there are none of apples, pears, plums, quinces, peaches, oranges, and grapes. There were no flowers, nor nutritious fruits, nor seeds, nor nuts, nor roots;—*and why was this?* The reason is simple and obvious. There were then absolutely *no animals breathing the air on the land,* and there was consequently no use whatever for a vegetation such as we *now* have in every portion of the globe. Would not Moses have made a fatal blunder, if, with the knowledge we now have of the coal vegetation, he had inadvertently introduced *air-breathing animals* into the picture, as some of our geologists, with less foresight, have ventured to do? But, instead of the fruits and flowers and waving grain that now surround us, the earth then produced trees like those of the lofty pines, and it was from *these* that the coal and various other substances allied to it were mainly extracted. There was then no immediate *use* for the vegetation itself; but the *resinous juices* which it secreted were to be entombed in the crust of the *earth for the future purposes of man;*—to subserve the designs of those great future eras, when, owing to the changed physical character of the earth, their production, distillation, and deposition would have been utterly impossible. The far-seeing sagacity of the great Author of the world is thus continually manifested; and we perceive at every step how uniformly consistent, great, and harmonious, are all his decrees.

There can be no doubt whatever that the forests of the carboniferous period largely predominated in coniferous trees, and that our beds of coal have been derived from their resinous secretions. I propose to give my reasons for this inference in due season; but in the mean time it is necessary to understand as nearly as we can the nature of the trees in question. This may be arrived at, in some measure, by studying the features of those now existing which the ancient trees most resembled.

The Coniferæ belong chiefly to the class *monœcia* and *polyandria* of Linnæus, and the gymnosperm phanerogamiæ of Jussieu. The existing family has been divided into thirteen genera, each containing a large number of species. The genera consist of, 1. *Pinus*, or the fir; 2. *Abies*, the spruce; 3. *Larix*, the larch; 4. *Shubertia*, deciduous cypress; 5. *Cupressis*, or cypress; 6. *Thuga*, or arbor vitæ; 7. *Juniperus*, or juniper; 8. *Auraucaria*, or New Holland pine; 9. *Belis*, or javelin-shaped; 10. *Agathus*, or dammer pine; 11. *Exocarpus*, or cypress-like; 12. *Podocarpus*, or Chinese pine; 13. *Taxus*, or the yew.

The distinct species of the *pinus* enumerated by botanists are upwards of twenty. None of these bear flat leaves, but a sort of spines, which, however, are true leaves. They are mostly evergreens; but the appearance of the tree, as well as the quality of the timber, varies with the species, as also with the situation in which it grows. Generally speaking, the timber is the more hard and durable the colder the situation and the slower the tree grows; and in peculiar situations it is not uncommon to find the northern half of a pine hard and red, while the southern half, though considerably thicker from the pith to the bark, is white, soft, and spongy.

In the peat-bogs of Scotland, the remains of pine trees are very abundant; and such is their durability, in consequence of the turpentine they contain, that, where the birch is reduced to a pulp, and the oak cracks into splinters as it dries, the heart of the pine remains fresh, and, embalmed in its own turpentine, is quite elastic, and used by the country people in place of candles.

<small>The wild pine of Scotland (*pinus silvestris*) is widely diffused; and is found growing in a state of nature in many situations. It is indigenous in the Alps, in the north of Germany, in Sweden and Norway, and in Russia. In favorable situations, it attains a height of eighty feet, and from four to five feet in diameter. The trunk is covered with a thick and deeply-furrowed bark; the leaves are in pairs, of a pale-green color, stiff, twisted, and about three inches long; the flowers are of a yellowish tint, and the cones are grayish, of a middling thickness, and a little shorter than the leaves. Each scale is surmounted by a retorted spine. There are several varieties of this pine. The *pinus silvestris* is that which yields the red wood; even young trees of this sort become red in their wood,</small>

and full of resin very soon. Pines generally are found *growing in forests, or clustered together*. In this position they grow tall and upright, with few lateral branches, except near the top. This pine very often, though not in trees completely matured, contains sap-wood next the bark, and toward the pith is a little spongy. The pines generally occur in much more *extensive* forests, and with a far less *admixture of other trees*, than any other genus whatever. Though it is not the last timber met with on the confines of the snow, as we ascend high mountains, or at the verge of vegetation as we approach the pole, yet, after a certain elevation, and north of the latitude of about fifty degrees, it is by far the most abundant timber in Europe, America, and Asia. Along the St. Lawrence, and in the British possessions north, large quantities of tar have for many years been distilled from it for the European market.

The other European species of the pine are: the *Corsican (p. laricio)*, which is nearly allied to the Scotch pine. Prof. Thonia considers it equally hardy with the Scotch pine; but its wood is more weighty and resinous. It grows wild on the summits of the highest mountains in Corsica. The *Cluster pine (p. pinaster)* is a grand and picturesque tree, and is a great favorite with the Roman and Florentine painters. The Stone pine *(p. pinea)* is very common in the south of Italy. The seeds of this and the cluster pine are eaten in Italy, both by the poor and rich. They are as sweet as almonds, but partake slightly of a turpentine flavor. The wood is not so resinous as most of the other species. The Siberian pine *(p. cerubra)*, the tennebaum of Byron's Childe Harold, grows higher in the Alps than any other tree; and is found in elevations where the larch will not grow. The peasants of the Tyrol make various carved works with the wood, and sell them in Switzerland, where the common people are fond of the resinous smell which it exhales. The Canary pine *(p. canariensis)* grows in the mountains of the Canary Islands. The wood is resinous and highly inflammable.

Of American species of the pine, Michaux enumerates ten. Of these, the Red Pine *(p. rubra)* is found in Canada and the northern parts of the United States. It occupies small tracts of a few hundred acres, either alone or mingled with the white pine. The wood has a fine grain, and is very resinous. It is largely produced in Maine, and along the shores of Lake Champlain. The Yellow pine *(p. mitis)* is very widely diffused in North America. It is a beautiful and symmetrical tree, the branches forming a pyramid at the summit. The concentric circles of the wood are six times as numerous in a given space as those of the pitch or loblolly pines. The heart is fine-grained, and moderately resinous. The Long-leaved pine *(p. australis)* is also known as the yellow pitch, broom, and Georgia pine. It is first seen near Norfolk, in Virginia, where the pine barrens begin; and it extends over the lower part of the Carolinas and the States of Georgia and Florida. Its mean stature is about sixty feet, with a uniform diameter of eighteen inches for two-thirds of its stem. The leaves are a foot long, of a beautiful brilliant green. The cones are also very large; the seeds are generally abundant, the kernel being of an agreeable taste, and is voraciously eaten by wild turkeys, squirrels, and swine. In some years, however, whole forests for hundreds of miles will not yield a single cone. The wood is compact, fine-grained, durable, and susceptible of a fine polish. It is from this tree that the principal supply of pitch, resin, and turpentine is obtained; the pine barrens, being of vast extent, afford an abundant supply of those materials, both for home and foreign consumption. I shall speak of the processes for obtaining these substances very shortly.

The Pitch pine (*p. rigida*) is another very resinous species, very common all over the United States, but particularly abundant along the Atlantic coast. It is a very branchy tree, and the wood is consequently knotty. The bark is thick, of a dark color, and deeply furrowed. The concentric circles are far apart, and *three-fourths* of the larger stocks consist of sap.

The White Pine (*p. strobus*) is one of the most abundant and valuable trees in America, and derives its name from the perfect whiteness of the wood. It grows extensively between the parallels of forty-three and forty-seven degrees, in almost all varieties of soil; but attains its greatest dimensions in New Hampshire, Vermont, and near the source of the St. Lawrence. This ancient and majestic inhabitant of the North American forests is still the loftiest and most valuable of their productions; and its summit is seen waving at an immense distance toward heaven, far above the heads of the surrounding trees. It is the foremost in taking possession of barren districts, and the most hardy in resisting the impetuous gales from the ocean. On young stocks, not exceeding forty feet in height, the bark of the trunk and branches is smooth and polished; but as the tree advances in age, it splits and becomes rugged, but does not fall off in scales like that of other pines. The wood is soft and light, and is extensively used in the United States for architectural purposes, as also in Great Britain. It is not resinous enough to furnish turpentine for commerce; nor would the labor of extracting it be easy, because of its diffusion in small tracts, and its admixture with other forest trees.

The Firs or Spruces (*abies*) form another genus of the *Coniferæ*, differing from the pines in the form and position of the leaves, as well as in the general aspect of the trees. In the firs, the leaves are generally shorter than in the pines, and placed solitary instead of in pairs. The Norway Spruce Fir (*abies communis*) is a beautiful and stately tree. It is one of the tallest of European firs. The leaves are solitary, slightly arched, and of a dark green color, which gives the tree a sombre aspect.

THE FAMILY OF RESINOUS PINES. 113

The cones are cylindrical, five or six inches in length, and contain small winged seeds. By incision it yields resin and pitch. The tops or young sprouts give the flavor to "Spruce beer." The white, black, and red spruces are natives of America, and nearly resemble those of Europe.

The Silver Fir (*a. picea*) is one of the most beautiful of this family. When standing alone developing itself naturally, its branches, which are numerous and thickly garnished with leaves, diminish in length as they approach the top, and thus form a pyramid of perfect regularity. The upper surface of the leaves is of a vivid green; and the under surface has two white lines running lengthwise on each side of the midrib, giving the leaves that silvery look from whence the common name is derived. The wood is light and slightly resinous, and inferior to that of the common pine. The resin of the tree is sold in England and the United States, under the names of Balsam, or balm of Gilead, although the true balm of Gilead is produced from an entirely different tree, the *amyris Gileadensis*.

Pinus Douglasii. This tree grows to the height of two hundred and thirty feet, and is fifty feet in circumference at the base. It has a rough corky bark, from an inch to twelve inches thick. The leaves resemble those of the spruce, and the cones are small. The timber is good and heavy. This pine abounds in Oregon, California, and Washington Territory, where it forms extensive forests, extending along the shores of the Pacific Ocean to the Rocky Mountains. It is impregnated with resin.

Pinus Lambertiana. This tree abounds in California, where it is dispersed over the country, but not in large forests. Like the other, it attains the most extraordinary dimensions—often exceeding two hundred and twenty feet in length, and sixty feet in circumference at the base. The cones average sixteen inches in length. The seeds are eaten, roasted or pounded into cakes. The tree bears much resemblance to the spruces; and like them, its turpentine is of a pure amber color, and the timber soft and white. One singular property of this tree is, that when the timber is partly burned, the turpentine loses its peculiar flavor, and acquires a sweetish taste. The Indians use it instead of sugar.

The Larch (*larix communis*) is, after the common pine, probably the most valuable of the tribe. The name seems to be derived from the Celtic, in allusion to the resinous juice which it exudes. Dioscorides remarks that *larix* is the Gallic name for resin. Though a native of the mountains of more northern regions, it thrives extremely well in Great Britain. The bark of the larch is more than half as valuable as that of oak in tannin, and the tree yields turpentine by incision. The black larch of America, (*l. pendula*) called by the Indians *tamaracke*, resembles the European species both in appearance and the excellent quality of the wood and bark.

The Cedar of Lebanon (*l. cedrus*). This celebrated tree is a native of the mountains of Libanus, Amanus, and Taurus; but it is not now to be found in great numbers. The forest of Lebanon never seems to have recovered the havoc made by Solomon's fortyscore thousand hewers, so that there are now probably more cedars in England than in all Palestine. Its resistance to wear is not equal to that of the oak; but it is so bitter that no insect whatever will touch it, and it seems to be proof against time himself. The timber in the temple of Apollo at Utica was found undecayed after the lapse of two thousand years. Some of the most celebrated structures of antiquity were made of this tree. "Solomon raised a levy of thirty thousand men out of all Israel; and he sent them to Lebanon, ten thousand a month, by courses; and he had threescore and ten thousand that bore burdens, and fourscore thousand hewers in the mountains. And he covered the temple with beams and boards of cedar. And he built chambers against it, which rested on the house, with timber of cedar. And the cedar of the house within was carved with knobs and flowers; all was cedar, there was no stone seen." Thus writes the sacred historian, who mentions that the same monarch had a palace of cedar in the forest of Lebanon. Ancient writers notice that the ships of Sesostris, the Egyptian conqueror, were formed of this timber; as was also the gigantic statue of Diana in the temple of Ephesus. The description of the cedar of Lebanon by the prophet Ezekiel is fine and true: "Behold the Assyrian was a cedar in Lebanon, with fair branches, and of an high stature; and his top was among the thick boughs. His boughs were multiplied, and his branches became long. The fir trees were not like his boughs, nor the chestnut trees like his branches; nor any tree in the garden of God like unto him in beauty."

The Yew Tree (*taxus baccata*) is a native of Europe, of North America, and the Japanese isles. The trunk and branches grow very straight; the bark is cast annually; the wood is red and veined; it is compact, hard, and elastic. The yew tree was also consecrated—one or more being in every church-yard, and they were held sacred. In former times, in funeral processions, the branches were carried over the dead by the mourners. Being an evergreen, it was thus made typical of the immortality of the soul.

The Cypress (*cypressus sempervirens*) obtains its name from the Island of Cyprus, where it grows in great abundance. Of all timber, that of the cypress is the most durable, superior even to that of cedar itself. The doors of St. Peter's Church, in Rome, which had been formed of this material in the time of Constantine, showed no sign of decay when, after the lapse of eleven hundred years, Pope Eugenius IV. took them down to replace them by gates of brass. In order to preserve the remains of their heroes, the Athenians buried them in coffins of cypress; and the coffins in which the Egyptian mummies are found are usually of the same timber. Like the yew, it was carried in funeral processions, and strewn over the graves of the dead. The White Cedar is a native of America; its growth is slow; but it is hardy, and forms a good variety in clumps of evergreens. *Arbor vitæ (thuja occidentalis)*, when burnt, gives out an agreeable odor, and was used by the ancients at their sacrifices. It is a native of Canada. It grows well in swamps and marshes. A Chinese species (*c. orientalis*) resembles it, and both are readily propagated by cuttings, seeds, or layers.

Norfolk Island Pine (*auraucaria excelsa*) attains a gigantic size, often measuring two hundred and twenty feet in height. It is a native of Australia, and presents a magnificent object, with its bright evergreen foliage, and innumerable waving branches. The leaves are closely imbricated, inflexed, and pointless. The longitudinal section of the wood, with all the distinctive marks of the Coniferæ, exhibits the peculiarity of three rows of oval disks. From this circumstance, the fossil trees of Craigleith quarry (previously referred to by me) have been identified with the auraucaria of Norfolk Island. Other fossil trees occurring in coal beds have likewise been identified with it. Sir J. Bank's auraucaria (*a. imbricata*) is also a beautiful variety of this species.

The Juniper (*juniperis communis*) is common in all the northern parts of Europe. It flourishes everywhere, but grass will not grow under it. Wood is hard and durable; the bark is so tenacious that it may be formed into ropes, and the berries are used for imparting flavor to gins. A gum oozes spontaneously from the trunks of old plants, which forms the gum sandarack, and in its powdered form is known as pounce. The berries and tops, by distillation, are largely used for medicines. Bermuda Cedar Wood is the product of a West Indian species of Juniper. The Red Cedar (*j. virginearia*) is one of the highest timber trees in Jamaica. The wood is bitter, and hence avoided by insects. Common Savin (*j. sabina*) is a plant which only attains the size of a few feet in England, but is found as a tree in some of the Greek Islands. The leaves and tops have a disagreeable odor, and a bitter, hot taste. These qualities are owing to an essential oil, which is obtained in large quantity by distillation. *Gum Olibanum*, supposed to be the incense of the ancients, and the substance now used in the Catholic churches, is the product of the juniper *licia*. Allied to the Coniferæ is the family of plants, Myrica, or candleberry myrtle. One of these, the sweet gale, is very abundant in bogs and marshes of Scotland. It is a small shrub, with leaves like the myrtle or willow, of a fragrant odor and bitter taste, and yielding an essential oil by distillation. The cones, boiled in water, throw up a scum resembling bees'-wax, which, collected in sufficient quantity, serve for candles. *Myrica Conifera*, or Tallow Shrub, is common in North America, where candles are made from a decoction of the berry. It grows in wet soils, or near the sea. A soap is also made from it, and in the Carolinas it is used for sealing-wax.

The Tallow Tree (*Croton Sebiferum*) yields a substance very much like tallow, and in China, where it grows abundantly, candles are extensively produced from it. The Piney Tree (*vateria Indica*) growing on the east coast of Malabar, yields a substance very similar to the foregoing, and is also very extensively employed in the production of candles, being superior, in many respects, to animal tallow. A resin, very similar to copal varnish, exudes from the same tree, and furnishes a very durable varnish. This resin is often mixed with the tallow, and applied as a substitute for tar in smearing the bottoms of boats.*

Such, in brief, are the *existing* coniferous trees, and there is abundant reason to believe that they are analogous to those of the coal-bearing period. The leading characteristic of the whole order, it will be observed (aside from their structural features), is their *secretion of resins, oils, tallows,* and *turpentines,* in varied qualities. These exudations, by a distillation presently to be described, have furnished the various beds of coal, mineral bitumen, asphalt, and anthracite, distributed over the earth. As I remarked before, I do not think that the ferns contributed materially to the formation of the coal, inasmuch as they secreted no resinous or inflammable juices. It is a curious fact, that immediately over the coal veins of the Alleghany mountains, which extend over eight hundred miles in length, the trees which abound most largely in these resinous secretions, are now *found growing in native strength and vigor*, and constitute the prevailing species of the forests; while underneath their tall and overarching tops *many species of the fossil Fern* are also found in extraordinary abundance! Of the two hundred species of fossilized Fern in my collection, there are many that can

---

* I abridge my description of the living Coniferæ mainly from *Rhind's History of the Vegetable Kingdom*, and the *Library of Entertaining Knowledge*, London editions.

DIVERSITY OF VEGETATION EXPLAINED. 117

be found growing immediately over the rocks in which they were imbedded! This is a singular fact, and suggests the idea obscurely intimated by Moses, that the seeds of vegetation are *within themselves upon the earth*—that is, the seeds of vegetation of a previous age, may be buried in the soil to germinate *anew* at subsequent periods! The comparative absence of the trunks of pine trees in the coal, may be accounted for on the supposition of their enormous dimensions as well as *distance* from the scene of resinous accumulation; while, on the other hand, we can trace a close alliance between them and the extinct Lepidodendria, the Sigillaria, and the Stigmaria. I have found many cones in the coal, of the most perfect and dissimilar structure—showing that, like the cone-bearing trees now living, there were originally many different varieties. But the extraordinary abundance and variety of the Ferns now growing in the coal regions, and their *absolute identity* with the fossil specimens, leads me to believe that, wherever the strata of the coal or Devonian measures have been uptilted, or *brought to the surface*, and the soil preserved in its native condition, free from obstructions or cultivation, the *original seeds of the ancient vegetation have again germinated*. The very fact that the words of Moses seem to authorize such an inference, leads me to give it paramount weight. How else could these seeds have been diffused? It will not serve our purpose to suppose that birds and animals could, by any possible means, disseminate seeds in such profusion, and in localities so *exactly* corresponding with the same plants and trees imbedded in the shales below! This would have been next to impossible; especially as every geological formation appears to have had originally, and *has still*, when unobstructed by cultivation, a *vegetation peculiar to itself.* Wherever there is coal, or, rather, wherever the *coal measures outcrop*, the ferns and resinous pines abound; and are frequently found to be similar to the fossil specimens. These, however, may have been removed some distance from the place of their growth, and due allowances should be made in such cases; yet the great fact still stands forth in its integrity, and seems to defy any other process of interpretation. Wherever the pines or coniferæ and ferns now abound, as *natives of the soil*, coal, in some form or other, is morally certain to exist in close proximity.

Upon the basis of this hypothesis, we can readily account for all the diversity which exists in vegetation—for primarily, every geological era has furnished in its rocks and shales the seeds for future generations. The Almighty Creator scattered the seeds of vegetation *in the beginning*, and every subsequent era brings forth its exhaustless crops—exhaustless, because the soil itself is exhaustless. It was only at the creation of man that he planted a garden in Eden, and *then*, for the first time, *introduced fruits and grains*, and the varied sorts of *nuts, melons,* and *esculents*, so essential to man and the animal creation. *Previous* to that time there was no necessity for fruits, and the domestic vegetables; land animals, properly so understood, *having had no existence*. With the creation of man, however, (or in anticipation of his creation,) an entire new order of vegetation was introduced, embracing every fruit and garden and field product now known, and including all the flowers that bloom and dispense their fragrance over the earth.

But it may be urged against this hypothesis, that the fossilization of the seeds would utterly have destroyed their fructifying principle. This, at first thought, would appear likely; but in the case of the coal vegetation, it has little force. In the first place, the shales overlying the coal are not *always* hard and indurated; and when they *are*, a brief exposure to the air decomposes them.

All the coal shales are, in fact, nothing but *baked mud*, but it is mud composed of the very finest particles of earth. When the vegetation grew, the seeds of the pines, concealed in the resinous cones, were scattered in this soft mud, and the turpentine and tar with which they were surrounded served to coat them, and to place around them an air-tight envelope which would preserve them as effectually for *one hundred million* of years as for *one year*. The soft mud was afterward baked by the heat below and the pressure from above, and thus became, in process of time, compact slate. The same process will apply equally to sandstones mixed with argillaceous clay and vegetable mould; and as every geological formation abounds in these rocks, the preservation of the smaller seeds, by an air-tight oleaginous coating, is rendered as probable in one geological formation as in another. In the case of those trees which bear nuts, as the oaks, the chestnuts, etc., they had *no existence* in the Paleozoic periods, and it would therefore be useless to assume that they had been similarly preserved. Wheat, it is well known, has been preserved in the catacombs of Egypt for several hundred years, and upon being planted, has brought forth prolific crops.

Tar, for local use, is produced in all the coal regions; but in North and South Carolina it forms, with turpentine, pitch or resin, an article of very extensive export to foreign countries. In the pine forests of those States, the sap or turpentine begins to circulate in the tree during the month of March, and the accumulation proceeds and increases with the warm weather, generally attaining the the maximum in the month of August. When the sap manifests itself, incisions are cut in the base of the tree, beneath which boxes are placed to receive it as it exudes. Sometimes three or four incisions are made, of variable depth, and at different spaces—from all of which the tree freely bleeds. The ground around the tree has to be carefully cleared of all dry weeds and brush, to prevent any liability of fire from the inflammable character of the liquid, which is often scattered around, through the carelessness of those in attendance. At the commencement it usually requires about two weeks to fill the boxes with turpentine—each of them holding from one to two quarts of the liquor. When the exterior sap is extracted, deeper incisions are again made, penetrating through at least four of the annual rings of the wood, and thus tapping the more vital parts of the tree, which, however, will continue to yield sap for five or six years afterward. When the receiving boxes are filled, the sap is transferred by wooden ladles into barrels; and this completes the process. The turpentine thus obtained is of the purest quality, from which the oil or spirit of turpentine is afterward distilled.

The process for extracting tar from the yellow pine or the long-leaved pines of the Carolinas, is exactly similar to that of coking bituminous coal. A round space is cleared in the forest, gradually sloping to one side. The space thus cleared is formed into a concave basin, and the ground beaten down with mallets to render it hard and compact. From the centre of the cavity a ditch is dug in the direction of the outward slope. Billets of pine wood, stumps, knots, roots, and branches, are now arranged in circular form, layer after layer, around the space thus prepared. The sticks are set up on end, and the pile of wood and branches terminates in a gradually sloping dome. The dome thus erected is covered over by leaves, branches, and loose material, and the whole then inclosed with a layer of moist clay. A few holes are left around it for the admission of air. The combustible material inside is now ignited, and it burns with a slow, smouldering heat—never being allowed to burst into a flame. As the combustion proceeds, the tar is liberated from the cavities of the wood, and draining into the centre of the conical pit, thence issues in a continuous stream through the ditch or trough running to the outside. Here it is immediately taken up and placed in barrels, and is thus ready for the market. In the Carolinas, tar is principally extracted from dead wood that has fallen by accident, and from the tops of the trees that have otherwise no value. The whole process is extremely simple; and we may add, was in practice among the most ancient nations of the earth, as well as among the more recent. But long before the Greeks, or Romans, or Egyptians, applied it in their forests, Nature had exemplified it in her great coal basins. And it is a singular fact, that all the varied contrivances of man, for extracting, elaborating, and compounding different elementary substances in minerals and vegetation are, after all, but the primary lessons which he has learned in the great school of Nature. He has done nothing in the arts of design, in mechanism, dynamics, hydrostatics, or the crucible, in which he has not been anticipated. Nature furnishes all his models; and he is a mere apprentice in *copying*. But his efforts, although necessarily local and experimental, are sometimes noble and even godlike; but those of Nature, the great teacher, are always infallible and universal, and she has entire globes for her laboratory.

Let us now return to the Devonian basins, which we described some time ago, and which we left fully prepared to receive the veins of coal so soon as we could elaborate them from the ancient vegetation. It was stated that these basins were in many respects similar to the great *inland seas* of the northwest; that the anthracite basin was much the deepest, and stood at the head of all the others, somewhat like that of Lake Superior. These coal basins, although they did not extend in a direct line, nevertheless communicated with each other in a manner precisely similar to those Lakes. Superior is connected with Lake Huron by the St. Mary's river, which is about sixty miles long, and generally not over half a mile wide. The descent of the stream is perhaps more than thirty feet, or six inches to the mile. Lake Huron is connected with Lake Michigan by the straits of Mackinaw, which are of greater width, but interspersed with numerous islands and rocky promontories. Lake Huron thus receives the waters of two great Lakes, and then passes them into a little shallow Lake, not over twenty feet deep at any place, nor over thirty miles wide, by means of the St. Clair river—a stream some forty miles in length, but a few hundred yards in width, and perhaps forty or fifty feet deep. From Lake St. Clair, the waters are passed through the Detroit river into the basin of Lake Erie. Here they are thrown over the falls of Niagara, and then, by another very narrow river, not over twelve hundred feet in width and fourteen miles in length, they are emptied into Lake Ontario. They are now again discharged into a river, (the St. Lawrence), and after expanding somewhat into the form of lakes, at intervals, are finally emptied into the ocean —describing another great basin or gulf before finally mingling with its saline waters. The distance thus traversed, from the head of Lake Superior to the gulf of the St. Lawrence, is sixteen hundred and fifty miles. The distance from the head of the anthracite basins by the route *originally pursued* by the primitive lakes and rivers, to the Missouri river, was nearly the same, and when they reached this point, they encountered the waters of the ocean, which *then formed* a gulf over portions of the States of Missouri, Arkansas, Mississippi, Louisiana, Texas, and Nebraska, very similar to that of Mexico or the St Lawrence.

After the Silurian seas had been withdrawn, the great lakes or basins left behind, began gradually to fill up by the deposition of the sediment drained from the adjacent rocks—comprising, at some places, limestone and silt, and at others, sandstone or conglomerate and silt. The *prairies were now covered with immense forests of coniferous trees*, the woody cells and fibres of which, like those of our existing pines and firs, consisted mainly of resinous and oily secretions. These forests in all probability extended hundreds of miles around the sloping plains of the lakes; and were liable to the same contingencies of ultimate decay and destruction as existing forests. It is perhaps hardly worth while to remark, that they were, in every respect, the most enormous fields of vegetation which have ever yet flourished upon the face of the globe. While those of the humid plains of Central American and Brazil may convey an *idea* of their extent, they certainly could make no pretensions as rivals. If some of the pines we have described, can *now* attain the height of two hundred and fifty feet, there is no absolute reason why they should not, at this particular era, have soared still higher in the air, because all the circumstances that surrounded them were in the highest degree favorable to the most extraordinary development. The forests of South America are described as absolutely impenetrable by man; while in California trees have recently been found of four hundred and fifty feet in height—or nearly twice the height of our loftiest steeples, and fully equal to the tower of Babel. While the trees themselves thus tower hundreds of feet in the air, the trunks are surrounded by younger shoots and weeds, which stand so close together that even the wild animals have difficulty in traversing them—some, indeed, more skillful than others, retreat to the thickets to escape from their enemies. The coal vegetation, in addition to tropical prolificacy, was not depredated upon by prowling animals. It grew in undisturbed luxuriance, and attained such development that those only who have witnessed the wild and unchecked profusion of nature amid the tropics, can form any conception whatever of its enormous extent. The atmosphere under which the vegetation flourished was in many essentials different from that which we now breathe. The proportion of carbonic acid was enormous, and while this *served to stimulate vegetable growth*, it rendered the existence of animal life impossible. It may not always have been materially warmer, as is generally supposed by geologists; but it certainly was more *humid*, and perhaps enveloped for the most part, in fogs and mists, such as prevail along the coasts of Newfoundland, where *the yellow pines even now attain their greatest and highest development*. But whatever heat existed, must have been mainly derived from the *earth itself;* as we shall hereafter demonstrate that *solar* heat had as yet scarcely established itself upon the earth. The radiated heat of the *interior* would envelope the surface in vapors, and these, we have every reason to infer, contributed largely to the growth of the vegetation. Growing under circumstances of extraordinary favor, the forests would at times yield to that unsparing law which levels every thing with the dust. Whether by tempests or the overpowering gravity of their elevated tops, or the prostration of one upon the other—it is certain that entire forests would finally bend to the ground, to give place to a new crop. Accumulating thus on the sloping prairies, constantly moist and wet with the atmospheric exhalation and condensation, the prostrate vegetable material would be *exposed to fermentation and distillation similar to that of the tar pits*. Trunks and fragments of trees, covered over by their branches and leaves, and the accumulating rubbish of the forest, under the smouldering fermentation thus evolved by the interior heat of the earth, would part with their resinous and oily juices, while the atmosphere would be blackened with the smoke and gas. The whole earth was thus enveloped in the fermenting process. The gases ascending from the smouldering vegetation, would be arrested by the fogs and vapors of the atmosphere, and thrown down upon the earth in the form of soot and lampblack. The soot would accumulate like layers of snow; and uniting with the oily liquids issuing from the vegetable mass, would thus be borne off to the waters of the adjacent lakes. All the streams, springs, rivers, and lakes were discharging *coagulated carbonaceous ink*. In the absence of solar evaporation, nothing was lost. As the vegetable material went on accumulating, its resinous juices were liberated by spontaneous fermentation, and, both in the form of liquids and gases, the elements of the vegetation would be drained down into the lakes or basins. We can thus imagine the ground which supported these vasts forests of pine to be literally moist, spongy, and miry with the escaping tar, and oils, and smoky soot; and that in every direction, for hundreds of miles around the sloping plains drained by the lakes, the pyroligneous liquid oozed out of the ground in constant springs and streams. The whole earth, wherever the dry land had yet appeared, was thus covered with stupendous tar-pits, while the atmosphere, already humid with the vapors of radiated heat, was blackened with ascending smokes or enveloped in *snows* of black carbonaceous soot. The surface of the ground, in every direction, having been thus periodically, if not almost constantly under the influence of the resinous fermentation, distillation, and combustion, there was, of course, but little sand and sediment. It was only occasionally that the stratum of vegetable mould would be removed, and the underlying sand and clay exposed. In such cases the mud and sand would be carried into the lake, and scattered over the accumulating coal seam, or distributed in irregular heaps or layers; while the mould resulting from the decomposition of trees, and charged with fragments of half-decomposed stems and branches, would in like manner be sometimes removed. As a general thing, the fine mud and vegetable mould was scattered uniformly and evenly over the bottom of the lake, during the occasional pauses in the supply of resinous matter. The mud thus distributed is now *the parting slate between the benches of the pure coal;* while the *faults* are derived from the layers of sand and silt, and the deposits of half resinous mould and mud into which the pure coal often degenerates. The *rock faults,* however, in many cases, originally existed in the *bottom of the lake*—the resinous material merely collecting around and accommodating itself to them. The resinous matter, as a general thing, was evenly distributed over the bottom; but there were, of course, occasional exceptions. Sometimes it would thin out, and give place to the original clay or sand of the bottom; while again it would expand into twice its regular thickness. These deviations, intrusions, and irregularities comprise what are now termed the *faults* of the coal veins—features which are entirely overlooked in all the other theories of the coal formation.

Now, after the process here described had gone on until a thick layer of resinous material accumulated (somewhat similar to the pitch lake of Trinidad), the outlets or narrow straits of the lakes became clogged, and the result was an *unusual accumulation of water.* The straits connecting the coal lakes were essentially similar to those connecting the lakes of the northwest; and it is easy to conceive *how* these could be temporarily choked up so as to temporarily impede the passage of the water. The formation of a sand-bar, rendering the water shallow, as in the case of the Lake St. Clair; or the drifting of logs and trees into the narrow perpendicular necks of the river (like that of Niagara), would readily suffice. This

ORIGIN OF COAL EXPLAINED.

is a phenomenon of such frequent occurrence in all our mountain streams, that I take it for granted it will appear self-evident in this connection. The waters of the Mississippi and the Missouri, by freshets or temporary obstructions, are sometimes diverted from their course, and overflow the surrounding plains for forty and fifty miles. The passages of the coal lakes, thus obstructed in the narrow rivers, the waters would at once *overflow the whole surrounding forests.* The effect of this is readily perceptible—all the rubbish of the forests, with the leaves, branches, stems, and logs, and the great bulk of the mud and mould of decomposed vegetation, would be *removed* and *borne off into the lake.* Upon the subsidence of the water, they would settle over the vein of coal; and the debris thus collected now forms the *top slate in which are found all the fossil impressions known of the coal vegetation.* The inquiry of the geologists for *large trees,* and their surprise at not finding them *in* the coal, is thus easily explained. The trees that exuded the resin were really but seldom displaced, while the overflowing of the water did little injury to the forests themselves beyond the removal of their loose and *scattered trees, limbs, leaves, and vegetable mould.*

As the waters of the lake forced their way through the connecting straits, and subsided to their customary level, they again began to wear down the adjacent shore, and received the debris of the mud and sand exposed in consequence of the removal of the decomposed vegetable mould. This process continued with activity until the forests had accumulated *another* layer of resinous material. In the mean time, however, the low flats immediately adjoining the lake brought forward *their* crops of soft and succulent vegetation, the most conspicuous of which was the plant called *Stigmaria.* Recent investigations have led some geologists to suppose, as I have already remarked, that this plant is the root of the Sigillaria. But I am of a different opinion. I have every reason to believe that it was a species of vine, which extended itself over the low and *half-submerged flats* along the margins of the coal lakes; and that it was borne off in the water and deposited *before* the regular supplies of resinous matter reached the lake—hence it is almost invariably found *under the veins of coal.* The phenomenon of finding trees in an erect position, in this under clay or shale, may be explained in the same way. They grew on these marshy flats, like the trees on the flats of St. Clair; and the *wearing away* of the shore on which they stood, during times of high water, caused their removal into the lake, where they would natually settle to the bottom in *an erect position.* This is daily exhibited in the Mississippi, and the Amazon river in Brazil. Trees are *undermined* by the water, and they fall down and are borne off by the stream, their foliage maintaining them in an erect position. Sometimes their roots find a lodgment in the bottom of the river, and they are thus supported until sufficient sand has gathered around them to enable them to stand erect after the subsidence of the water. The whole phenomena of finding trees penetrating *through* the coal vein, and of lying *over* and *under* them, is to be *explained in this way.* They are the results of *accident,* not of geological law. All the trees ever found in these situations, had they been converted *into* coal (*which they never are*), would not have made a seam as thick as a sheet of paper in the great basins in which they occur. But the fact that they never furnish coal at all, except where they have been *flattened by pressure,* and their resinous sap thus *squeezed out,* is conclusive that their solid woody fibre contributed *nothing* directly to its formation.

It is hardly necessary to observe that the process here described was repeated again and again, upon the deposition of every subsequent layer of coal. Indeed, it was often partially carried out without the *interposition of the coal*—for the alternating strata show many little veins of slate and leaders of impure coal which have been derived *solely from the vegetable mould of the forests.* Again: The overflowing of the forest was *not* an absolute essential to every seam of coal. By no means. These overflowings were *irregular*, and generally terminated, for the time being, the flow of the oils and resins. For this reason, the seams of coal vary in thickness from a few inches to forty feet—the latter, however, are invariably separated into benches or laminæ of coal, varying from one inch to three feet. These benches may be regarded as *separate veins*, because they are *parted by layers* of slate, mud, or sand, which sometimes run into very thick strata. We have already alluded to this fact, and mention it again only to show that no regularity is claimed for the floods. They sometimes occurred during the deposition of the largest veins; but whenever they *did* occur, the debris of the forests and the surrounding rocks was invariably brought into the lake. Again: Some of the veins of coal were deposited without the occurrence of floods at all. In these cases there is, of course, a comparative absence of *top slates and of fossils.* The vein of coal is then often overlaid by *sandstone* — or limestone — the immediate debris of the basin terraces. The liability of the submergence of the forests was, however, very great. The sloping prairies were in no instance elevated more than from five to ten feet above the level of the lakes. The Mississippi river, for more than fifteen hundred miles, does not descend at an average of over half an inch to the mile. Even the Ohio, which emerges from and traverses a great mountain slope, has a fall of only four or five inches to the mile. A flood of ten or fifteen feet on the lower Mississippi, will inundate the surrounding country for a distance of twenty or thirty miles. The adjacent plantations are protected by means of levees; but when a crevasse occurs, the mighty river extends itself for many miles over the level prairies, and carries off fences, logs, and all the loose rubbish it encounters. The river is often diverted from its regular channel by the deposition of sand-bars or other obstructions; and so it was with those of the coal lakes. Their liability to such obstructions, as well as to great freshets, is sufficiently apparent by the known circumstances which involve, to a greater or less extent, all existing streams. These we can see and clearly comprehend; and if similar contingencies be allowed for the primitive rivers and lakes, we have no further difficulty whatever in accounting for the varied strata *alternating* with the seams of coal, nor for all the other phenomena associated with them. We can thus dispense with terrific earthquakes, volcanoes, upheavals, and depressions of the land, and satisfactorily explain all the circumstances of the origin and deposition of the coal, according to existing principles of natural causes and effects.

But I have thus far been describing only those basins which are of *fresh-water origin*—as the anthracite regions on the eastern slope of the Alleghanies. All the bituminous coal on the *western* slopes of those mountains contain marine as well as fresh-water fossils. I have already observed that, at the beginning of the Devonian period, the estuaries of the sea extended over the whole region of country now comprising these mountains, and that they were gradually receding westward, to the great gulf which overflowed Nebraska, Texas, Louisiana, and several other southwestern states. When the coal forests attained their vigor, these arms of the sea were met by the drainage of the land. The basins began to be separated by shoal water, then by sand or concretionary limestone bars; then the waters of the sea were wholly withdrawn, and the bars cut through by connecting *rivers*. The high tides may still have penetrated far eastward, especially as the basins were yet on a comparative level with the sea; but when the coal began to accumulate in the *upper basins*, its influx was separated by longer intervals, and the quantities of marine fossils show a consequent diminution and an ultimate thinning out in that direction. The veins of fossiliferous lime in the coal measures of McKean, are seldom over two feet in thickness; on the Monongahela, and at Cumberland and Broad Top, they vary from two to twenty feet; while at Wheeling they expand to forty, fifty, and sixty feet in thickness. There are, however, local variations; sometimes the limestone does not occur at all in the eastern basins, while it may appear at other places three or four feet thick. The coal is deposited in thin seams, and was perhaps cut off by the influx of the sea, since the fossiliferous limestone occurs directly *over* some of the veins, without the interposition of the carbonaceous shale. This is a very common feature in all the bituminous basins, not only in the United States, but elsewhere throughout the world. The coal itself, in Missouri and Illinois, is penetrated by cubical laminæ of silex, thus showing its presence in the water of the basins wherever the coal was deposited.

In going westward, we find the sea lingering for long periods in the coal basins. Indeed, in many cases, the coal was deposited in calcareous or silicious waters, somewhat modified by the drainage of the land. The absence of fish and crustaceous animals in the *coal*, can be accounted for from the fact of its waters having been impregnated with the prevailing *tar* and *pyroligneous juices* of the vegetation. Upon the destruction of the vegetation, and the *purification of the waters*, all these animals again made their appearance—though never in great abundance. That the waters of the sea, during the coal measures, never *reached so far up* as the anthracite basins, is evident from the absence of limestone, and of all the marine fossils which are found in those below. The veins in the upper bituminous basins are few and thin. The coal material, instead of being deposited in the immediate basins, on the margins of which it grew, was borne *downward by the current*, and helped to form the larger veins at Pittsburg, Cumberland, and the Monongahela. Below these points, a thinning out again occurred, which subsequently formed the separating axis or rim between two basins; and thus all the region from Pittsburg, Wheeling, Pomeroy, Kanawha, and Alabama was cut up and divided into *numerous basins*, connected by narrow straits, precisely like the fresh-water lakes of the Northwest. The estuaries and tides of the sea were gradually withdrawn as the upper basins filled up; and the number of distinct veins they contain is in proportion to their depth and the time occupied in the filling up. The anthracite basin was very deep. When the coal began to accumulate, it could not have been less than fifteen or eighteen hundred feet. The depth of Lake Superior is one thousand; and notwithstanding the extraordinary purity of its waters, and their freedom from sediment, it must originally have been at least twice the present depth. The depth of all the other lakes is considerably less—that of St. Clair not being over twenty feet, while the surrounding flats, embracing millions of acres of surface, have apparently just emerged from the water. And it was thus with the lower and some of the intermediate coal lakes; but while many of them, like St. Clair, filled up at an early day, the anthracite basin continued deep, like Superior, and it went on quietly accumulating its coal, layer after layer.

The numerous deposits and veins of asphalt, chapapote, bitumen, petroleum, pitch, condidum, and other liquid and solid combustibles, occurring in various quarters of the globe, may here be briefly referred to in further illustrations of the formation of coal. The celebrated pitch lake of Trinidad, lying upon one of the West Indian Islands of that name, is said to be three miles in circumference; but its thickness or depth is unknown, from the difficulty of measuring it. It occupies the highest land in the island, and emits a strong resinous odor, sensible at a distance of ten miles. Its first appearance is that of a lake of water; but when viewed at a nearer point, it seems to be a surface of glass. In hot weather, it liquifies to the depth of an inch or more, and cannot then be walked upon. The geological data in the vicinity exhibit traces of volcanic action; and not only in the lake itself, but in the neighborhood, are seen holes and fissures, sometimes containing liquid bitumen or petrol oil. Fissures of great length, from four to six feet wide, traverse the surface of this lake, in every direction, and are generally filled with water. The consistence and general appearance of the pitch or bitumen, when hard, is similar to that of coal, only the color is rather greyer. It is very brittle, and breaks into small cellular glassy fragments. Some of the more elevated parts of the surface are covered with thin brittle scoriæ. The pitch is used for coating ships, and thereby protecting them from that pest of the West Indian seas, the *teredo*, or *borer;* it is also applied as an ordinary varnish, and in some other minor uses. Not far from this lake, and near the sea shore, both coal and schistose plumbago are found in considerable abundance. Lignite, or brown coal, also exists. Near the same island, south of Cape de la Brea, is a submarine volcano, which occasionally boils up and discharges a quantity of petroleum. Another occurs on the east side of the island, which throws up on the shore masses of bitumen, black and brilliant as jet. It would appear that the island is underlaid with seams of petroleum oils and gas, which are thus injected to the surface by the expansive force of the latter. The pitch of the lake evaporates the carburetted hydrogen, thus parting with a portion of the oil with which it was originally associated. This is sufficiently manifest in the fact that efforts long since made to use it as ordinary vegetable tar or pitch, have failed to make it available or profitable, because it requires too much oil to be mixed with it. The substance is passing, by gradual transitions, from its original condition of petroleum into that of bituminous coal.

The Chapapote of Cuba, commonly called coal, is mined in the same manner as the latter mineral, and appears in several positions in the rocks in the vicinity of Havana and Matanzas, in enormous deposits. It occurs in the fissures of stratified rocks, in wedge-shaped veins, enlarging from the surface downward, thereby indicating its origin from below, among magnesian and metamorphic rocks of serpentine, diorites, and euphotides, accompanied with quartz and chalcedony, and sometimes copper. The heat which metamorphosed the rocks also expelled the petroleum, which thus solidified in the caverns and fissures of the upper strata. Chemical essays of this chapapote show: of carbon, 34.97; volatile matter, 63.00; ashes or cinders, 2.03 = 100. A mine situated six miles from Havana, and which was described by M. Castales, in 1842, was found to contain a deposit forty-eight yards deep, perpendicularly, and more than one hundred and eighty feet in horizontal extent. The bottom, however, had never been reached, but the explorations made indicated one of the greatest deposits of mineral asphalt or bitumen ever found in the world. While the chapapote exists in many places on the island, whenever a disturbance of the strata has occurred, *flowing springs* of petroleum are no less abundant. Some of these springs have been known for more than two centuries. Indeed, the whole island is penetrated with bituminous matter to a most surprising extent. Even the solid quartz, the serpentine rocks, and the veins of chalcedony, have cells and cavities filled with liquid pitch; and the air is scented with it when these rocks are broken by the blows of a hammer. In this respect it resembles the mineral pitch found filling the cavities of chalcedony and calc-spar in Russia.* Even in the bay of Havana, the shore, at low water, abounds with asphalt and bituminous shale in sufficient quantity for the paying of vessels as a substitute for tar. It is stated that, in buccaneering times, signals used to be made by firing masses of this chapapote, whose dense columns of smoke could be recognized at a great distance, and served as signals to vessels at sea. It is a matter of history that Havana was originally named by the early visitors and settlers, *Carine*,—"for there we careened our ships, and we pitched them with the natural tar which we found lying in abundance upon the shores of this beautiful bay."† Petroleum leaks out in numberless places, in this delightful island, and it is astonishing that it has thus far excited no particular notice, except as a natural phenomenon. M. Bousingault, in a dissertation on the bitumens of France, remarks that the only contradictory fact opposed to his conclusion that the geological position of mineral pitch is in formations referable to the super-cretaceous group, is that given by Alexander Von Humboldt, who, in his travels in South America, saw at Punta d'Acaya, on the coast of Caraccas, petroleum issuing from mica slate, and extending far out into the sea. To these exceptions might be added many more, for all the springs of petroleum and of mineral pitch in the West Indian Islands, and in South America, are associated with metamorphic rocks, or rocks very nearly as old as mica slate. But it does not necessarily follow that the oil itself is of cotemporary age, although it might be assumed that the unctuous

* Allen's Manual of Mineralogy. † Early History of Cuba.

touch of mica and felspar may proceed from contact with such substances. But coal is frequently found reposing on granite, as near Richmond, in Virginia, while anthracite is found among metamorphic rocks, as in Sweden and Rhode Island; yet those facts do not prove the coal to be of the same age as granite or of mica slate. It only proves, in short, that vegetation existed at an earlier epoch than the geologists have hitherto allowed—that such vegetation, or oily remains of vegetation, lodged on and in the stratified seams of primitive rocks, and that by the heat or sublimation which decomposed it, the expansive force of the resultant gases injected the oil into the fissures and cracks of the overlying and adjacent strata.

Compact mineral pitch, like that of Cuba, and copious streams of petroleum, also occur opposite the city of Maracaybo, in Venezuela, and on the borders of the lake. The petroleum is employed here, as in Havana, for paying the sides and bottoms of vessels. Toward the north-east margin of this lake, which is two hundred and fifty miles in circumference, is a remarkable mine of asphaltum, the bituminous vapors of which are so inflammable that, during the night, phosphoric fires are continually seen, which, in their effect, resemble lightning. They are more frequent during times of great heat, than in cool weather, and go by the name of the "lanterns of Maracaybo," because they serve both for lighthouse and compass to the Spaniards and Indians, who, without the assistance of either, navigate the lake.[*]

The bitumen of Murindo, in New Grenada, is of a brownish black color, soft, and has an earthy fracture. It has an acrid taste, burns freely, with a smell of vanilla, and is said to contain a large quantity of benzoic acid. This arises, apparently, from the decomposition of trees which contained benzoin,[†] their decomposition precipitating the

---

[*] McCullough's Geographical Gazeteer.   [†] Ures' Dictionary of the Arts.

secretions of which the trees were composed. Coal is found in this state at an elevation of over six thousand six hundred feet—being about the same as that of New Mexico and Upper California. The coal mines are worked extensively by English and American companies.

In Mexico, on the Salado river, near Reveilla, situated about one hundred and twenty-five miles above Camargo, bituminous coal exists in quantity, and has been worked by an American company. A coal formation, fifty miles in breadth, very likely a continuation of that of the Rio Salado, crosses the Rio Grande from Texas into Mexico at Loredo. Coal is also found in the provinces of Oajuca, San Louis Potosi, and Vera Cruz. In the villages of Sayultepec and Muloacan are fountains of petroleum, which discharge their contents over a wide extent of country. The oil called "Mexican Mustang Liniment," formerly used for sprains and rheumatism, is derived from these and similar springs. In the interior of Mexico, according to a writer in Hunt's Merchants' Magazine, are lakes of fresh water, where chapapote is found bubbling up to the surface. When washed upon the borders, it is gathered and used as a varnish and for the bottoms of canoes. It has a pungent smell, like that of liquid asphalt, and possesses many of its qualities.

In Texas, about one hundred miles from Houston, there is a small lake of petroleum that closely resembles the pitch lake of Trinidad. A description was given of this lake, in 1844, in a report to the War Department. It is said to be filled with bitumen or asphalt, and is about a quarter of a mile in circumference. During the cool weather of winter, its surface is hard, and is capable of sustaining a person. From November to March it is generally covered with water, which is acid to the taste, from which cause it has been commonly called the *Sour Pond*. In the summer months a spring occurs near the centre of the lake, from which an oily liquid like petroleum continually boils up. This liquid gradually hardens on exposure to the air, and forms a black, pitchy substance, similar to that which coats the sides of the lake. It is said to be precisely similar to the bitumen of Trinidad, and promises to be of great value for the production of gas. It burns with a very clear bright light, but gives out a pungent odor. This lake, in losing its volative oils, is evidently hastening into the incipient stages of coal. Coal also abounds in various quarters of Texas—the main formation having been already described.

On the falls of the Wallamette river, in Oregon, fossil copal or resin has been found, as also on the shores of the Pacific, north of the Columbia river. Beds of imperfect coal were also discovered by General Fremont, in his explorations in 1844, near the cascades of the Columbia river. One stratum consisted of coal and forest trees, imbedded in strata of alluvial clay, containing the remains of vegetables, the leaves of which indicated that they were of the dicotyledonous order. A very significant fact is mentioned by Fremont, viz., that the stems of the ferns were *not* mineralized, but merely charred, retaining still their vegetable structure and substance; and in this condition, also, a portion of the trees remained. But some portions of the coal precisely resemble the cannel-coal of England; and, with the accompanying fossils, have been referred to the tertiary period. The *pure* coal, it is plain to see, resulted from the resinous secretions of the trees; but as the ferns contained none, of course they were incapable of being transformed into lignite.

Wood and brown coal, of very recent origin, has been found in Kansas, and was described by Lieutenant Johnson in 1845. It occurs on the escarpment of a bluff fifty feet in height, in which are various seams of wood and lignite, intermingled with iron pyrites, and on the surface of the bluff alum crystalizes in considerable quantities. Permanent springs flow from the base, and taste strongly of alum. Seams of wood and sandstone alternate, and the formation, which is described as of the postdiluvial era, has been traced for several miles, at an elevation of one hundred feet above the Red river. On the False Washita river, towards the Wishetaw mountains, the same gentleman met with a dark sandstone having a vertical dip, out of which, throughout its course, a great quantity of bitumen has flowed. A specimen of the liquid bitumen has the consistence and appearance of common tar. It occurs as a mineral oil or petroleum on the surface of a spring near that place. This spring is in the vicinity of granite, upon which the oil doubtless rests.

In the lead-bearing magnesian limestones of Wisconsin are occasionally observed thin seams, or lamina, of a buff-colored shale, which, on being placed on a fire, burn for a while with a moderate flame; after which the residue presents a preponderance of earthy ashes. This asphaltic shale is calcareous, and frequently fossiliferous. It has been, in the absence of other fuel, economically employed in lime burning, as it contains inflammable matter in sufficient quantity to calcine the limestone without additional combustibles.*

At the Albert mines of New Brunswick there is a bituminous substance which, for many years, has been ranked alternately as coal and asphalt. A lawsuit once depended on its being pronounced one or the other, and after hearing the opinions of several of the most distinguished geologists and mineralogists of England and the United States, it was finally determined by the court to be coal. It is a beautiful mineral, very black and glossy, burns freely, and makes an abundance of gas. A gentleman long connected with the mine has shown us specimens of a peculiar white

---

* R. C. Taylor's Statistics of Coal.

resinous wax, which is obtained from the coal by distillation, and which sufficiently indicates the resinous and oily material of which it is composed. Oil is now obtained from it in large quantities, and is very extensively used, under the name of Portland oil, throughout the United States and British Provinces. The earthy shales in the vicinity of the mine are also impregnated with the oil—thus showing its former liquid condition, and its previous existence in the form of oil springs.

The existence of petroleum, or rock oil, in various parts of the United States, was known to the Indians, and to many of the early explorers. Father Hennepin speaks of it in his missionary explorations among the Indians of the North-west lakes, and of the Upper Mississippi, over two hundred years ago. Its presence along the shores of Oil creek, in Venango county, Pennsylvania, gave that stream the name by which it is distinguished. In all the borings for salt, of which there are a great number on the Kiskeminitas, the Conemaugh, and other tributary streams of the Alleghany and Ohio, oil was the usual accompaniment of the saline waters. Whenever these waters oozed out, along the slopes of the mountains, the oil would collect as a thin scum on the surface of the springs, and swamps, and morasses in the vicinity. The Indians used to collect it and use it as a medicine; and it is said to possess peculiar healing properties. The same substance has long since been sold, under the name of "Rock Oil," and "Seneca Oil," as a quack remedy for sprains and rheumatism, as also, more recently, under the appellation of "Mexican Mustang Liniment." The quantity of oil found in many of the salt borings was so great that the wells were abandoned as worthless—the oil giving to the crystallized salt an odor, and an unctuous feel, which destroyed its value in the market. There are many of these old borings which will soon be found valuable for oil, now that the nature of the substance is better comprehended; while it is not unlikely that both oil and salt may, by a little chemical skill, be rendered available to the use and requirements of our domestic economy.

At one of the old salt mines on the Kanawha river, in Virginia, when the borings were being made, a reservoir of gas was struck, the explosive force of which hurled the augur and the surface machinery into the air. The gas was finally tubed, and used for fuel to evaporate the salt obtained from adjacent borings. About the same time, if not in the same well, a vein of oil was struck, which issued up in great force, and diffused itself over the surface of the ground, and thence into the adjacent river. The ignition of the gas extended itself to the oil, and the flames followed the latter to the river, setting fire to the boats along the shore, and illuminating and covering the river in a sheet of flame for many miles below. The extraordinary spectacle of a *river on fire* was thus presented, for the first time in the history of the world. Since the traffic in oil began on the Alleghany river, this spectacle has frequently occurred, both on that river and Oil creek. The destruction of property which such accidents effect has at times been immense. The burning oil spreads over the water, and attacks every thing that it encounters—steamboats, flat-boats, rafts of lumber, and all the inflammable material of the shore, share the infection and the conflagration. The oil was formerly floated down the river *en masse* in wooden scows to Pittsburg, where it was refined and barrelled. An improvement on these wooden structures was the introduction of sheet-iron compartments, or entire boats of sheet-iron, which, in case of fire, are less liable to ignite. Nevertheless the danger is still imminent, and measures have latterly been taken to prevent its transportation from the wells in open tanks, unless accompanied with additional safe-guards.

As soon as the true nature and inherent qualities of coal oil were ascertained, in the United States, systematic efforts were instituted to obtain it by distillation from coal. Several establishments for distilling and refining it from cannel-coal were erected, the most prominent of which was from the coal of Breckenridge county, in Kentucky, and the coals of Coal river, and the Kanawha in Virginia. The oil thus obtained from coal served to furnish the distinguishing name for all the liquids of that character now in the market, while the true coal oil itself has disappeared. So soon as it was ascertained that the oil obtained from coal was identical, or nearly so, with that found in a liquid state in saline wells, and in the natural springs abounding along the margin of the Alleghany coal measures, borings were made into the earth to obtain it. These borings, at first, were confined to the upper layers of the oil strata, and furnished only the oil which, in the course of ages, had been ejected by springs, and subsequently drained into the fissures of the overlying rocks. While the true oil-bearing strata occupy a position at and near those of the saline rocks, and generally occur from five hundred to one thousand feet below the surface of the ground, the oil and saline water, aided by the gases which are always associated with them, have driven them upwards, and diffused them not merely over the surface of the earth, but into all the pores and fissures of the statified measures. In some instances the amount of oil thus held in the cavities of the overlying rocks, and in the seams and cracks of the alternate lamina, has been immense, and yielded extraordinary and unlooked-for results. Nevertheless, the deposit was merely superficial, and was in time exhausted. The exhaustion of particular wells intimidated others from similar enterprises, while, in the mean time, the oil itself became so abundant as to overstock the market. For several years its value at the wells was less than one dollar per barrel, and scarcely seemed to pay the cost of transportation. But the extraordinary cheapness and abundance of the substance stimulated inquiry into its character and qualities, and uses were soon created for it which suddenly enlarged the field of consumption, and greatly enhanced its value. The first use to which it was applied was for light; but it required a long time to develop its qualities. The usual prejudice with which every new and untried experiment is assailed by the public, had to be overcome and lived down. Lamps to properly burn it had to be invented; dangers which attended its combustion had to be corrected; and every precaution taken in its preparation to prove its value, its safety, and its convenience. When the prejudice of the people was thus finally conquered, the oil found its way into every household in the land, and the demand for it rapidly increased. In the meanwhile, certain qualities of the oil were found to be adapted for lubricating machinery, either alone or by combination with other oils. Its cheapness, compared with mineral oils, was so great, that an extraordinary demand was created for this purpose —the railways and machinists all over the country becoming the principal consumers. At the same time, it was found that turpentine could be distilled from it—that, in fact, the oil contained the same elements as were formerly distilled from the vegetable resins of the long-leaved pines of the Carolinas and Georgia; and no sooner was this fact ascertained than it found its way into many new departments of art and manufactures, and became an article of export to France, England, and other portions of Europe. Its uses are now almost innumerable. Containing the bases which we extract from vegetation—because it is itself derived from vegetation—it would be difficult to predict what it is *not* capable of being applied to. The writer of these pages, in a lecture which he delivered on the subject some five years ago, and which was subsequently published in a pamphlet, predicted that it would ultimately be used as the base for alcoholic or spirituous liquors. "We thus find," if I may be permitted to quote from the pamphlet which is now before me, "we thus find in our coals and oils the ingredients that we have for ages been accustomed to look for in vegetation, in animals, and in fish. We can obtain from them the means to warm our houses, the gases to illuminate them, the oils and tallow to lubricate machinery, the turpentine and other spirits to supply the demands of the arts; and no doubt we can, and will distil from them very *good brandy, excellent camphene-gin,* and old, oily, unctuous, *Monongahela whiskey!*" It is perhaps needless to assure the reader that, when making this prediction nearly six years ago, it seemed too preposterous and absurd to give it the dignity of serious language. While I ventured to throw it out as a playful suggestion, I nevertheless had solid grounds to believe in its ultimate realization. A parallel case was in a prediction which I made, after the war broke out, and the supplies of pitch, resin, tar, and turpentine, which came formerly from certain of the Southern States, were suddenly cut off by the blockade, would all be obtained by extracting them from our coals and petroleums. But, singular as the fact may appear, *brandy is now actually distilled,* and that, too, on a somewhat extensive scale, from the coal oils of Pennsylvania! And why not? Do they not furnish the alcoholic base of spirituous liquors? What, then, is to prevent their conversion into liquors of any desired quality? Whiskey and brandy obtain *oil* by age—but in this case they can be made to obtain age by *oil!*

But perhaps a more important use reserved for coal oil in the future is its introduction as a fuel. The day may, indeed, be not far distant, when it will be economically used for heating houses, for driving locomotives and steamships, if not for smelting iron. And why not? Oil constitutes the inflammable principle of coal, as well as of vegetation. Coal and vegetation are alike associated with more or less earthy matter, which is precipitated as a slag or ashes during combustion. The ashes contribute nothing to the heat which the fuel creates, but are rather an incumbrance. Why not, then, employ the inflammable material direct? To do this with economic advantage, it would seem to require only a mode of combustion adapted to the object; for it is plain that a fire-hearth that will burn coal and wood would not answer to burn coal oil or the tarry bitumen containing coal oil. Experiments recently made in United States war steamers, under direction of the Navy Department, have demonstrated the practicability of using these oils as a fuel; but as yet there appears to be no economical advantage, owing to the high price of the article. So soon, however, as we learn *how* to burn it, the cost saved in transportation over coal, and the probable greater convenience of using it, will ultimately introduce it into fields of usefulness not now contemplated in our philosophy.

The almost universal use to which the illuminating properties of petroleum are now applied, with the constantly increasing demand for lubricating running machinery, has suddenly awakened new interest in the subject, and led to the investment of capital, under the stimulus of speculation, to a most stupendous extent. While I write this, it is the all-absorbing subject of conversation. Nearly every man you encounter has his pockets stuffed with oil stocks, with leases of oil lands, or with the bonds involving the fee simple. All the machine shops are busy making pumps and boring apparatus. The railway cars are filled with travellers, hurrying forth and back from the land overflowing with oil and saline water. The hotels of our large cities and towns buzz with oil speculators, like a hive of busy bees. The newspapers are full of prospectuses for oil companies. Prominent merchants—dignified judges and senators—railway presidents—learned lawyers, doctors, preachers—all are anointed with it, and shine and glisten with it, like Moses when he returned from the Mount. Oil has burst forth as a great speculative flame, and it now rules the "court, the camp, and the field."

And why is this? Because we have found out that it is no temporary or ephemeral thing. When the first superficial borings were made, and some of the wells failed, although a certain few knew better, the great mass of our people thought the quantity was exhausted, or at least exhaustible. Many believed, and still believe, that it is eliminated from the veins of coal—that it is secreted and confined to the caverns of the upper rocks in close proximity to the coal, and that permanent supplies cannot be looked for.

This theory has been partially dispelled by the deeper borings which have been made during the last few years. It is now well ascertained that there are two, and perhaps more, distinct seams of the oil, running along the western slope of the Alleghany, with all the regularity of the upper coal veins. But in boring, the same contingencies of success prevail as in mining coal. The veins of oil, like those of coal, have their *faults*. They abound in rocky, and slaty, and aluminous barriers, which cut off and intercept the liquid seam. If the augur happens to pass through one of these, little or no oil can be obtained. If the same augur, however, should go down fifteen or twenty feet in either direction from the first well, it would avoid the fault, and tap the oil. This is our daily experience in mines, and especially in coal mines; and this fact explains the reason, otherwise unaccountable, that of two wells of equal or nearly equal depth, one will obtain more oil than the other, or one will obtain oil in quantity, while the other yields none at all. This singular fact has happened time and again, and is likely to happen so long as we are ignorant of the measures beneath.

But the deep borings of the salt wells of the Ohio, and the numerous streams emptying into it, prove that the strata of oil extends all along the slope of the Alleghanies, and extend, in many-places, far beneath the overlying group of rocks which cover the adjacent States of New York, Ohio, Indiana, Kentucky, Illinois, Missouri, Kansas, Texas, and many others. Oil exists in more or less abundance in all these States; it only requires deeper shafts to obtain it.

To establish this fact, we have now only to recur to a consideration of the circumstances under which it was formed. The reader will bear in mind that the chain of mountains which constitutes the Alleghany, is nearly one thousand miles long and more than one hundred miles wide. During the early Devonian era, this entire mountain system was submerged—it consisted of several distinct basins or lakes, not particularly dissimilar to those of Huron, Erie, or Superior. I have already described the leading characteristics of these basins or estuacies of the sea, and wish merely to recall my observations here, so as to enable the reader to understand the modus operandi of the formation of the beds of oil. The adjacent prairies I have previously described as resembling the St. Clair flats—flats that were impenetrable with rank and luxuriant vegetation. The resinous secretions of the coal plants were strewn into the earth, and drained into these lakes, which, up to this time, were filled with the saline waters, and the concomitant infusoria and mollusca of the sea. The petroleum collected into these lakes formed a thick tarry crust on the surface, and sank down to the bottom beneath an inundation or overflow of sedimentary water. The overflows were sudden, and the vegetable oils were buried and confined amid the rubbish of the forests and the sand and mud which they held in suspension. When the waters drained off, another crop of vegetation furnished supplies for another stratum of liquid bitumen, and again it was submerged, and a strata of sand and mud and debris deposited over it. Now these rocks were all porous; but over them, after the waters became more clear, was deposited a thick seam of plastic clay, having in it a large content of kaolin, or decomposed felspar. This seam of clay is from ten to forty feet thick, and is so compact that neither water nor oil can penetrate it. It overlies the oil; it shuts it down; it imprisons it in the earth; it can only escape along the margins of the clay seam, or where it has been fractured or worn away. The decomposition of the oil by a sublimation which is always going on in the bowels of the earth, forces the oil and water to seek crevices, for escape. They are accordingly driven toward the central axis of the Alleghany, where they have again been driven into the upper measures, which have a gradual receding pitch from the centre. After the oil had been thus imprisoned, the process of accumulating alternate layers of sandstone, slate, and mud, and the vegetable material of the forests, continued for long intervals, in the manner previously described—but with this important difference: as the earth and the vegetation attained a more reciprocal action, or equilibrium to each other, the vegetable material contained less liquid oil, and parted with more of its volatile gases while exposed to the atmosphere on the surface of the lake. We notice that on the pitch lake of Trinidad, as also on that of Texas, the oil is evaporated by long exposure, although oil springs are constantly emptying into these lakes. Now the veins which made the coal were exposed to exactly similar influence. There was not so much evaporation, because the atmosphere was not then adapted to its absorption; but there was abundant time to allow the mass to solidify, which was *not* the case with the previous veins of oil. *They* were imprisoned suddenly, while the veins of coal were accumulated slowly. There was also a very material difference in the vegetation itself. The first was richer in liquid oils—the latter furnished secretions more resinous in their character, and better adapted to coalesce and solidify under the pressure of sedimentary waters. That there was a very considerable variation or alternation in the character of the vegetation is very certain, and the variation was governed then, as it is now, by a positive law of nature. We know that a law of alternation of crops exists in our forests now— that as pine trees disappear, a crop of oaks, or chestnuts, or hickories, succeeds, and *vice versa*. This fact has not escaped the experience and observation of man, and we have a right to assume that the law which exists now was in full force then. It is probable, therefore, that if a crop of calamites prevailed for a certain interval, a crop of Lepidodendria or Sigellaria succeeded, and that such crops varied in their relative contents of oil, as the existing pine trees vary in their relative richness in resinous secretions— the yellow pine exceeding the white pine, and the white pine in turn exceeding that of the spruce, or the hemlock, but all alike belonging to the great family of coniferous trees. During the coal period several hundred of such trees flourished—perhaps in all not less than four or five hundred distinct species. I mentioned before that I had collected some seven hundred and fifty fossil specimens of this vegetation, all of which are now in the Academy of Natural Sciences, in Philadelphia, where they may be seen, under the regulations of that institution, by those curious to learn something of the paleontology of the coal measures.

The phenomena attending the upheaval of the Alleghany mountains have already been discussed. It will suffice here to say, that when the coal and coal oil basins arose, the measures on the western slope were comparatively undisturbed. They became indurated—but the content of oil and coal was not transformed into anthracite, as on the eastern slope, where the greatest amount of heat prevailed. The eastern slope was traversed by trap dykes and volcanic action, metamorphosing the rock, and heaving up ranges of rocks belonging to the previous Silurian era. But to the west, the upheaval was gentle, and was attended with but few instances of contortion. The coal oil measures were thus brought within a few hundred feet of the surface, with the coal measures reposing still nearer the surface, but with a gradual increasing dip towards the north, where, after forming a basin, they again arise, allowing the gases to escape in Western Canada, and in Western New York and Ohio; and thus forming a true but extensive anti-clinal axis is of both coal oil and coal.

But that there was an alternation of the resinous vegetation, is established in the fact of the difference in the veins of coal oil and the coal themselves. We know that some veins of coal are richer in oils than other veins in the same basins. Nor was this alternation in the vegetation universal at any one time. The variation appears to have been local, as indeed it is now. We do not find *all* our forests abounding in any one particular species of tree at the same time. Different geographical sections furnish different shades of vegetation. And so it was during the coal and coal oil period—and hence the local differences in the quality of the oils. Now, on the Kanawha, in Virginia, on the Big Sandy, in Kentucky, and generally on the head waters of the Monongahela, we find richer and more fatty oils than we find on the Alleghany, the Clarion, or on Oil creek. And why is this? Because the coals, themselves, are richer in oils, and because a species of vegetation flourished in these points, which secreted more oils than these of Pennsylvania. But it is nevertheless notorious, that *between* the layers of fat, bituminous coals, which exist in those regions are other veins which are comparatively worthless—veins not only full of bituminous shales, and iron pyrites, but extremely lean in the carburetted hydrogen, which constitutes the true value of mineral coal. But while we have these lean veins, we have also others that surpass even the ordinary bituminous veins in richness of oils—veins that, to all intents and purposes, consist entirely of solidified coal oil or petroleum. Such a vein, eight feet thick, is found on the Kanawha. If you take a lump, and expose it to boiling water, it will dissolve into a scum, and precipitate the oil, of which it is in great part composed ; or, if you throw it on a red hot stove, the gas and oil will be immediately evolved. It will thus be perceived that there is the same difference between these coals and oils, in their constituent elements, that there is between yellow pine and white pine wood—between white pine and chestnut or hickory. They include coals that are lean and barely inflammable, and coals that are fatty, oily, and full of the gases by which we light our houses.

Conceding, then, that the difference in the qualities of the oil and coal proceeds from the difference in the vegetation from which they are derived, we find a positive chemical identity in their origin. And while there is an intimate relation between the coals and the various resinous oils, they occur, as we have already shown, in nearly every quarter of the globe, in one form or another. They seldom are in direct proximity to anthracite, and this for reasons which are sufficiently manifest. The great family of mineral combustibles comprehends naphtha, petroleum, elastic bitumen, mineral caoutchouc, compact bitumen, asphaltum, mineral pitch, bituminous candidum, mineral oil, and the Seneca oil of New York—many of which, in liquid form, are now obtained in enormous abundance in Pennsylvania, Virginia, Ohio, and Kentucky, from the lower coal measures along the western slope of the Alleghanies. There are, at this time, perhaps not less than three hundred wells in operation, and more than that number under way. If we can suppose five hundred to be productive during the present year, each averaging twenty barrels per day, (which is extremely moderate), the gross aggregate would be about 3,640,000 barrels per year, which, at twenty dollars per barrel, on the average, would yield $72,800,000 in money. We conceive this to be a really low estimate. We believe the actual result will far exceed it; but even if it should only approximate this result, it will be seen that it exceeds the entire value of the wheat crop of the several States mentioned—that it exceeds the entire coal and iron product; and that, as an article of export, it *surpasses all our other staples combined*—including cotton and tobacco. It might be worth while to inquire, however, how long we will be called on to furnish supplies for foreign nations, when we know that the same oil exists in abundance in China, the East Indies, in Sweden, in Norway, in Russia, in South America, California, and Mexico, and in nearly all the West India Islands. There *is* a time, but how far distant every one may estimate for himself, when coal oil will be everywhere produced and consumed, as coal now is, and its value as an article of commercial traffic will be restricted. Any one may have a well of his own, as the farmers in the West now have coal mines for their own domestic use. The value, as a commercial staple, will be regulated by the cost of transportation and the cost of barrelling and refining it.

While we have a family of solidified and liquid combustibles, it is easy to pass from their varieties, by gentle gradations, to a corresponding assemblage of the *true coals*—as the coal-asphalt of New Brunswick, the cannel-coal of Kanawha and Breckenridge, the fat or oily bituminous coal of the Monongahela, the tar coal of North Carolina and Virginia, the semi-anthracite coal of Broad Top and Cumberland, the free-burning anthracite of Trevorton and Lyken's Valley, the medium anthracite of Pottsville, the more compact anthracite of Tamaqua, and the hard, stony anthracite of the Lehigh. *These all belong to one great family of combustibles*, and, of course, have a common vegetable origin. The anthracites may be regarded as the patriarchs—the venerable *heads* of the family group; while the bituminous stand intermediate between them and the more immature or youthful offspring of the ancient forests.

But the question now suggests itself as to the physical circumstances—local, chemical, and mechanical—under which these vegetable resins were converted into so many different mineral substances. It appears from the investigations of the celebrated Baron von Liebeg, and other eminent chemists, that wood, and every kind of vegetable matter, when buried in the earth, exposed to moisture, and partially or entirely excluded from the air, decomposes slowly, and evolves carbonic acid gas, thus parting with a portion of its original oxygen. By this means, it becomes gradually converted into lignite or *wood-coal*, which contains a larger proportion of *hydrogen* than wood does. A continuance of decomposition changes this lignite into common or bituminous coal, chiefly by the discharge of carburetted hydrogen, or the gas by which we illuminate our streets and houses. According to Bischoff, the inflammable gases which are always escaping from mineral coal, and are so often the cause of fatal accidents in mines, invariably contain carbonic acid, carburetted hydrogen, nitrogen, and olefiant gas. The disengagement of all these, it has been inferred, gradually transforms bituminous into coke or anthracite coal.

The accuracy of the chemical changes here enumerated has never been questioned; they are universally recognized as strictly true. It was, indeed, owing to the *decomposition of the wood*, and the *fermentation* or heat thereby produced, under circumstances of moisture and pressure, that the *resinous properties of the coal forests were eliminated*. Yet the authority of Liebeg and other distinguished chemists, is always referred to, as in the foregoing paragraph, to prove the conversion of the *solid wood* of trees into *coal*. Even Sir Charles Lyell, in his "Elements of Geology," leads us to such an inference. It strikes me that chemistry can sanction no such conclusion; and the paragraph above, if it means any thing, means quite the reverse of what Mr. Lyell and his numerous professional satellites have inferred. Liebeg says that the "*partial decomposition of the wood*, exposed to moisture and fermentation, converts it into wood-coal or lignite." Very well. What, then, *is* wood-coal or lignite? It is wood in the state of *decomposition*, expelling or precipitating its *resinous* and oily *juices* into the form of *bitumen*. It is wood undergoing *combustion*, or *parting* with its pyroligneous oils. Combustion is of various kinds and degrees. When you throw a billet of wood on the fire, its resinous juices, in the form of gas, are *consumed*, while the *woody structure* is converted into *charcoal*. A continuance of combustion, reduces the charcoal to *ashes*, and the ashes show that the woody fibre was composed of *earth; viz., iron, lime, silex, clay, magnesia*, etc. Another kind of combustion is that which we see in the open air. It is slow, and occurs without any visible heat, except the *insensible oxygen of the atmosphere*. A fallen tree will slowly decompose—its gases are volatilized and mingle with the air, while the *woody fibre* crumbles away, and mingles with the *earth*—its ashes or mould being precisely similar to the ashes of wood exposed to a flame. Now, combustion is always the same in its ultimate results; when it occurs under the ground, it is by *fermentation*, and the gases, being *unable to escape*, form *compounds*, which compounds are *coal;* and this coal increases in purity with the *extinction* of the woody fibre. In lignite, this woody fibre still exists; but in pure coal it does *not;*—hence the ashes of lignite contain more than *four times* the quantity of *earthy material* as the ashes of pure coal. But *in consequence* of its partial combustion, as Liebeg observes, lignite contains more *hydrogen* than the original wood. The reason is, that the juices in forming into solids, are first decomposed, and thus part with their *water*. The lignite mined in the immediate vicinity of Giessen, in Germany (the very spot from which the great chemist wrote his *Letters*) contains from forty-five to fifty per cent. of hydrogen as it comes from the bowels of the earth. The heat generated in the mine by the *decomposition* of the lignite, when exposed to the air, is so intense that the miners are compelled to disrobe when they enter their breasts. It is this decomposition (or combustion) that discharges the carburetted hydrogen gas to which the Professor alludes, and which finally terminates in the *extinction of the wood*—leaving behind pure bituminous coal, derived from the previously eliminated resins.

The fermentation produced in the vegetable detritus of the forests was sometimes so great and intense, that it amounted to absolute combustion, but combustion without flame, as in the case of charcoal pits. This is proved from the *abundance of mineral charcoal found in the coal measures*. This charcoal, however, is not always charred *wood*, but not unfrequently is charred *resin*, or bituminous coke. Now, to have produced such coke, the escape of illuminating gas was essential, and we can suppose that the atmosphere of the coal period was at times suffused with these gases, which, in consequence of its humidity and density, would precipitate them upon the earth in the form of *smoky soot or lamp-black*. The city of Pittsburg affords an idea how bituminous smoke can thus evolve lamp-black. When the atmosphere is heavy, that city is involved in the sulphurous fumes, dense vapors, and floating soot of its manufacturing establishments, and the houses are blackened with it, and the streets incrusted with it. During the coal period, the dense black smoke, sometimes overspreading the forests, would *precipitate soot in the form of black carbonaceous snow*, and this, mingling with the waters and liquid oils, would be borne off to the coal lakes as a black, viscous, and tarry ink. The entire lakes would, in time, be resolved into lakes of oleaginous tar; which would finally settle down, like the pitch of Trinidad, into a solid and compact body.

But, it may be remarked, the coal, after all, is *derived* from the vegetation. And so, indeed, it is. But the geologists would have us infer that it was derived *directly from the wood*, without the interposition of resin or bitumen; that, in fact, the wood was not *decomposed*, but only *changed* from one condition to another. It is against this empirical theory that I object, for there is no sanction for it in all the laws of chemistry. It is a matter of astonishment to me that the decomposition of the wood, and its final expulsion in carburetted hydrogen (or, rather, its new chemical combination, for I do not think it was *expelled*, except under the circumstances already mentioned), should so long have escaped observation in treating the chemical phenomena of the coal. For this reason, among several others, woody fibre cannot be detected *in* the coal—the microscopists to the contrary notwithstanding. And had the fact been investigated on independent premises,—had Buckland, Brongniart, Bowman, Lyell, Hitchcock, Rogers, or any one of the host of geologists who have remarked and freely speculated upon the absence of such trees in the *solid coal*, investigated for themselves, instead of adopting the ill-digested surmises of each other, there never could have been any disagreement or difficulty in reconciling *all* the phenomena of its origin and deposition. It was mainly through their misapprehension, extensively promulgated in their writings, that the idea of alternate elevation and depression of the land, of earthquakes and floods, periodical submergence of the forests, and their *direct* conversion into bituminous, and thence by other earthquakes, into anthracite coal, has been so long and so generally entertained ; and this, too, in the face of the fact (which they never have been able to explain), that, wherever trees, and limbs, or leaves *were* found in the solid coal, they were invariably converted into non-combustible earths. There is, unfortunately, too great a willingness, among the professors of the Natural Sciences, for one writer to tacitly adopt or quote or give currency to the visions of another. The books are full of theories which have to be changed or modified with every progressive step ; and these theories are transferred from one book to another, with occasional emendations and additions. One-half of the geologists whose names are mentioned with respect, and who receive credit for scientific acumen, are really but mere tinkers and peddlers in the small facts which the experience of practical miners and workers in the rocks has disclosed. And such, too, rank with the *great men* of the earth—beneath whose severe and awful frown, even our Bible must be read with stealth, and its sacred truths believed in with mental reservations or overshadowing clouds of dark suspicion. The world, indeed, is governed too much by its so-called "great men." Popular opinion is too often based upon the flimsy shams of science; professional reputation is too often created by the "dummies" of Bob Sawyers and Ben Allens (late Nockemorfs)! When plain, sober, practical men once obtain a hearing in the earth, it will grow wiser and better, and the number of empirical quacks will be diminished.

Lignite occurs in a formation more recent than that of the coal measures, and, on strict geological principles, ought not to be considered in this place. I may remark, however, that it generally occurs in small and shallow basins—little lakes, into which were deposited the vegetable resins of Tertiary forests, and the solid trunks and branches of trees. The liquid oils penetrated the pores of the wood as they yielded to decomposition; and in time the whole mass of vegetable material, by overlying pressure, settled down into a compact deposit of *impure bitumen*. The lignite has therefore some resemblance to coal. When relieved of its large content of water (which sometimes amounts to one-half its weight), seventy pounds evolve as much heat as fifty pounds of anthracite. The ashes it leaves behind vary from twelve to seventeen per cent., while that of pure anthracite rarely exceeds four per cent. These ashes are composed of quartz, clay, iron, lime, and mere traces of some other primary substances. Now, the ashes of wood consist almost entirely of *potash*, and hence we are entitled to infer that, with the increase of bitumen, the wood disappears, and that, even in lignite, such is the preponderance of the bituminous principle, that very little wood, properly so-called, is left behind. And as showing how far the decomposition of the wood has proceeded, it may be observed that the ashes of the lignite contain exactly the same ingredients as the surrounding shales and slates that overlay the coal

ORIGIN OF COAL EXPLAINED. 159

beds. The same remark applies to *all* coals; and I may here mention again; that the remains of the ancient coal forests—that is, the decomposed trunks of trees, grasses, and weeds—constitute the rich *vegetable mould of the western prairies.* The black soil, often from six to ten feet deep, which comprises the extraordinary fertility of those regions, is the result of the dissolution of the stupendous and wide-spread forests of the coal era. The indurated shales that accompany the veins of coal are somewhat similar; but they have passed through different chemical changes, and their original character has been considerably modified. But wherever a fossil tree or branch is found, the solid trunk—the woody tissue of the interior—is invariably represented by such shale, while the outside is as uniformly surrounded by a *thin coating of coal.* These facts are curious, but they are plain and overwhelming: they are not isolated cases, selected to establish an empirical law; but they are *universal, undeviating,* and *irresistible.* When a tree decomposes in the forest, we see it gradually crumble into dust; and in time, no one could separate it from the earth of which it forms a part. It was so with the coal vegetation, with this difference: that before *entire decomposition* ensued, pressure and fermentation occurred, the result of which was that the resinous juices of the tree were *squeezed out,* and while the heat converted *them* into a thin crust of coal, it resolved the *woody tissue* into the silex, alumina, and iron of which it was primarily composed. It is not unusual to see the limbs of trees thus changed into quartz, sandstone, limestone, slate, or iron pyrites; *but no human being has ever seen them converted into pure coal.*

Amber is a species of lignite; and it is not only valuable in medicine, chemistry, and the arts, but is highly prized as a beautiful ornament in jewelry. While it was yet soft, and exuding from the tree in the form of a *gum,*

11

amber formed a complete trap for the forest insects, and held them firmly in its sticky clasp. As the gum received successive additions, these insects became buried in the substance. The subsequent hardening and transparency of the mineral exhibits their fossils very beautifully, as also the form of leaves and flowers collected in a similar manner. The stone is very valuable in some countries, especially in the East Indies; but gems are by no means common. Nodules, varying in size from a chestnut to a pine-apple, are often found in the lignite; but their degree of transparency and beauty is irregular.

Lignite can only be converted into coal by separating the woody tissue from the resinous matter that fills its pores. As the woody tissue is volatilized by decomposition, the fermentation operates upon the remaining resins, and converts *them* into bituminous coal. This process can only be accomplished by heat, or very slowly and partially by the fermentation caused by overlying pressure. But it may be assumed that, so long as the lignite remains unaffected by heat, the chemical changes will be so slow as never materially to affect it, since the tar and water, and the oils permeating the woody structure, act as a preservative. It is known that railway sleepers are impregnated with oil of tar, and other bituminous substances, to prevent their decay; and this process is copied from nature in the preservation of lignite. The amount of resinous liquid injected into railway sleepers, by means of hydrostatic pumps, averages about *ten pounds* per cubic foot. The sleepers thus prepared are, in many respects, similar to lignite; and were the process of injection continued as the wood itself decomposes, there would ultimately be nothing left in the sleepers but *compact vegetable resin*. If this resin were now exposed to slow heat, in an air-tight retort, the result would be bituminous coal. If the heat were continued, so as to *char* and

ORIGIN OF COAL EXPLAINED. 161

partially *consume* the volatile gases, the result would be anthracite coal. Here, then, we have the whole process of the chemical transformations which the different vegetable substances embedded in the earth undergo. Were the pitch lake of Trinidad, or the chapapote of Cuba, or the asphalt of New Brunswick, subjected to *slow heat*, under the pressure of superimposed rocks, they would be flattened down into a thin seam; and the *imprisonment* of their gases (not their escape) would result in a combination which would be bituminous coal. Increased heat would *consume* or *decompose* the inflammable gases, and the next result would be *anthracite*. The heat again increased, would transform the anthracite into unctuous plumbago—increasing the *weight* and *compactness*, in every case, in proportion to the decrease of bulk.

Before concluding my observations on the coal formation, I will notice a phenomenon which Prof. Rogers claims to have originally pointed out in explanation of the gradual transition of bituminous into anthracite coal. His remarks have been copied and tacitly indorsed by Lyell and others; and while they are certainly more plausible than his numerous theories generally are, they cannot stand the test of critical examination. "It is invariably found," says Mr. Rogers, "that the coal of the Alleghany is most bituminous toward its western terminus, where the veins are level and unbroken, and that it becomes progressively debituminized as we travel southeastward toward the more bent and disturbed rocks. Thus, on the Ohio, the proportion of hydrogen, oxygen, and other volatile matters, ranges from forty to fifty per cent. Eastward of this line, on the Monongahela, it still approaches forty per cent, where the strata begin to experience some gentle flexures. On entering the Alleghany Mountains, where the distinct anticlinal axes begin to show themselves, but before the dislocations are considerable, the volatile matter is generally in the proportion of eighteen or twenty per cent. At length, when we arrive at some insulated coal fields, associated with the boldest flexures of the Appalachian chain, where the strata have been actually turned over, as at Pottsville and Summit Hill, we find the coal to contain only from six to twelve per cent of bitumen, thus becoming a genuine anthracite."

It would thus appear that the bituminous or non-bituminous character of the coal is entirely due to the amount and nature of the disturbance which the inclosing measures have undergone. Lyell says, "the coincidence of these phenomena may be attributed, partly to the greater facility afforded for the escape of volatile matter, where the fracturing of the rocks had produced an indefinite number of cracks and crevices, and also to the heat of the gases and water penetrating these cracks, when the great movement took place which have rent and folded the Appalachian strata. It is well known that, at the present period, thermal waters and hot vapors burst out from the earth during earthquakes, and these would not fail to promote the disengagement of volatile matter from the carboniferous rocks."

Now, in the anthracite basin, there are from forty to fifty different veins of coal, varying from one to thirty feet in thickness, and separated from each other by rocks from ten to one hundred yards in thickness. The disturbance which permitted the gases to escape must have occurred *simultaneously* throughout all the measures; if it did not so occur, then the *escape* of the gases must have been irregular as to time, and a material difference in the *fixed contents* of the coal of the lower and the upper veins would unavoidably have ensued. But if these gases escaped *simultaneously* from all the veins of coal, where have we any evidence of the fact? The lower white ash veins in the southern part of the Schuylkill basin are *two*

ORIGIN OF COAL EXPLAINED. 163

thousand feet below the surface;—is it to be inferred that *their* volatile gases escaped through the pores of the rocks *overlying them?* If so, where is the evidence of the fact? Do the rocks disclose any traces of such escape? Again: In the Wilkesbarre region, the mammoth vein lies within forty feet of the surface, and often outcrops, and on the Lehigh summit, sixty feet of coal were long worked in *open day.* This coal was placed there by the disturbance which twisted and folded the measures;—if there was any *escape of gas*, it would certainly have been *greater in the coal thus exposed* to the surface than in that *two thousand feet below;*—yet there is no *difference* in the fixed elements of the coal. It is *all* anthracite alike, in all the veins, and in all the regions. But the measures at some places are *not* disturbed—not nearly so much as the *bituminous measures* often are; and yet the character of the coal is invariably maintained. The theory of the escape of gas is therefore absolutely preposterous.

The most probable explanation of the phenomenon is, not that the gases *escaped*, but that they were all retained and formed *new chemical compounds*, at the time or before the measures were folded. The heat that elevated the strata, with the concomitant friction and distortion, converted the coal into anthracite by the *combustion* of the volatile gases—*not* by their escape through the fissures and pores of the rocks. For if the coal had obtained access to the air, under the heat that was then metamorphosing it, it would have burned into cinders wherever so exposed, and in no other way and at no other *time* could the hydrogen have escaped. The evidences of heat are found in the scattered fragments of *charred* coal, often found in the very heart of the solid coal, and somewhat abundantly in the adjacent slates. With the coal thus on fire, nothing would have checked its reduction to ashes if carburetted hydrogen had been expelled during the elevation of the measures. No: the coal was anthracite *before* that time, and combustion had been maintained from the *beginning*, the result of which was the final extinction of volatile gases, and the solidification of their resultant *soot* or *lamp-black* into mineral anthracite.

A few years ago, in England, specimens of bituminous coal were subjected to microscopical examinations; the result of which was the supposed recognition of the vegetable structure of the coal. This, although contradicting the idea of the Lyell geologists, that the woody tissue had been *reduced to pulp* as a necessary preliminary for its conversion into coal, was hailed as an overwhelming demonstration of its vegetable or arborescent structure. Having examined coal under the most powerful microscopes in America, I may here give *my* opinion of the value of the test. A slice of coal has to be prepared so thin that it will be transparent. The slice is not thicker than a sheet of ordinary writing paper, and not larger than the superficies of a ten cent coin. It is almost impossible to prepare such a slice from pure bituminous or anthracite coal;—hence the experiments that have been reported were originally made *with brown coal or lignite*, the woody structure of which *nobody has ever questioned*. Coarse and earthy bituminous coal exhibits cells similar to that of pine wood. *These cells are filled up with resinous matter;* and the porous structure of the wood is maintained after the woody reticulation is obliterated. Earthy resin, therefore, has the same structure as the coal; and its porosity only diminishes with the degree of compactness. The porosity of coal and resin is always after the original porosity of the wood—and the supposed woody structure is altogether due to the *instincts* of the mineral (if I may use such a word) to crystallize in that form. Sulphuret of lead and of iron always crystallize in the cubic system, and it may be assumed that every mineral has its peculiar crystalline form. Crystals of acetate of lead, dissolved in water, upon the introduction of metallic zinc, will accumulate upon its surface, as a thin coating, and branch out in a manner exactly resembling the branches and foliage of a tree. The crystals thus obtained are metallic spangles of pure lead. Even native copper, silver, and metallic antimony often assume a branching and arborescent structure. In view of the vegetable origin of coal, no one can be surprised at its cellular, fibrous, reticulated, or medullar structure, under the microscope; but inasmuch as the trees found in the coal are *usually squeezed down to the thinness of pasteboard*, it would be folly to look for the original structure of the trees themselves! And since we know that all coal was soft, and a glutinous liquid, as a necessary preliminary to its chemical transformation, it is idle to maintain that the arborescent structure it reveals is in reality that of trees.*

---

* In drawing the affairs of the Third Day to a close, I may perhaps add a few words, for the sake of affording completeness to its principal phenomena, rather than from any pertinency which they may bear to the general subject. We have alluded to the escape of carbonic acid and other gases as the cause of explosions in coal mines. Such explosions are similar to those of powder, except that they are generally more violent and destructive. The atmosphere is converted into a cloud of fire, and every thing is dashed to atoms that falls within its grasp. The fiery tempest seizes the timbers of the mine, the rubbish, and fragments of coal, and dashes them against the side-rocks;—the men, if they elude the blast, have their ears, mouths, and nostrils filled with sand and mud, and sustain more or less bodily injury from the mere atmospheric concussion. They often avoid the fire by falling down on their faces, and letting the terrific demon ride over them. But, unless the ventilating currents in the mine are very strong, the choke damp immediately ensues, which is even more formidable in its effects. The atmospheric air being destroyed by the explosion, for a time there is left nothing to breathe but poisonous mephitic vapor—hence death by suffocation often follows.

> This is thy work, fell Tyrant!—this the miner's common lot.
> In danger's darkling den he toils, and dies lamented not!

To revert, in conclusion, to the antiquity of the earth, which was briefly considered among the phenomena of the First Day: No one can fail to perceive, in the alternation of so many distinct seams of coal and earthy strata, the utter impossibility of the creation having occurred simultaneously with the first effort of the Almighty volition. It will be observed, on the contrary, that Moses, in assigning *six distinct* and separate epochs, and giving to *each* their respective features, was perfectly correct, and that his narrative is entirely corroborated by all the known laws and facts of Science, and practical experience and observation. The work of the great coal era occupied

> The army hath its pensioners—the sons of ocean rest
> When battle's crimson flag is furled, on bounty's downy breast—
> But who regards the mining slave, that for his country's wealth
> Resigns his sleep, his pleasures, home, freedom, and his health?
> From the glad skies and fragrant fields he cheerfully descends,
> And eats his bread in stenchy caves, where his existence ends!

In England, on one occasion, out of two hundred men in the mine during an explosion, one hundred and ninety-six were instantly killed. In France, on a Monday morning, when the miners, one after the other, were descending the shaft, the first fell dead in a paroxysm of asphyxia. The next one, attempting to aid him, came within the stratum of carbonic acid, and also fell dead. The third, fourth, and fifth shared the same fate, in the effort to extend aid; and there is no telling where it would have stopped, had not the sixth man turned round, and forced the others to return up the ladder. The number of victims to these dreadful casualties in England, some years ago, caused the government to institute measures for the better security of life in the coal districts; for although the Davy lamp, which was then introduced, enables one to penetrate the fiery mixture with impunity, and to point out its presence wherever it exists in dangerous combination, it cannot be conveniently or economically used for the practical purposes of mining. The lamp is surrounded by thin wire gauze, like the delicate net-work of a bolting-cloth, and the discovery that the noxious gases did not penetrate through it, so as to produce explosions, constitutes its great merit and beauty. In this respect, it is one of the greatest achievements of modern science on record. The lamp is called after its distinguished inventor, Sir Humphrey Davy.

many successive ages—ages coextensive in duration with the stupendous magnitude of the work itself.

The universal distribution of coal over the surface of the earth, in cold as well as in warm climates (but more particularly in cold ones !), points to the universal climate that must have prevailed during that era as one of the necessities of the growth of the vegetation. But, as already intimated, the climate was affected more by the radiated heat of the interior earth, than by the solar rays. This is manifest from the fact that volcanic eruptions continued, at intervals, until the close of the Tertiary; and their effects are exhibited everywhere in dykes and the upheaval of vast mountain chains. The Creator invariably placed coal in all those situations where the climate now demands it. Dr. Kane brought with him specimens obtained in the frozen regions of the Arctic circle, where the vegetation that at all resembles that of the coal period is now dwarfed and stunted The same species in tropical regions, however, attains a prolific development—thus leaving us to infer that, during the coal period, the climate was everywhere warm, humid, and similar to our vernal seasons. So thoroughly was the whole Paleozoic atmosphere adapted to vegetation—so completely and exclusively carboniferous in its qualities, that it was, in fact, totally unfit for the support of the higher sorts of animal nature, and we can find no traces of any other creatures in it than those belonging to the class of Radiata and Mollusca, the two lowest divisions of animal life, with a few obscure and singular creatures partaking of the nature of fish and reptiles, but far beneath both in physical organization.

Agreeably, therefore, to the Mosaic revelation, to geological evidence, and to all rational, practical, and philosophical deduction, we are bound to recognize this great period (from the Metamorphic to the Carboniferous) as the *age of Vegetation*—an age *beginning* with vegetation, and *closing* with it, and occupying, for the time being, the whole surface of the earth;—and having thus fulfilled, in every respect, the grand purposes primarily contemplated in the divine scheme, a series of new and somewhat different phenomena were now to be developed. Moses, therefore, in devoting the Third Day exclusively to vegetation, was essentially correct, although, for thousands of years, the world was ignorant of the fact that what were regarded as *black stones*, were in reality the fossilized remains of vast primitive forests! The discovery of this fact was reserved for modern Geology—a science still in its infancy, and wholly unknown to the earlier races of mankind. And the simple fact that Moses pointed it out, in the true order of geological position, shows conclusively that he was endowed with an intelligence amounting to absolute inspiration!

I will again conclude with a quotation from Milton, who briefly recounts the phenomena of the Third Day in classic measure—well worthy of the angel Raphael, who is supposed to be enlightening the mind of Adam in the mysteries of the earth and of his own creation:

> The earth was formed, but, in the womb as yet
> Of waters embryon immature involved,
> Appeared not: over all the face of earth
> Main ocean flowed, not idle, but with warm
> Prolific humor soft'ning all her globe
> Fermented the great mother to conceive,
> Satiate with genial moisture, when God said,
> Be gathered now, ye waters under heaven,
> Into one place, and let dry land appear.
> Immediately the mountains huge appear
> Emergent and their broad bare backs upheave
> Into the clouds, their tops ascend the sky.
> So high as heaved the tumid hills, so low
> Down sunk a hollow bottom, broad and deep,

Capacious bed of waters: thither they
Hasted with glad precipitance, uprolled
As drops on dust conglobing from the dry;
Part rise in crystal wall, or ridge direct,
For haste; such flight the great command imprest
On the swift floods: as armies at the call
Of trumpet (for of armies thou hast heard),
Troop to their standard, so the wat'ry throng,
Wave rolling after wave, where way they found:
If steep, with torrent rapture, if through plain,
Soft-ebbing: nor withstood them rock or hill,
But they, or under ground, or circuit wide,
With serpent error wand'ring, found their way,
And on the washy ooze deep channels wore,
Easy, ere God had bid the ground be dry,
All but within those banks, where rivers now
Stream, and perpetual draw their hurried train.
The dry land Earth, and the great receptacle
Of congregated waters he called Seas;
And saw that it was good, and said, Let the earth
Put forth the verdant grass, herb yielding seed,
And fruit tree yielding fruit after her kind;
Whose seed is in herself upon the earth.
He scarce had said, when the bare earth, till then
Desert and bare, unsightly, unadorned,
Brought forth the tender grass, whose verdure clad
Her universal face with pleasant green;
Then herbs of every leaf, that sudden flowered
Opening their various colors, and made gay
Her bosom smelling sweet; and these scarce blown,
Forth flourished thick the clustering vine, forth crept
The swelling gourd, up stood the corny reed
Embattled in her field; and the humble shrub,
And bush with frizzled hair implicit: last
Rose, as in dance, the stately trees, and spread
Their branches hung with copious fruit, or gemmed
Their blossoms: with high woods the hills were crowned
With tufts the valleys and each fountain side
With borders long the rivers; that earth now
Seemed like to heaven, a seat where gods might dwell,
Or wander with delight, and love to haunt
Her sacred shades: though God had yet not rained
Upon the earth, and man to till the ground

None was; but from the earth a dewy mist
Went up, and watered all the ground, and each
Plant of the field; which, ere it was in the earth,
God made, and every herb, before it grew
On the green stem: God saw that it was good,
So ev'n and morn recorded the Third Day.—*Milton.*

THE FOURTH DAY—ASTRO-GEOLOGICAL.

14 And God said, Let there be lights in the firmament of the heaven to divide the day from the night; and let them be for signs, and for seasons, and for days, and years: 15 And let them be for lights in the firmament of the heaven to give light upon the earth: and it was so. 16 And God made two great lights: the greater light to rule the day, and the lesser light to rule the night: he made the stars also. 17 And God set them in the firmament of the heaven to give light upon the earth. 18 And to rule over the day and over the night, and to divide the light from the darkness; and God saw that it was good. 19 And the evening and the morning were the fourth day.

THERE is no portion of the Old Testament which has created more difficulty in the reconciliation of Revelation and physical Cosmogony, than the lines here quoted; and yet I venture to say there is no portion of the holy record the integrity of which is more readily susceptible of vindication. Moses addresses himself to the people of all climes, and nations, and tongues; and while his language has all the simplicity to commend it to the meanest intellect, it has the extraordinary peculiarity of embodying the most wonderful scientific phenomena. While his words and facts are distinctly comprehended by the weak and lowly, they defy the closest scrutiny of the learned. He is plain to the plain; but doubly fortified against the wise. To the one, he presents the fixed and unalterable quality of numerals; to the other, he occupies the loftiest heights of natural philosophy, and seemingly anticipates all the assaults of human speculation, reason, and experience.

We have already remarked, that during the coal period, the atmosphere was highly charged with carbonic acid.

The interior heat of the earth was still felt upon the surface; and its radiation had the effect of generating moisture and mists from the shallow shores of seas and lakes. The whole surface of the land may be described as somewhat similar to the coasts of Newfoundland. The seashore for twenty, thirty, and fifty miles from the land, is very shallow, but precipitately falls into the basin of the ocean. The shallowness of the water generates mists and fogs, which often extend several hundred miles into the ocean, or over the nearest land. The steamers crossing the Atlantic are frequently surrounded by these fogs for two or three days at a time; and, notwithstanding the usual precautions, the most terrible accidents and loss of human life in the records of nautical experience, have resulted from them. They are sometimes so dense, that the most brilliant light cannot be perceived at a distance of thirty or forty yards.

The vapor radiated from the surface of the earth may not have been as dense as these fogs, but its geographical diffusion was considerably greater. By far the largest portion of the surface of the earth was still occupied by water, and land occurred, for the most part, in vast islands or marshy plains and peninsulas. The atmosphere was consequently warm, humid, and not unlike that which encircles a volcanic crater previous to an eruption—it was full of carbonic acid. We often experience something like it immediately before a summer thunder shower. The air gradually becomes sultry, and the sky is murky and strewn with dark clouds. During the spring, such weather often continues several weeks without interruption; and while it *awakens*, and greatly promotes the *growth of vegetation*, the effect on the *animal spirits* is in the highest degree depressing. If such weather continued for years, instead of days and weeks, it would render vegetation very nearly as prolific as that of the coal period, and perhaps materially change its character; but it would at the same time prove deleterious to animal life, if it did not ultimately extinguish many air-breathing species.

During the carboniferous era, although the light of the sun may have prevailed to some extent on the surface of the earth, and had regularly and gradually increased in force over the previous periods, yet it had not thoroughly *penetrated or dispelled* the zone of vapor which then hung over it. Its light must have been subdued, mellow, and bronze-like. Had it shone with full brilliance and intensity, vegetation could not have attained so prolific a development—its leaves and succulent stems would have prematurely withered and decomposed, and the juices of the forests been evaporated. But the earth was, in fact, a *vast hot-house*, surrounded by a zone of carbonic acid, which the feeble rays of the sun in some measure rendered luminous; but which it was necessary completely to *dispel*, before the atmosphere could be made *transparent*. This was finally accomplished mainly by the prolific vegetation itself, which, in absorbing the carbonic acid, and transferring it in fixed carbon to the bottom of the coal lakes, removed the principal obstruction. But after the deposition of the innumerable layers of coal and limestone, a general and universal expansion of the "dry land" again occurred, and continued with force only diminished at particular localities, throughout the succeeding Secondary and Tertiary formations. Thousands of volcanoes were at this time in active operation in South America, Asia, and Africa; and they were constantly enlarging the base and increasing the elevation of mountains, while raising up new ones out of the sea or on the level plains. In the United States there were but few volcanic *eruptions*, but the expansion, vibration, and upward movement of the crust, produced the series of great wave-like flexures which distinguish the Alleghany mountains. The work of elevating this vast chain or system was, in all probability, gradual; and the up-lifting or expansive movement must have commenced soon after the coal which they contain was deposited. Similar movements were begun at the same time, in various other quarters of the earth, where coal had been thus formed. The main Alleghany, or central axis, was the first to emerge; and its rising necessarily produced vibrations in the soft matter of the crust, on either side. When a submerged log is raised, or while emerging from the water, a wave-like movement is produced on the surface on both sides, which gradually diminishes with the distance from the disturbing object. It was thus with these and other mountains; while we find them lying parallel to each other, they gradually diminish in altitude from the central axis, which may be distinguished by its uninterrupted continuity, or by its not being cut down by the great chasms or gaps which characterize all the subordinate ones. As the central axis arose, the waters of the lakes and sea rolled away with violence; but were, in a measure, again arrested by the emergence of the secondaries, and thus confined, for a time, in the intervening valleys which they helped to originate. Mountain after mountain thus arose, upon a plane gradually sloping from the central or primary axis. The retrogression of the sea occupied successively the valleys thus formed, and deposited, during its brief sojourn, the beds of new red sandstone and fossiliferous limestones which we find in them in various places, and which, of course, lie in unconformable order to, but not *over* the coal strata. The new red was derived from the beds of the old red sandstone, which, in the anthracite regions, was deposited by fresh water; but its subsequent attrition in the sea, rendered the new red layers a marine deposit, and they consequently exhibit sparingly the remains of marine life. The mountains were finally drained by the sea, in its gradual withdrawal in a direction toward the southwestern terminus; but the valleys were for a long period afterward occupied by great fresh-water lakes. These finally found vent through the gaps of the mountains, and then pursued the natural slope of the Alleghany plane to the present basin of the sea. The continuance of elevation filled these valleys with the water permeating the strata of the mountains, and these constitute the sources of all the rivers and lakes now flowing from them. And it is worthy of notice that, from the highest peak of the Alleghany, in a spot not over two square miles in area, one may drink from the crystal waters gushing up from springs which are discharged respectively into the Gulf of the St. Lawrence, the Gulf of Mexico, and the Chesapeake bay. The waters thus emerging from the Alleghany mountain, in Pennsylvania, mingle with the waters of nearly every State in the Union, and finally with those of every nation on the earth.

Now, while the climate of the coal and of the preceding periods was very nearly universal all over the land, the elevation of the mountains, from time to time, and the consequent disturbance of the waters of the sea, produced a *very material refrigeration and diversity of atmospheric temperature.* The currents of air and water, reciprocally generated by mountains and intervening valleys, as well as by surface drainage, soon changed it into many varieties, varying from the icy coldness of the frigid zone to the melting heat of the torrid. The atmosphere thus purified, was not long in revealing, in all their glory and beauty, those heavenly lamps, the light of which separates forever the day from the night, and prepared the earth for the still higher creative scenes that were to ensue.

God said, "Let there be light in the firmament of heaven to divide the day from the night; and let them be for signs, and for seasons, and for days and years." Although, so far as animal life was concerned, there could then have been no practical *use* for signs, or seasons, or days, we have the best grounds to infer that they did not and could not previously have existed in their present form. The climate was a *perpetual tropical spring*—vegetation was bursting forth at all times, and was only checked by its own excessive gravity to proceed further. The obscurity of the sun and moon and stars rendered it a perpetual day, or a day combining the features of night and day; and there was, consequently, no regular astronomical movement by which "signs, seasons, days, or years," could have been determined. It was, in short, an embryonic condition immediately preceding the *regular* operation of the law of gravitation, of vital force, and organic life; for although we had light in the first day, as an element of the *nebulosity* of the earth, and primarily of all the other planets;—yet, in this case, we require the lights in the firmament of heaven expressly "to *rule* over the earth," and to be "for signs and seasons." God, therefore, "made two great lights—the greater light to *rule* over the day, and the lesser light to *rule* over the night. He made the stars also."

Now, it is *not* to be supposed (as it almost always *is* supposed) that the lights here spoken of were *made* for the first time—that is, created. The lights were now "made to *rule* over the day and night," and God had set them *previously* in the firmament to afford such light to the earth. The action, in some measure, is expressed in the present tense;—but the true meaning is unquestionably retrospective upon the previous days The natural and logical inference then is, (for no one can doubt but that Moses fully comprehended all the phenomena upon which his scenes are based), that the light of the sun, moon, and stars had not *yet penetrated* through the vapors surrounding the earth, and could therefore have exercised but little astronomical influence; and it was to permanently *estab-*

lish that influence, and to *rule* over the earth, that their light finally predominated. In support of this proposition, it is sufficient to know three great facts : *first*, that in the earlier eras volcanic eruptions were constantly occurring, the hot steam and vapor of which were enormous and uninterrupted ; *second,* that upon the partial subsidence of volcanoes, a prolific vegetation was nourished, which never could have withstood the scorching rays of the sun, in *addition* to the radiated heat of the earth ; and *third,* that no *land animals* are found in any of these strata, and that they could, under no circumstances, have *breathed* the mephitic atmosphere that then prevailed.

And it is a fact of no ordinary significance, as betraying the solid and enduring basis upon which the holy record is founded, that while Moses makes the sun and moon *rule over the earth,* the system of Hipparchus and Ptolemy, which prevailed in Greece and throughout the world for nearly fourteen centuries, and was as firmly and universally established as any religious dogma, made the earth rule over *them!* It is true that different and more correct views had previously been entertained, not only by the earlier Grecian philosophers, but by the Chaldeans, Egyptians, and the Chinese. But Ptolemy, in the second century after Christ, was the first astronomer who wrote and promulgated a *complete system;* and the theory upon which it was based, was in direct contradiction to that indicated by Moses. Regarding the earth as the centre of the universe, the Ptolemaists found great difficulty in accounting for the irregularities of the planets in their supposed revolutions around it. They sought to overcome this barrier by supposing an individual, holding a light, and describing a waltzing movement around a room;—to a spectator in the centre, the light would appear only *alternately.* Thus it was ingeniously assumed that the centres only of the planets revolved regularly around the earth, while their diurnal movements produced the singular gyrations for which they could not otherwise account. These absurd theories were overthrown, early in the sixteenth century, by the celebrated Nicholas Copernicus, who again placed the sun in the centre of the universe, and resolved a great many complicated phenomena through the discovery of the movements of the earth, and how far those movements affected our observations of other sidereal bodies. Under his theories and mathematical deductions, Astronomy assumed a precision which it had never known before; and the result was, that it soon became invaluable for "signs, and for seasons, and for days and years." Although the discoveries of Copernicus did not directly lead to the subsequent brilliant achievements of the seventeenth century, his *fundamental* views served to indicate to "theoretical astronomy paths which could not fail to lead to sure results, and to the solution of problems which of necessity demanded and led to a greater degree of perfection in the analytic calculus."* An opinion has prevailed that Copernicus was intimidated in the expression of his theories by the fear of priestly persecution; but Humboldt dispels this impression, and observes "that the founder of our present system of the universe (for to him incontestably belong the most important parts of it, and the grandest features of the design,) was almost more distinguished, if possible, by the intrepidity and confidence with which he expressed his opinions, than for the knowledge to which they owed their origin." In describing, in his dedication to Pope Paul III., the origin of his work (a work only printed at his death, 1543, and which he merely saw and touched on his dying bed), he does not scruple to term the opinion generally expressed among theologians, of the immobility

---

* Humboldt's Cosmos, vol. ii. p. 305.

and central position of the earth, "an absurd acroama," and to attack the stupidity of those who adhere to so erroneous a doctrine. "If ever," he writes, "any empty-headed babblers, ignorant of all mathematical science, should take upon themselves to pronounce judgment on his work through an intentional distortion of any passage in the Holy Scriptures, he should despise so presumptuous an attack." In order to show that, deeply penetrated with the truth of his own deductions, he had no cause to fear the judgment that might be passed upon him, he turned his prayers from a remote corner of the earth to the head of the Church, begging that he would "protect him from the assaults of calumny, since the *Church itself would derive advantage from his investigations on the length of the year and the movements of the moon.*" Astrology and improvements in the calendar long procured protection for astronomy from the secular and ecclesiastical powers, as chemistry and botany were long esteemed as purely subservient auxiliaries to the science of medicine.*

Notwithstanding the light which the great mind of Copernicus had thrown upon the mechanism of the universe, most astronomers still adhered to the main features of the Ptolemaic theory, and persisted in regarding the earth as the common centre, around which all the other planets, including the sun, revolved. At the head of these, and at the head of astronomical science, stood Tycho Brahe, whose investigations, for more than a quarter of a century, were munificently supported by the king of Denmark. Toward the close of the sixteenth century, when the doctrines of Copernicus had almost been forgotten, Brahe received at his observatory on the Island of Huen, an enthusiastic young German, named Kepler. He had pre-

---

* Humboldt's Cosmos, vol ii. p. 307.

viously elaborated an ingenious but fallacious hypothesis on the cosmogony and morphology of worlds, which excited the admiration of his future instructor, as betraying the depth, elasticity, and clearness of his mind. Up to this time, it was not only generally thought that the earth was the centre of the planetary universe, but it was also held, contrary to the exemplifications of Copernicus, that their revolutions around their primaries described *true circles* and a *uniform movement*. Copernicus had so far modified this hypothesis as to suppose that the earth, in revolving around the *sun*, gradually moved *outside* and then as gradually *inside* of the circle which its revolution described. This may be familiarly illustrated thus: Place an orange on the top of a hoop, and another on the bottom, both on the *outside;* then place two other oranges on each side of the hoop, but *inside* of the rim. The hoop thus divided into four equal parts by the oranges, represents a true circle; but the *oranges*, it will be observed, are alternately *outside* and *inside* of the rim—consequently, in revolving around the hoop, they will be in the track of a circle, but in point of fact, they describe an *ellipse*. This was a very important and interesting invention, and, although founded in error, had the effect of unlocking the door to subsequent research. The long-continued investigations of Tycho Brahe, on the orbit of Mars, attracted the attention of Kepler, and the data accumulated by the former, led him to the conception that the varying motions of that planet could be explained on no other hypothesis than that of its *revolving around the sun in the form of an ellipse*—the sun itself being *not* in the centre, but in one of the two *foci* of such ellipse. This proposition assumed, the *irregularity* of its velocity had yet to be determined. After an amount of labor which few persons are capable of appreciating, Kepler accomplished the task by demonstrating that the quantity of space *between* the radiating vectors of a planet, and the sun around which it moves, is always in *proportion* to its velocity. If a wagon-wheel were elongated or compressed at the sides, so as to describe an ellipse *instead* of a circle, this proposition might be illustrated thus: The hub, or radiating focus, would be placed on a line drawn through the elongated centre, about one-fourth of its diameter either to the right or the left of the true centre. From this focus or hub, we would extend the several spokes, (the radius vectors of the astronomers), and in proportion as their *length increased*, they would be drawn closer together at the tire, or circumference line. While the space *between* the short spokes would thus be *wide*, that *between* the long spokes would be proportionally *narrow*, and the actual quantity of space or superficial area remains the same in all. Now, it was ascertained that the *velocity* of planets, in their movements around their primaries, *varied* in exact proportion to the radius vectors constituting their orbitual lines. When their distance from the focus *increased*, their velocity was *reduced;* when they approached *near*, it was *accelerated*. This law was subsequently extended by the assumption, *first*, that all the planets moved by similar laws, and were combined as one apparatus; and *second*, that the differences in their revolving movements depended on their distance from the sun. These propositions of Kepler subsequently formed the basis for the discovery of the law of universal gravitation. In the mean time, however, new light had been shed on astronomy by the discovery of the telescope. The merit of this discovery has been claimed by several individuals, all natives of Holland; but priority seems to belong to a spectacle-maker named Hans Lippershey, or to another spectacle-maker named Metius, both of whom, in the year 1608, offered to sell them to the government. The telescope, however, for several years after its invention, was only applied to distant terrestrial objects. In 1609, while in Venice, Galileo first heard of the instrument, and from the description given, he conjectured at once what were its essential features, and produced one for his own use at Padua. By its aid, (though it magnified but seven diameters), he was the first man on the earth to peer into the physical mysteries and configuration of the heavenly bodies. He discovered the satellites of Jupiter, the varied phases of Venus, a multitude of stars in the Milky Way, invisible to the naked eye, and all the mountains, craters, and valleys of the Moon. The transition from natural to telescopic vision, which characterizes the first ten years of the seventeenth century (from Brahe and Kepler, to Galileo) was more important to astronomy than the year 1492, in respect to our knowledge of terrestrial space—(geography). "It not only infinitely *extended our insight into creation*," says the learned Baron von Humboldt,* "but also, besides enriching the sphere of human ideas, raised mathematical science to a previously unattained splendor, by the exposition of new and complicated problems. Thus the increased power of the organs of perception reacts on the *world of thought*, to the strengthening of *intellectual force*, and the *ennoblement of humanity*. To the telescope alone we owe the discovery, in less than two and a half centuries, of thirteen new planets, of four satellite systems (the four moons of Jupiter, eight satellites of Saturn, four, perhaps six of Uranus, and one of Neptune), of the sun's spots and faculæ, the phases of Venus, the form and height of the lunar mountains, the wintery polar zones of Mars, the belts of Jupiter and Saturn, the rings of the latter, the interior planetary comets of short periods of revolution, together with many other phenomena which otherwise escape the naked eye.

* Cosmos, vol. iii, p. 70.

While our own solar system, which so long seemed limited to six planets and one moon, has been enriched, in the space of two hundred and forty years, with the discoveries to which we have alluded, our knowledge regarding *successive strata* of the region of the fixed stars, has, unexpectedly, been still more increased. Thousands of nebulæ, stellar swarms, and double stars, have been observed. The changing position of the double stars which revolve round one common centre of gravity has proved, like the proper motion of all fixed stars, that forces of gravitation are operating in these distant regions of space, as in our own limited mutually-disturbed planetary spheres. Since Moria. and Gascoigne (not indeed till twenty-five or thirty years after the invention of the telescope) combined optical arrangements with measuring instruments, we have been enabled to obtain more accurate observations of the change of position of the stars. By this means we are enabled to calculate with the greatest precision every change in the position of the planetary bodies, the ellipses of aberration of the fixed stars and their parallaxes, and to measure the relative distances of the double stars, even when amounting to only a few tenths of a seconds-arc. The astronomical knowledge of the solar system has gradually extended to that of a system of the *universe.*" This expansion has kept pace with the increased development of the telescope itself, which, from the instrument used by Galileo, has attained a space-penetrating power, in that of Lord Rosse, several thousand times exceeding the original grasp. The Rosse' instrument has an aperture six feet in diameter, and a tube fifty-three feet in length. It is suspended between two towers; and many nebulæ and other objects, previously out of range of the telescope, have been resolved under its gigantic eye. The visual and sensuous domain of man has thus literally and absolutely been extended into the heavens, and his own ideas of law, order, and harmony, derive fresh vigor from an understanding of those which govern *all* the worlds of space.

The discoveries which Galileo made by means of the telescope necessarily placed him, for some time, at the head of astronomical science, which, under the laws of Kepler, now began at once to assume a new aspect. But under his telescopic observations, Galileo also adduced several very important philosophical laws, which still further accelerated the progress of our knowledge of the planetary universe. He laid down, as a fixed law, that every body receiving an impulse to move in space, would move forever in a perfectly *straight line*, provided it were not interrupted or disturbed by any other force. This law, in connection with that previously evolved by Kepler, viz., that all planets invariably moved in *elliptic curves*, and that their radius vectors pass over equal areas in equal times, led the illustrious Newton to the investigation of the *cause of motion*, and the reason *why* the planets, in their orbitual revolutions, departed *from* a straight line. Up to this point it was conceded, on the basis of Kepler's laws, that the sun attracted the planets *to* it, and that its attractive powers varied with the *distance:* or, as it is technically expressed, the sun attracts or deflects the planets by a force which decreases as the squares of the distance increase. The square of a number is that number multiplied by itself, consequently the square of 2 is 4, that of 3 is 9, and of 4—16; if the sun, therefore, is *twice* as far off from a planet at one time as it is at another, its power of attraction is *four* times less, or *vice versa*. The quantity of matter of the sun being considerably greater than that of *all* the other planets combined, and being the centre of the system to which they belong, and from which they all primarily emanated, they have an *inherent* and constant disposition to *return home*. Their *natural* movement, therefore, would be in a *straight line* toward the centre of the sun, precisely as a ball in the air will fall toward the centre of the earth; but before they reach the sun, they encounter streams of *opposing force* (if I may use the word) resulting from *his own motions on his axis,* into which the planets are borne, and then, like a cork in a whirlpool, are whirled around his circumference. These solar currents or zones occur at various distances from the sun, and may be compared (for illustration) to the successive rinds of an onion, with vast intermediate spaces between them. It is in these zones that planets revolve. Every planet describes a movement in such a zone, around the primary orbs; but while their natural impulse is always to fall to the *centre,* they are constantly prevented and thrown off by the swift revolution of the central and parent body. The sun, therefore, alternately attracts and repels the planets; and it is this that not only makes them revolve around him, but also compels them to revolve constantly on their own axes— the periodic times of such revolution increasing with their *distance* from the sun.

The principle of universal gravitation, it is said, was first suggested to the mind of Newton by the observation of falling bodies. When he saw an apple in his orchard, falling to the ground, he was instructively led to reflect on the *cause* of such a phenomenon. Other eyes had witnessed the same thing, times without number. Other eyes observe it now, every day;—but what of that? Men have eyes, and yet cannot always *see;*—have ears, but do not always hear. For many thousand years, people had eaten fruits, and observed them fall, ripe and luscious, into the lap of Autumn; and yet it required many thousand years, and many millions of human beings, before *one* could be found sufficiently god-like to fathom so *familiar a mystery,* and to place it on the broad basis of

Universal Law. Newton not only demonstrated that all bodies, suspended in space, evince a constant tendency to fall to the centre of the earth; but that all the planets were in *like manner* attracted to the great luminous centre of the system. While the amount of such attraction varies with the distance, and with individual planetary circumstances, the universality of its operation is manifest in all nature and to all process of reason. The sun and the earth both attract the moon; but while the latter is compelled to revolve around the earth in consequence of its *nearer proximity*, the other is compelled to drag her along in her annual journeys around the sun. It is thus with all the other planets that have satellites or moons. They are, as it were, *servants* of the superior body, and perform certain functions assigned them by the sun. The moon of our earth, besides furnishing light during the night, exercises a potential influence on the tides, the "seasons, and for signs." Her services are absolutely indispensable; and it is not too much to infer, that were any accident or casuality to occur, by which our lunar influence would be materially impaired or destroyed, the earth would again relapse into that half-chaotic aspect which characterized it during the earlier Silurian and Metamorphic periods. The principle of universal gravitation thus extends throughout all nature, and to every atom; and is perhaps the parent of galvanic attraction and chemical affinity, since it always evinces a disposition to return into *Unity*—to become *One*—to nestle in the bosom of *God!*

"The immortal author of the *Philosophiæ Naturalis Principia Mathematica*," observes the no less immortal Alexander von Humboldt, in his introduction to the third volume of Cosmos; "Newton succeeded in embracing the whole uranological portion of the *Cosmos* in the casual connection of its phenomena, by the assumption of one all-controlling, fundamental, moving force. He first applied physical astronomy to solve a great problem in mechanics, and elevated it to the rank of a mathematical science. The quantity of matter in every celestial body gives the amount of its attracting force; a force which acts in an inverse ratio to the square of the distance, and determines the amount of the disturbances, which not only the planets, but all the bodies in celestial space, exercise on each other. But the Newtonian theory of gravitation, so worthy of our admiration from its simplicity and generality, is not limited in its cosmical application to the uranological sphere, but comprises also telluric phenomena, in directions not yet fully investigated; it affords the clew to the periodic movements in the ocean and the atmosphere, and solves the problems of capillarity, of endosmosis, and of many chemical, electro-magnetic, and organic processes. Newton even distinguished the *attraction of masses*, as manifested in the motion of cosmical bodies and in the phenomena of the tides, from *molecular attraction*, which acts at infinitely small distances and in the closest contact."

"Bodies," says Newton himself, "act one upon another by the attraction of gravity, magnetism, and electricity; and it is not improbable that there *may be* more attractive powers than these. How these attractions may be performed, I do not here consider. What I call attraction may be performed by *impulse,* or by some other means unknown to me. I use that word here to signify only in general any force by which bodies tend toward one another, whatsoever be the cause."*

The seventeenth century, beginning with the telescope, was essentially the age of modern astrononomical, mathematical, and philosophical discovery. Such names as

---

* Principia Phil. Nat., p. 351.

Tycho Brahe, Kepler, Galileo, Bacon, Huygens, Newton, and Leibnitz, stand forth like marble statues, or like the brilliant stars of the firmament in a dark and cloudy night. The laws which govern the fall of bodies were understood for the first time. The rotundity of the earth—to believe which had lately been a high crime against the church (or rather, against the snapping, barking curs that, in all ages, have guarded its portals, and prevented "miserable sinners" from entering!)—had now been established. "The pressure of the atmosphere—the propagation of light, and its refraction and polarization were investigated. Mathematical physics were created, and based on a firm foundation. The invention of the infinitesimal calculus characterizes the close of the century; and, strengthened by its aid, human understanding has been enabled, during the succeeding century and a half, successfully to venture on the solution of the problems presented by the perturbations of the heavenly bodies; by the polarization and interference of the waves of light; by the radiation of heat; by electro-magnetic re-entering currents; by vibrating chords and surfaces; by the capillary attraction of narrow tubes, and by many other natural phenomena."* In addition to the discoveries of astronomy, geology has established the antiquity of the earth, and vastly extended our knowledge of its history and formation, and the cosmical laws which govern it.

But the introduction of the light and heat of the sun, for the first time and in full effulgence, not only produced the most important changes on the dry land, but also in the atmosphere and in the waters of the sea. The ocean, it is known, is traversed by many well-defined and powerful currents, varying in temperature, in geographical direction, and in volume, precisely as the great rivers and

---

* Cosmos, vol. ii., p. 303.

lakes which traverse the interior dry land. The Gulf Stream is one of the most stupendous and marvelous features of the globe. In its physical aspect and relations to the terrestrial economy, Lieut. Maury has happily compared it to a vast *steam heater*, such as are now used for warming houses during the winter. The heating-furnace is the torrid zone; the great boiling caldrons are the Gulf of Mexico and the Caribbean sea; while the Gulf Stream itself, traveling through the waters of the ocean with accelerated speed, and hemmed in on both sides by walls of cold water, constitutes the *great heat-conducting pipe.* The heat of the torrid zone is thus borne off and diffused *en route* to the grand Banks of Newfoundland, and thence, by the trade winds, to the opposite shores of Europe, and finally to the polar regions, where the *cold* waters and vapors are again returned to the caldron to take the place of the out-going hot waters; they are thus again heated, and then again sent through the conducting-pipe. Not only is the climate of the ocean thus diversified, so as to adapt it for the countless creatures that inhabit its "dark unfathomed caves," but the drainage of the land upon which they feed, and from which many of them elaborate their tiny and sculptured shells, is thus distributed and diffused; at the same time that the vapors are caught up by the winds, borne over the earth, mingled with the atmosphere, and thus made to moderate the rigor of the climate in the torrid, the temperate, and the frigid zones. "The quantity of heat daily carried off by the Gulf Stream from these regions, and discharged over the Atlantic, is sufficient to raise mountains of iron from zero to the melting point, and to keep in flow from them *a molten stream of metal* greater in volume than the waters daily discharged from the Mississippi river. Who, therefore, can calculate the benign influence of this wonderful current upon the climate of the south? In the pursuit of this subject, the mind is led from nature up to the Great Architect of nature; and what mind will the study of this subject not fill with profitable emotions? Unchanged and unchanging alone, of all created things, the ocean is the great emblem of its everlasting Creator. 'He treadeth upon the waves of the sea,' is seen in the wonders of the deep. Yea, 'He calleth for its waters, and poureth them out upon the face of the earth.'"*

It was generally supposed, until Prof. Maury proved the contrary, that the Gulf Stream was caused by the influx of waters from the Mississippi and other rivers emptying into the Gulf of Mexico, and backing up into the Caribbean sea. The altitude thus attained, it was thought, was sufficient to produce the current, as a river descending an inclined plane; but the true solution of the origin, as suggested by Maury, must be sought in the *atmosphere*—for the Gulf Stream, in volume, is 'greater than a thousand Mississippi rivers combined, and the supplies from such sources would be wholly inadequate. The atmosphere, aided by the sun, exerts a power inconceivably great; and in lifting water from the earth, transporting it from one place to another, and letting it down again, performs the functions of a great steam engine. "The south seas, in all their vast intertropical extent, are the boiler for it, and the northern hemisphere is its condenser. What is the horse-power of the Niagara, falling a few steps, in comparison with the horse-power that is required to lift up as high as the clouds and then let down again, all the water that is discharged into the sea, not only by the Mississippi or the Amazon, but by all the other rivers in the world." In speaking of the currents of the Pacific Ocean, page 167,* Prof. Maury remarks: "The better to appreciate the operation of such agencies in producing currents in the sea,

---

* Physical Geography of the Sea, M. F. Maury, LL.D., U. S. N.

let us imagine a district two hundred and fifty-five square miles in extent, to be set apart, in the midst of the Pacific ocean, as the scene of operations for *one day*. We must now conceive a machine capable of pumping up, in the twenty-four hours, all the water to the depth of *one mile* in this district. The machine must not only pump up and bear off this immense quantity of water, but it must discharge it again into the sea, on the *same day*, but at some *other place*. Now here is a force for *creating currents* that is equivalent in its results to the effects that would be produced by bailing up, in twenty-four hours, two hundred and fifty-five cubic miles of water from one part of the Pacific ocean, and emptying it out again upon another part. The currents that would be created by such an operation would overwhelm navigation and desolate the sea; and, happily for the human race, the great atmospherical machine which actually does perform, every day, on the average, all this lifting up, transporting, and letting down of water upon the face of the grand ocean, does not *confine itself* to an area of two hundred and fifty-five square miles, but to an area *three hundred thousand times as great;* yet, nevertheless, the same quantity of water is *kept in motion*, and the currents, in the aggregate, transport as much water to restore the equilibrium as they would have do were all the disturbance to take place upon our hypothetical area of one mile deep over the space of two hundred and fifty-five square miles. Now, when we come to recollect that evaporation is lifting up, that the winds are transporting, and that the clouds are letting down, every day, actually such a body of water, we are reminded that it is done by little and little at a place, and by hair's breadths at a time, not by parallelopipedons one mile thick—that the evaporation is most rapid and the rains most copious, not always at the same place, but now here, now there."

But the currents of the sea, as well as the varying temperature of aqueous belts of climate, are also powerfully influenced by the influx of fresh water from inland lakes and rivers. The water thus drained into the ocean, as the common sewer of the dry land, is of course charged with every variety of mineral, vegetable, and animal matter. These substances, being held in suspension and solution, constitute the food of the fauna and flora of the sea, as well as affording the material from which many of them build their sculptured houses, and elaborate their weapons of warfare. These creatures, moreover, are as much dependent on the laws of climate, and the arbitrary failure of supplies of food, as are those of *terra firma*. But after the solid substances are extracted from the fresh water; after they have been absorbed by millions of polyps and molluscan animals, and secreted in vast coral reefs or beds of shells and sponges, the water becomes a thin lixiviate or ordinary sea-water. The thin water, by the law of gravitation, will be displaced by the heavier water, and as the latter sinks, the other is lifted up by the atmosphere to be returned to the land in genial dews and rains. By this process, not only is the solid contents, formerly held in suspension or solution by the water, *left behind*, but also the salt of the sea with which it afterward becomes united. The water raised from the ocean by the clouds is neither *salt water* nor *sedimentary*—it is thin and attenuated rain-water, and, by the process it has undergone in the ocean, is again well adapted for dissolving the particles of earth, and of nourishing and invigorating vegetation, and once more emptying its loads of food for the animals of the ocean! The constant displacement of water in the ocean, as Prof. Maury has shown, is one of the primary causes which give motion to its waters; but, in addition to the mere displacement, the movement is accelerated by the temporary changes going on in the density and specific gravity of the waters—for as the atmosphere abstracts only the water, and leaves the sediment and the salt behind, it follows that the weight of the waters, for the time being, is increased; while their density would form a natural wall to inclose streams of water of less gravity, less density, and warmer temperature. Thus, reciprocal and corresponding currents are maintained in the sea and in the air, the disarrangement of which would involve the lives of all the inhabitants of both elements.

> The least confusion but in one, and not all
> That system only, but the whole must fall.

Well has Prof. Maury shown, in his admirable investigations of the physical wonders of the sea and air, that the "wind goeth toward the south, and turneth about unto the north; it whirleth about continually, and the wind *returneth again* according to his circuits." The same may be said of the sea; whose waters whirl about continually, but invariably return again according to their circuits. If, indeed, the proportions, density, and gravity of earth, sea, and air were not fixed upon a basis of reciprocal and harmonious action, why should we be told that the Creator "measured the waters in the hollow of his hand, and comprehended the dust in a measure, and weighed the mountains in scales, and the hills in a balance."

Now, in taking a retrospective glance at the Third Day, during which the coal was produced as its most prominent feature, who would venture to assert that the complicated and nicely-adjusted laws of aqueous and atmospheric currents, here indicated, were then in full force? If, indeed, they were in force during the Third Day, why not also in the Second, or even the First, when all was comparative unshaped nebulosity or chaos? On the Second Day, there was no land exposed whatever. All was yet under the sea. On the Third Day, the land had indeed emerged, but it was yet miry and swampy. There were no mountains, and consequently there could have been no such rivers as the Mississippi, the St. Lawrence, the Amazon, or the Nile. There were rivers and broad lakes; but they were rather arms of the sea than great inland streams two or three thousand miles in length. There were rivers and lakes—but they were rather ponds resulting from the drainage of the marshes and swamps of the land, than rivers deriving their supplies from the fountains in the bowels and snow-capped peaks of lofty mountains. There were rivers, I repeat;—but they were rivers left behind to carve out their passage to the sea as the land itself arose; they were *not* rivers dependent on rain-clouds and subterraneous caverns for their supplies. The exposed land then did not comprise one-twentieth part of the land now elevated above the ocean.

But that no such system of air and ocean currents prevailed during the Third Day, is overwhelmingly manifest in the fact that a climate prevailed which was *almost uniform all over the face of the earth.* The coal basins then produced are now found in every portion of the globe, in the hottest as well as in the coldest and most temperate zones. Coal exists everywhere, in irregular patches; and how could it have been thus diffused if the climate had been diversified by currents of air and ocean, as we now find it? The plants which produced the coal required *heat*—where could they now find it in the frigid zone? Would the Stigmaria, the Calamite, the Sigillaria, Lepidodendria, and the arborescent Fern grow in unsurpassed luxuriance on the icy slopes or the towering glaciers of the North pole? Are these cold and inhospitable regions the habitats of vegetation? Yet, coal is found there!

If, then, the climate were nearly uniform during the coal period, it follows that no such currents of air and ocean as give equilibrium to our present terrestrial economy could then have existed; and as it is indisputable that these currents originate and perform their functions through the *instrumentality of the sun*, we are bound to conclude that, up to that time, the sun had not yet cast his rays upon the earth in the manner that he now does; but that the atmosphere itself was humid and vapory, and that it only attained its transparency and diversity of temperature *after* the sun had dispelled the mists of the marshy plains, and set in motion the currents of the sea and air. And, in the absence of the sun, how could a system of storms and rains, and of land drainage and ocean currents, be maintained? And where was the necessity for it, when the land, expelling heat itself, was constantly giving rise to fogs, which fed the vegetation with moisture? There was no rain, because there was yet no necessity for it;—there was no direct action of the sun, because the radiated heat of the earth sufficed. Thus vegetation grew, as under the stimulus of a prolonged spring; until finally, toward the close of the Third Day, the Devonian mountains arose—the vapors were dispelled—fresh or sedimentary waters poured into the ocean—continental rivers began to flow, and, as a consequence, the sun illumined the atmosphere, and the present phenomena were set in motion! This movement was aided by volcanic action occurring during the day we are now considering. The whole aspect and qualities of the previous era were changed—the climate was at once refrigerated by the elevation of the mountains containing the coal, such as the Alleghanies; and it was *then*, for the first time, that the rays of the sun pierced the murky atmosphere, and awakened new life in the sea, air, and land!

That Moses, in introducing the sun, moon, and stars on the Fourth Day, contemplated a *radical change of the climate*, so as to inaugurate the scenes that were to follow in the subsequent days, is apparent from the fact that, in the *Fifth Day*, immediately succeeding, he orders the seas to bring forth *abundantly* the moving creatures that have life, including whales, and birds that might fly in the air. The atmosphere and circumstances of the earth, during the Third Day, were wholly unfavorable to the existence of winged animals, or to air-breathers of any kind; while, in the absence of belts of climate in the sea, it was utterly impossible for animals such as now fill its waters to exist to any considerable extent. The few specimens of animals afforded by the Silurian and Devonian rocks sufficiently indicate the fact that no great currents traversed the seas of those epochs, and that, to produce them *abundantly* and in *diversified species and genera*, it was absolutely requisite to create currents of hot and cold water, shallow and deep basins, and drainage of animal, vegetable, and mineral debris from the land. The climate, therefore, had to be *changed;* and this change was originated by the elevation of the coal basins, and the immediate appearance of the sun's beams.

There are few persons on the earth that may be said to be wholly insensible to its beauties, or the familiar working of its physical laws, as exemplified in everyday life. Poets are filled with emotion on beholding a rocky precipice or a frowning mountain pass; they dwell with ecstasy over a landscape, and point out secret beauties in every tree, shrub, or flower; they apostrophize teeming valleys, with their sloping banks of green verdure, and their streams glistening through vistas of bending foliage like sheets of burnished silver. Travelers write books descriptive of their journeyings, in which the familiar and characteristic scenery of the earth is depicted in every possible style and form. The varied zones of climate and physical configuration, aided by the operations of man, constantly open to view new scenes, and present the earth in new aspects. The subject, therefore, is sufficiently comprehensive and exhaustless;—yet, while so much has been written and painted of the earth, is it not singular that so little has been said, and that, consequently, so little is known, of the *sea and air?* Look down into the depths of the ocean;—look at its wondrous caverns, its huge mountains, its high rocky steeples, its vast plateaux of coral reefs, its forests of coral trees, pendant with silicious and calcareous jewels! Behold its swarms of animals—its minute polyps and its enormous whales! Observe its rivers and currents, moving in every direction, but in undisturbed harmony with each other! Observe the shades of human character impressed upon its fishes, its crustaceous, and its sauroid creatures;—see how they are armed for battle—how the weak elude the strong—how the strong prey upon the weak! Observe their social habits, their domestic instincts, their mechanical pursuits! Then consider the magnitude and mechanism of the ocean itself! The mind becomes bewildered and overpowered with a contemplation of the vast field. It is, as it were, looking down upon a *new world*—that which we inhabit, were it exposed to our view for the first time, could hardly excite emotions more profound or reverential. We turn tongue-tied from the view, to silently adore!

It has been computed that the salts of the sea, were they precipitated, and spread over the northern half of our continent, would make a stratum *one mile thick!* "What force," asks Lieut. Maury, "could move such a mass of matter on the dry land? Yet the machinery of the ocean, of which it forms a part, is so wisely, marvelously, and wonderfully compensated, that the most gentle breeze that plays on its bosom, the tiniest insect that secretes solid matter for its sea-shell, is capable of putting it instantly in motion!" Yet the contrivances of man, with all the steam-engines of the earth, could hardly move a stratum a foot thick, if exposed on a heap, during a thousand years!

Nor is the atmosphere less interesting, though still less of it is popularly known. It has a system of interior, local, and general currents precisely similar to that of the sea. It is at once the source and protection of life—without it, no organized beings could exist. We cannot see, though we constantly *feel* its presence. It presses on us with a load of fifteen pounds on every square inch of surface of our bodies, or from seventy to one hundred tons in all,—yet we cannot realize a sense of its weight. "Softer than the softest down—more impalpable than the finest gossamer—it leaves the cobweb undisturbed, and scarcely stirs the lightest flower that feeds on the dew it supplies; yet it bears the fleets of nations on its wings around the world, and crushes the most refractory substances with its weight. It is sufficient, when in motion, to level the most stately forests and stable buildings with the earth—to raise the waters of the ocean into ridges like mountains, and dash the strongest ships to pieces like toys. It warms and cools by turns the earth and the living creatures that inhabit it. It draws up vapors from the sea and land, and again throws them down in rain or dew. It bends the rays of the sun from their paths, to give us the twilight of evening and of dawn; it disperses and refracts their various tints, to beautify the approach and the retreat of the orb of day. But for it, sunshine would burst on us and fail us at once, and at once remove us from midnight darkness to the blaze of the noon."

While so little is known of the wonders of the air and

sea, it is a matter of astonishment to find that, *whenever the Bible has occasion to allude to either element, it does so with a perfect knowledge of all their laws and characteristics.* Things that *our* wise men have been for ages attempting to solve, are here frequently alluded to as facts understood, and hence the unceasing laudation of all God's works in the holy volume. "But where shall wisdom be found," exclaims the perfect man of Uz, "and where is the place of understanding? The depth saith, It is not in me; and the sea saith, It is not with me. It cannot be gotten for gold, neither shall silver be weighed for the price thereof. No mention shall be made of coral or of pearls, for the price of wisdom is above rubies. Whence, then, cometh wisdom? Destruction and Death say, We have heard the fame thereof with our ears. God understandeth the way thereof, and he knoweth the place thereof; for he looketh to the ends of the earth, and seeth under the whole heaven, *to make the weight of the winds;* and he *weigheth the waters by a measure.* When he made a decree for the rain, and a way for the lightning of the thunder; then did he see it, and declare it; he prepared it, yea, and searched it out."

"When the pump-maker came to ask Galileo to explain how it was that his pump would not lift water higher than thirty-two feet, the philosopher thought, but was afraid to say, it was *owing to the 'weight of* the winds;' and although the fact that the air has weight is here distinctly announced, philosophers never recognized the fact until within, comparatively, a recent period, and then it was proclaimed as a great *discovery!*"*

It has required more than five thousand years of hard study and patient investigation of the wisest men of different nations and periods, to elaborate the great truths

---

* Lieut. Maury—Physical Geography of the Sea.

of Nature which we have thus glanced at; and yet every one of them is *plainly anticipated in the prophetic vision of Moses*. Geology was then *utterly unknown* as a science; was utterly unknown even to the philosophic Greeks; was utterly unknown until man became a practical miner, and learned (by a simple contrivance) to raise water out of the pits he made in the earth, as he penetrated into its bowels. Astronomy had not been arranged even into an *imperfect* system until the time of Ptolemy. The fragmentary conjectures of previous ages had no solid basis for support; while the premises of Ptolemy having been founded in *absolute error*, of course all his conclusions were necessarily fallacious. And yet we see Moses, in the *earliest epoch of mankind*, without geological experience or telescopic aid, unfolding the whole scheme and process of creation, from stage to stage. His description stands the test of time. All the schemes of the wise men of old, fade away like the fantastic visions of a dream; but Moses looms up unchanged and unchangeable; and every new discovery only adds lustre to the brightness and beauty of his narrative. Every thought embodies a philosophical *law*—every word carries with it a volume of significant meaning, the *key* to which is found in the rocks, and the elements, and the creatures constituting the earth itself. In vain has reasoning man set up his subtle barriers against it—in vain all his dark innuendos, his pseudo-logic, and his nicely-wrought theories! If they do not lead him back subdued and repentant and wiser to the fountain of all truth, they sink into the earth with his forgotten bones; while the holy record, deriving new strength with every new discovery, and every progressive step in intelligence, still goes on "conquering and to conquer." "Beware," says the eloquent Paul to the Colossians, "Beware lest any man spoil you through philosophy and vain deceit, after the tradition of men, after the rudiments of the world, and not after Christ." In the 6th chapter of his second letter to Timothy, he exclaims: " O Timothy, keep that which is committed to thy trust, avoiding profane and vain babblings, and oppositions of science, *falsely* so called—which some professing, have erred concerning the faith. Grace be with thee." Again, in his second epistle to his brethren at Corinth, he exclaims with an eloquence which nothing less than the spirit of heaven could inspire : " Howbeit we speak wisdom among them that are perfect ; yet not the wisdom of this world, nor of the princes of this world, that come to nought : But *we* speak the wisdom of God in a mystery, even the hidden wisdom which God ordained before the world unto our glory ; which none of the princes of this world knew : for had they known it, they would not have crucified the Lord of glory. But as it is written, eye hath not seen, nor ear heard, neither have entered into the heart of man, the things which God hath prepared for them that love him. *But God hath revealed them unto us by his Spirit ;* for the Spirit searcheth all things, yea, the deep things of God ! For what *man* knoweth the things of a *man,* save the *spirit* of man which is *in* him ? Even so the things of God knoweth no man, but the *spirit of God !* Now, *we* have received, not the spirit of the *world,* but the spirit which is of *God ;* that we might know the things that are freely given to us of God ; and which things also we speak, not in the words which man's wisdom teacheth, but which the Holy Ghost teacheth : comparing spiritual things with spiritual. But the natural man receiveth not the things of the spirit of God, for they are foolishness unto him ; neither can he know them, because they are *spiritually discerned.*"

While the Bible was not written to promulgate a system of science, it is, in fact, the only book ever written that can stand the tests of scientific truths. The Veda, the

Shaster, and the Koran, can stand no such tests, but the Bible can; and even where it seems most beset with difficulty and mystery, those passages, by modern investigation, shine with brilliant light. "If, indeed," says Dr. Cumming, in an address before the London Bible Society, "if the Bible had been written by mere human hands, they might have indicated, here and there, something like a system of science. It speaks of flowers and trees, from the hyssop on the wall, to the cedar of Lebanon; but there is not a hint of a system of botany. It speaks of stars, and sun, and moon; but not a hint of a system of astronomy." And so, too, it may be added, it speaks of nearly every animal on the earth; but not a hint of a system of zoology. The investigator of Natural History is thus unimpeded in his work; but when the Bible *does* make a statement of fact, which involves the truths of science, it invariably turns out that the Bible is *correct*, and that the *science was false*. There are many examples of this, some of which have already been adverted to, while others remain to be noticed hereafter. Job, for example, speaks of himself as standing on the *circle* of the earth; while Isaiah speaks of the *circle of the sea*. It is a singular fact that the rotundity of the earth had not been established until the time of Newton; and that, in consequence of his inability to understand, or rather to measure its true form, he was *unable to promulgate his discovery of the law of universal gravitation*, which was withheld seven years longer from the world in consequence of his inability to satisfactorily demonstrate it, owing to *this very cause*. Even the Church of Rome, notwithstanding these expressions of the Bible, had long made it heresy for any one to believe that the surface of the earth was other than a flat plain, or that the stars were not really *riveted* to the ethereal vault—hence the origin of the term "fixed stars." "Hast thou an arm like God, or canst thou thunder with a voice like him? Gird up thy loins and declare! Canst thou bind the sweet influences of the Pleiades, or loosen the bands of Orion? Canst thou bring forth Maggaroth in his season, or bind Arcturus with his stars?" These words of Job were long a mystery. What *are* the sweet influences of the Pleiades? Astronomy tells us that the stars, the sun, the moon, and the earth, with their leading satellites, constitute *one* group which revolves round *another* central sun, and *that* central sun is but *one* of the mysterious Pleiades. The Bible thus indicates what astronomers, with the aid of the telescope, have only lately realized—that there are *other* systems of worlds planted in space, besides our own, and that these revolve around central suns still greater and grander than our own. Nothing short of an *inspired conception* of the Almighty Creator and his works, could have enabled the promulgators of his divine word to deal in cosmical truths like these—truths so far beyond the ordinary range of the human understanding.

The profound and erudite Alexander von Humboldt, in describing the literature of the Hebrews, and especially their poetry and descriptions of Nature, remarks of the 104th Psalm, that it almost represents, of itself, "*the image of the whole cosmos.*" "Who coverest thyself with light as with a garment: who stretchest out the heavens like a curtain: who layeth the beams of his chambers in the waters: who maketh the clouds his chariot: who walketh upon the wings of the wind: who laid the foundations of the earth, that it should not be removed forever! He sendeth the springs into the valleys, which run among the hills. They give drink to every beast of the field: the wild asses quench their thirst. By them shall the fowls of the heaven have their habitation, which sing among the branches. He causeth the grass to grow for the cattle, and herb for the service of man: that he may bring forth fruit out of the earth : and wine that maketh glad the heart of man, and oil to make his face shine, and bread which strengtheneth man's heart. The trees of the Lord are full of sap; the cedars of Lebanon which he hath planted; where the birds make their nests: as for the stork, the fir trees are her houses." "The great and wide sea" is then described, "wherein are things creeping innumerable, both small and great beasts. There go the ships: there is that leviathan, whom thou hast made to play therein." "The picture of the heavenly bodies," says Humboldt, "renders this picture complete. 'He appointed the moon for seasons: the sun knoweth his going down. Thou makest darkness, and it is night; wherein all the beasts of the forests do creep forth. The young lions roar after their prey, and seek their meat from God. The sun ariseth, they gather themselves together, and lay them down in their dens. Man goeth forth unto his work, and to his labor unto the evening.'"

"We are astonished to find," says the venerable Humboldt, "in a lyrical poem of such a limited compass, the *whole universe*—the heavens and the earth—*sketched in a few bold touches!* The calm and toilsome labor of man, from the rising of the sun to the setting of the same, when his daily work is done, is here contrasted with the moving life of the elements of nature. In the 37th chapter of Job, the meteorological processes which take place in the atmosphere, the formation and solution of vapor, according to the changing direction of the wind, the play of its colors, the generation of hail and of the rolling thunder, are described with individualizing accuracy; and many questions are propounded which we, in the present state of our physical knowledge, may indeed be able to express under more scientific definitions, but scarcely to answer satisfactorily." "The book of Job," observes Humboldt, "is alike picturesque in the delineation of individual phenomena, and artistically skillful in didactic arrangement of the whole work." "The Lord walketh on the *heights* of the waters, on the ridges of the waves towering high beneath the force of the wind." The morning red has colored the margins of the earth, and variously formed the covering of the clouds, as the hand of man moulds the yielding clay. The habits of animals are described, as those of the wild ass, the horse, the buffalo, the rhinoceros, the crocodile, the eagle, and the ostrich. We see "the pure ether spread, during the scorching heat of the south wind, as a melted mirror over the parched desert." "*Who shut up* the sea with doors, when it brake forth, as if it had issued out of the womb? When I made the cloud the garment thereof, and thick darkness a swaddling band for it; and brake up for it my decreed place, and set bars and doors, and said, Hitherto shalt thou come, but no further: and here shall thy proud waves be stayed!" It would not be difficult to construe, in these words, the phenomena of the origin of the world on the basis of the nebular hypothesis. Again, in the 77th Psalm, "The waters saw thee, O God, the waters saw thee: they were afraid: the depths also were troubled. The clouds poured out water: the skies sent out a sound: thine arrows also went abroad. The voice of thy thunder was in the heavens; the lightnings lightened the world: the *earth trembled and shook*. Thy way is in the sea, and thy path in the great waters, and thy footsteps are not known. Thou didst divide the sea in thy strength: thou didst cleave the fountain and the flood: thou *didst dry up mighty rivers*. The day is thine, the night also is thine: thou hast prepared the light and the sun," etc., etc.

I might multiply quotations, *ad infinitum*, for nearly the entire book of Job, which is of equal if not greater antiquity than that of Moses, together with a large portion of the Psalms, are each full of instructive and impressive meaning in relation to the physical phenomena of the earth. No one can read these beautiful portions of the Bible without being impressed with the accuracy of their descriptions, and the spirit of fervid sublimity which distinguishes them as poems.

In concluding my observations on this day, it is, perhaps, hardly necessary to remind the reader that, inasmuch as the light of the sun had only been introduced for the *first time*, during the *fourth day*, and was especially intended to mark the days and seasons, as well as to *rule* over them—no days, properly so understood, *could have existed, or been contemplated by Moses, in the previous epochs.* This proposition is so self-evident, that no candid mind can, for a moment, question it. Although Moses is constantly speaking of *days*, it is clear that he always contemplates *lengthened periods of time*, as we have already surmised in the earlier pages of this volume; for nothing is more certain than that the days of the coal and metamorphic periods (if they could exist at all as a diurnal measure of time) were wholly *dissimilar* to those which occurred *afterward*, and which occur *now*. This conclusion is made perfectly overwhelming in the second chapter of Genesis, where Moses observes (in allusion to this period) that "the Lord God had not caused it to rain upon the earth, and there was not a man to till the ground; but there went up a mist from the earth, and watered the whole face of the ground." Is not this wonderful? Does it not confirm, in the most remarkable manner, the peculiar phenomena of the coal period—but above all, its *singular climate*? If such were its general aspects, we at once perceive the propriety of specially introducing the light of the sun on the fourth day. It is spoken of before, among the phenomena of the first day, and therefore already existed; but its light was now required to *rule* over the days and seasons; and this was fully accomplished after the absorption of the carbonic acid and the elevation of the mountains. The atmosphere soon became more diversified, as well as beautifully clear and transparent, so that, on the following day, or era, the birds were enabled to spread their pinions in the "*open* firmament of heaven." The earth was thus again prepared for the still nobler, and grander scenes that were to follow:

> Again the Almighty spake: Let there be lights
> High in the expanse of heaven to divide
> The day from night; and let them be for signs,
> For seasons, and for days, and circling years;
> And let them be for lights, as I ordain
> Their office in the firmament of heaven
> To give light on the earth; and it was so.
> And God made two great lights, great for their use
> To man, the greater to have rule by day,
> The less by night, altern: and made the stars,
> And set them in the firmament of heaven,
> To illuminate the earth, and rule the day
> In their vicissitude, and rule the night,
> And light from darkness to divide. God saw,
> Surveying his great work, that it was good:
> For of celestial bodies first the sun,
> A mighty sphere, he framed, unlightsome first,
> Though of ethereal mould: then formed the moon
> Globose, and every magnitude of stars,
> And sowed with stars the heaven thick as a field.
> Of light by far the greater part he took,
> Transplanted from her cloudy shrine, and placed
> In the sun's orb, made porous to receive
> And drink the liquid light, firm to retain
> Her gathered beams, great palace now of light.
> Hither, as to a fountain, other stars
> Repairing, in their golden urns draw light,
> And hence the morning planet gilds her horns,
> By tincture or reflection they augment
> Their small peculiar, though from human sight
> So far remote, with diminution seen.
> First in his east the glorious lamp was seen,
> Regent of day, and all the horizon round
> Invested with bright rays, jocund to run

THE FOURTH DAY—ASTRO-GEOLOGICAL.

His longitude through heaven's high road: the gray
Dawn and the Pleiades before him danced,
Shedding sweet influence. Less bright the moon,
But opposite in leveled west was set
His mirror, with full face borrowing her light
From him, for other light she needed none
In that aspect: and still that distance keeps
Till night, then in the east her turn she shines,
Revolved on heaven's great axle, and her reign'
With thousand lesser lights dividual holds,
With thousand thousand stars, that then appeared
Spangling the hemisphere: then first adorned
With their bright luminaries, that set and rose,
Glad evening and glad morn crowned the Fourth Day.— *Milton.*

THE FIFTH DAY—GEOLOGICAL.

20 And God said, Let the waters bring forth abundantly the moving creature that hath life, and fowl that may fly above the earth in the open firmament of heaven. 21 And God created great whales, and every living creature that moveth, which the waters brought forth abundantly, after their kind, and every winged fowl after his kind; and God saw that it was good. 22 And God blessed them, saying, Be fruitful, and multiply, and fill the waters in the seas, and let fowl multiply in the earth. 23 And the evening and the morning were the Fifth Day.

ZOOLOGISTS classify animal species into four grand divisions, named respectively the *Radiata*, the *Mollusca*, the *Articulata*, and the *Vertebrata*. The first division (*Radiata*) is divided into five subordinate groups, consisting of *Spongiaria*, *Polypifera*, *Infusoria*, *Foramenifera*, and *Echinodermata*. These are the lowest and most minute species of animals in existence, and they are frequently only perceptible by means of the microscope. They exist in lakes and seas limited to a certain depth, beyond which the pressure of the water is perhaps too great for their delicate anatomy. The difference between some of them and vegetable structure is so slight as to render it difficult to establish a line of demarkation. The second division (*Mollusca*) comprises a soft and pulpy animal, most generally covered by a calcareous shell. This division includes a great variety of subordinate species, among which may be mentioned the *Brachiopoda*, the *Conchifera*, the *Gasteropoda*, the *Cephalopoda*, and the *Tentaculifera*. These are numerously represented in the Silurian strata by the beautiful fossil shells of the *Ammonite*, the *Nautilus*, etc. The third division (*Articulata*) comprehends six subordinate groups, the leading peculiarities of which, as a whole, are their jointed and ring-like structure. Of the subordinate types, the *Annelida* are worms; the Cirrhipoda a shell-worm, generally found adhering to the shells of other animals; the Crustacea are represented in part by the crabs and lobsters of our seas, though they also comprehend many other types essentially different in structure, among which may be mentioned the sculptured shells of the Trilobites, the fossils of which are often coiled like serpents; while the Arachnida include spiders, scorpions, and the several varieties of insects, butterflies, bees, grasshoppers, and beetles. The fourth and last great division (*Vertebrata*) includes all animals that have internal, articulated skeletons, and is divided into four leading groups — Fishes, Birds, Reptiles, and Mammifera.

Nearly all those comprised in the three divisions first named, flourished during the Paleozoic and the Secondary eras, and, with some changes or alternations of genera and species, many of them reappeared in the subsequent Tertiary and modern formations. Of the Vertebrata, the only animals that did *not* appear before the Tertiary, were those of the Mammifera, though these were represented in the *seas* by the class of Cetacea, comprising whales, dolphins, and seals. Inasmuch as all these animals have contributed largely to the changes wrought upon the earth, a brief description of them is essential; but this must necessarily be general rather than particular, and comprehend classes instead of individuals and species. Most of those belonging to the first division being microscopic animals, little or nothing is known of them except through the labors of original investigators. All we know of them, therefore, or of most of them, is embraced in methodical and technical language, and to adopt this in a work addressed to the popular mind, would be injudicious and foreign to its purpose. Having already derived aid from the valuable work of the late Dr. Richardson, entitled an "Introduction to Geology," (by far the best compendium of that science with which we are acquainted,) I shall compile mainly from it the information necessary to conduct the reader over this important branch of the subject—a subject which, under the name of *Conchology*, comprises in itself a vast and exhaustless field in the illimitable domain of Natural Science. "The Spongiaria," he observes, "are among the lowest forms of animal life. They are composed of a horny frame-work, invested with a simple gelatinous tissue, and furnished with vibratile cilia, for causing currents of water to flow through their porous structure; the horny net-work is consolidated with silicious or calcareous spiculæ. The Spongiaria remain rooted to rocks at the bottom of the sea, or hang like living stalactites from the vaulted arches of submarine caves; or their delicate vegetable forms droop in endless variety from the shelving edges of rocks exposed to the washing of the surge. They are reproduced by small gemmæ, covered with cilia, which are free organisms during the first period of their existence. We form three orders of this class: the first have *silicious* spiculæ, the second *calcareous* spiculæ, and the third have a *horny net-work* without either. Fossil sponges are found in most strata, either entire or decomposed into spiculæ. . . Like the bodies of recent Spongiaria, the silicious fossils contain Infusoria in the interior, which may be detected by the microscope." The spiculæ of fresh-water sponges are found in great profusion in lacustrine Tertiary beds, along with the shields of fossil Infusoria. Many of the *moss agates* are of spongious origin, and present some of the most beautiful gems that come under the eye of the lapidist.

"The skeleton of the *Polypifera* assumes a vast variety of forms, being horny or calcareous, globular or branched, solid or tubular, stellate, porous, or retiform. The gelatinous organized substance of the animal is inclosed in ramified tubular sheaths, or expanded over the surface of the calcareous skeleton which it incloses and secretes. The mouth of the polyp is surrounded with numerous filaments, which, in the highest groups, are furnished with vibratile cilia. Each polyp, or digestive sac, contributes a moiety to the nourishment of the compound body with which it is originally united. This physiological relation occasions remarkable associations; hence the stupendous results obtained from their operations in the seas of intertropical regions, by which the life of the individual is combined with the life of the whole, and the nutriment prepared by each organism is made to contribute to the nourishment of the community, as in the red coral." The calcareous skeletons of some *Anthozoa* are very abundant, and attain a great magnitude in the Pacific, where they contribute largely to the formation of islands and continents. Dr. Darwin, in his work on Coral Reefs, has shown that the zoophytic productions may be classified into three groups—Atolls, Barrier-reefs, and Fringing-reefs; that the vital operations of the animal are limited within a range of thirty fathoms, and that beyond that depth they cannot live; while the forms which the reefs assume depend upon the elevation or subsidence of the ocean's bed, on which the foundations of the zoophytic structure are laid. Coral-reefs stretch along the shores of New Caledonia to the length of four hundred miles, while they extend on the northeast coast of Australia for upward of a thousand miles. Hundreds of islands in the

Pacific are made up almost exclusively of their calcareous remains.

The class of Polypifera is divided into two orders, the *Bryozoa* and the *Anthozoa*. The first comprises five or six families, and the other more than a dozen, separated into two sub-orders. It is unnecessary here even to mention the names of these subordinate groups, because the distinctions between them are extremely delicate, and difficult of specific identity, while it is by no means essential to a proper understanding of the animals as a class. The Bryozoan order is the most highly organized coral of the class, and the fact that it existed during the Silurian era, disproves the idea of a progressive development even in the *first* and *lowest forms* of animal life ever created.

The class of *Infusoria* are so called because they originate and abound in infusions of decomposed vegetation in ponds, lakes, rivers, and seas. They are the animalculæ of vegetable juices, and are so inestimably minute, that they can often be detected only by the highest powers of the microscope. But, small as they are, they comprise many distinct genera, and many hundreds of species. Some of them are inclosed in silicious shells, marked with longitudinal, transverse, or oblique lines, or adorned with various other forms of minute sculpturing. Many of them, especially of the families *Bacillaridæ* and *Peridinidæ*, are found in a fossil state in the Tertiary beds of Europe and America.

Prof. Ehrenberg, of Germany, who has devoted particular attention to this branch of microscopic investigation, has ascertained that twenty-four thousand of these organisms, placed together, would not measure one inch in length. In some infusions, indeed, the creatures are so small, that ten thousand can swim in such a space; hence a cubic inch would contain more organized animalculæ than there are human beings on the surface of the earth! The substance known as Tripoli is composed wholly of the silicious shields of these animals. And Ehrenberg has estimated that a cubic line of this polishing stone contains, in round numbers, the remains of no less than 23,000,000 of individuals. But, as there are in a cubic inch 1,728 lines, therefore there would be in a cubic inch of Tripoli, as sold at the shops, no less than 39,744,000,000 of the fossil armor of the extinct animalculæ! "This overpowering force of numbers," says the late Alexander von Humboldt, in speaking of the propagation of the light of luminous cosmical bodies, "is as clearly manifested in the smallest organisms of animal life as in the Milky Way of these self-luminous suns which we call fixed stars. What masses of Polythalamiæ are inclosed, according to Ehrenberg, in one thin stratum of chalk! This eminent investigator of nature asserts that one cubic inch of the Bilin polishing slate, which constitutes a sort of mountain cap forty feet in height, contains forty-one thousand millions of the microscopic *Galionella distans;* while the same volume contains more than one billion seven hundred and fifty thousand millions of distinct individuals of *Galionella ferruginea*. Such estimates remind us of the treatise, named *Arenarius*, of Archimedes —of sand-grains which might fill the universe of space! If the starry heavens, by incalculable numbers, magnitude, space, duration, and length of periods, impress man with the conviction of his own insignificance, his physical weakness, and the ephemeral nature of his existence; he is, on the other hand, cheered and invigorated by the consciousness of having been enabled, by the application and development of intellect, to investigate very many important points in reference to the laws of Nature and the sidereal arrangement of the Universe."

In Lapland and Finland there are beds of fossil fauna, which the natives mix with flour, and eat. The microscope shows that this farina, which occurs as a comminuted powder, consists of the shields of infusoria. Their fossils are also found in opals and semi-opals, stones which often rank next to the diamond in value, and, in the play of colors, fully equal it in beauty.

The *Foramenifera* are microscopic animals of a simple, gelatinous, fleshy substance, without appreciable organization, which secrete a delicate, calcareous, and many-chambered shell, of extreme beauty, into the cells of which the body of the animal retires. The animal, by peculiar expansions and retractile movements, is enabled to crawl and swim. They are alike wonderful for the simplicity of their organization, and the variety and delicate structure of their shells. Plancus collected 6,000 shells from an ounce of sand, on the shores of the Adriatic, and D'Orbigny found 3,840,000 in the same quantity of sand on the shores of the Antilles. Saldami collected from less than an ounce and a half of rock, from the hills of Casciana, in Tuscany, 10,454 fossil shells of Foramenifera. Several of the species are so minute, that 500 weigh only one grain, and others, still more minute, would require double that number to make the same weight. Yet, so abundant are their remains, that they often form banks that blockade navigable channels, obstruct gulfs of the sea, or fill up harbors, and, aided by the polyps, form extensive islands in tropical seas, or along continents, hundreds of miles in length. Their shells occur in great abundance in the Tertiary strata, while about twenty species are found in the oolite, and two hundred and fifty in the chalk. But they are still more numerous in the modern era—D'Orbigny, the celebrated French Conchologist, having identified over nine hundred species now living in our seas.

"The *Echinodermata*," says Dr. Richardson, "forms the true type of the radiata division, and is composed of animals, fixed and free, with a highly-organized integument, for the most part armed with movable spines. In the rayed families, the organs of locomotion are disposed around a central axis (and are hence often called sea-stars). In the spherical forms, they are ranged in rows like the lines of longitude on a terrestrial globe, and the mouth and the arms are situate at the opposite poles. Each element of the body is in general repeated five times. Thus, the sea-lily has five primary arms; the sea-star, five rays; and the sea-urchin, five pairs of perforated and five pairs of imperforated plates in its shell. The external surface of the skeleton supports a series of movable spines, and the perforated portion gives passage to thousands of tubular feet for gliding over the bed of the shallow shores they inhabit. The three higher orders of this class, are free, ambulatory animals, while the crenoids are generally fixed by a calcareous stem, like the polyps. The higher forms also possess visual organs—a feature in which all the other radiata are deficient. The *Echinodermata* have a distinct system of vessels for the circulation of the blood, and some of them a tree-like organ for respiration. The class comprises four orders—*Holothurida, Echinida, Asterida,* and *Crinoida.*

Such, in brief, are the varied and minute aqueous creatures whose delicate secretions have, in the course of ages, wrought the most stupendous and wonderful changes upon the earth. Inhabiting the more shallow bottoms of the sea, their vegetable instincts have covered it with magnificent coral orchards, that sparkle in the water like the ice-spangled foliage of winter, and elaborating, as it were, pebbled fruits of agates, opals, and emeralds. While thousands of millions of them, as we have seen, could occupy a cubic inch of water, and then not feel any thing like the ambition "for enlarging the area of freedom" which has ever distinguished filibustering man, they have yet reared up monuments of their untiring industry and combined power, far greater and more enduring than the pyramids of Egypt, or the marble temples of Greece and Rome—temples which, in fact, are often erected with rocks composed of their fossil remains. Their islands stretched, and even now extend hundreds of miles amid the watery fields of ocean, and their peninsulas have brought whole continents into friendly union. Such are the creatures—the unseen workers which the great Creator employs to carry out his architectural designs.

The *Mollusca*, which form the second great division of animal life, are generally distinguished for a body inarticulate, soft, and pulpy, and usually inclosed in a calcareous shell. The primary division of the class is based on the development of the nervous system, and on the presence or absence of the ganglia that represent the brain. The first form the *Encephalous*, and the second the *Acephalous* orders. The encephalous possess organs of sense, and their blood circulates in a system of arteries and veins, aided by the contraction of a two-chambered heart. The terrestrial, and most of the lacustrine species, breathe by an air-sac or lung; and all the marine, and most of the lacustrine species, have branchiæ for respiration. Their shells are composed of a cellular, albuminous membrane, indurated by carbonate of lime, and secreted by a portion of the tegumentary system or mantle. The shell is generally external, and presents a great variety of forms. Sometimes it is internal, and appears like a rudimentary bone. They are for the most part marine animals; but many inhabit fresh-water lakes, and a few live on the land. The character of the shells varies with their habitat—those of the marine species are large and heavy, while in the others they are usually light and delicate. The *Acephalous* are all aquatic, and embrace three sub-classes: the *Tunicata* have no shell, but are inclosed in an elastic muscular sac, with two openings; some are solitary, others social, and organically united in groups like polyps. The *Brachiopoda* are inclosed in a bivalve shell. They have two long spiral arms, developed from the sides of the mouth, and fixed to an internal frame work. The *Conchifera* have also a bivalve shell, and respire by laminated branches attached to the mantle. Most of them have a fleshy foot for locomotion.

The *Encephalous* class are divided into sub-classes according to the modifications of their organs of locomotion. The *Gasteropoda* are generally inclosed in a univalve shell. They creep by means of muscular discs, situated under the body. The *Pteropoda* swim by two wing-like membranes, situated at the sides of the neck. When they have a shell, it is thin, fragile, and univalve. The *Cephalopoda* have their locomotive organs arranged round the head, in the form of eight or more arms, with or without sucking discs. Some have internal bones, as the *Lepia* and *Laligo*. Others have an external many-chambered shell.

The classes of Mollusca, thus enumerated, embrace together many hundred species, some of which, from their relation to the ancient earth, deserve special consideration. Of the Brachiopoda, the species *Terebiatula* were very numerous and constituted nearly one-fourth of the Mollusca of the primary eras. The *calceola, chonetes, leptœna, productus,* etc., are found only in the paleozoic rocks. The *Conchifera* embrace nearly a thousand species, many of which are found in the most ancient rocks, and although their generic forms have slightly varied during the long periods which have elapsed, nothing is observed in their history to justify the supposition that there is any process of development from lower to higher forms. Very remarkable changes, however, have taken place in species at different periods. Of nearly a thousand species obtained from the

Tertiary, more than a seventh part of them were found to be identical with *living* species of the same or of distant latitudes. While, in the same genera, the number of species found in the Tertiary, often exceeds that of the species now known in a living state. It appears that there has been a constant oscillation in the *number* of species in each genus in the Tertiary, as compared with the modern epoch; but there has been no gradual perfectioning of the same.*

The remains of the *Gasteropoda* are important to the geologist, as affording unequivocal evidence of the fluviatile, lacustrine, and marine conditions under which strata were formed. "The species of gasteropoda," says Prof. Grant, "are much less abundant in the ancient grauwacke (Silurian) limestones than those of bivalved mollusca; only about seventy species of the former having been yet identified in the strata of that epoch, and a quarter of these belong to the extinct genus *euomphalus*, which ceases with the carboniferous rocks. Most of the species, however, observed in these ancient grauwacke formations, are referred to existing genera, as *turritella*, of which about ten species occur; *turbo*, six species; *buccinum, patella, delphinula*, five each; *nerita, pileopsis, trochus*, and *phasianella*, three species each."

The *Pteropoda*, it has already been remarked, swim by muscular membranous expansions of the mantle, which project from the sides of the head. Their body is naked, or sometimes protected by a delicate shell. They are small animals that float on the surface of the ocean far away from shore. In the North Seas, the *clio* and *liruacia* swarm in such abundance that they are said to constitute the food of the whale. The *clio* is provided with a singularly complex apparatus, which has recently

* Richardson's Introduction to Geology, p. 237.

been described by Eschricht of Copenhagen. The head is furnished with six retractile appendages, which have a reddish tint, from the number of distinct red spots distributed over the surface, amounting in each to about three thousand. When viewed with a microscope, each speck is seen to be the orifice of a sheath which contains about twenty pedunculated sucking discs, that are capable of protrusion for the prehension of prey; so that the head of the *clio borealis* is armed with three hundred and sixty thousand microscopic suckers—an instrument which, for complexity, is quite unique in the animal series.

The *Cephalopida* have a thick, soft, fleshy body, sometimes protected by a shell, and sometimes naked. The mantle is a musculo-membranous sheath, inclosing the digestive, respiratory, circulating, and generative organs. The head is distinct from the trunk; is of large size, and round form. It contains the organs of the five senses, and those for mastication and deglutition. It is surrounded by a circle of fleshy processes, or feet, from whence the name of the class is derived. The eyes are two in number, of large size, and highly organized. The mouth is armed with a pair of vertical, horny or calcareous jaws, resembling the bill of a parrot, and inclose a fleshy tongue. The class comprises two orders, the principal one being the *Tentaculifera*, of which the Nautilus Pompilius is the type. They have large extended univalve shells, symmetrical in form, and divided internally into a series of chambers, the last being very capacious, for lodging the body of the animal. A tube passes through all the chambers, and opens into a muscular sac surrounding the breast. This apparatus is intended to facilitate the ascent and descent of the animal in water, by determining an increase or diminution in the specific gravity of the shell—the reservoir and siphon can be distended with water, thereby augmenting the weight of the shell, or emptied by the

ANIMALS OF THE ANCIENT SEAS. 221

contraction of its muscular walls, and thus enabling the animal to float. The weight of the sea-water is the ballast by which they thus ascend or descend. Their eyes are more simple than those of the *Acetabuliferæ*, which compose the other class of the order.

The *Tentaculifera* comprise three families—the Nautilidæ, the Clymenidæ, and the Ammonitidæ. The first have the siphon in the middle of the septa, a spiral or straight shell, and a septa simple or sinuous. The family contains but one living genus, known as the nautilus. This has a spiral shell, rolled on the same plane, and volutions at all ages contiguous, apparent, or concealed. It contains more than one hundred and twelve species, which made their first appearance in the Devonian rocks, and attained their highest development during the coal period. After appearing in all the subsequent eras, but two living species now remain. The *Clymenidæ* have the siphon in the internal part of the septa; and the shell is spiral, arched or straight. The genera *melia*, *cameroceras*, *phragmoceras*, and *clymenia* are extinct, and belong to the paleozoic era. These shells are very beautiful. The *Ammonitidæ* have the siphon at the external dorsal part of the septa; and the shell is spiral or straight, arched or bent in various forms. The genera *oncoceras*, *cyrtoceras*, *gyoceras*, *cryptoceras*, and *stenoceras* are all found in the paleozoic era. Goniatites are Devonian and carboniferous, and ceratites are triassic forms. The G. Ammonites form a regular spiral, rolled on the same plane, with the turns contiguous. Five hundred and thirty species of this genus have been identified. They comprise some of the most ornate and magnificent shells of the ocean. In each of the eighteen geological stages in which ammonites are found, certain groups of specific forms are found to characterize the different beds.

The ten-armed Cephalopods, with internal shells, comprise the families of *Spiculidæ, Loligidæ, Teuthidæ,* and *Belemnitidæ.* The first have an internal calcareous shell, with a series of air-chambers, traversed by a siphon. It includes living and fossil genera. The second family have an internal horny plate of a feather-like form, but no air-chambers. Fossils are found in the oolite. The third have an internal blade, like an arrow, also without air-chambers. Fossils of these genera occur in the oolite. The Belemnites have an internal horny skeleton, and a testaceous shell, formed of air-chambers, piled on each other in a straight line, and traversed by a lateral and marginal siphon. The genera are all extinct. They occur in the oolite and chalk, and distinguish the strata by their specific forms.

The *Tentaculifera* were among the first animal forms that appeared on the earth, being found in the Silurian rocks. There were twenty-two genera in the Paleozoic period; seven in the triassic; the same number in the oolite; fourteen in the cretaceous, and but one genus in the modern seas. The *Acetabulifera* appeared in the oolite with twelve genera, and in the Tertiary with four; five, however, still survive.*

The annexed table, originally compiled by the great French conchologist, Alcide d'Orbigny, presents the number of species of Radiata and Mollusca belonging to each geological or stratigraphical stage of the earth. It exhibits at a glance the whole extent and distribution of their fossil remains, and of their specialty to particular eras. The species now living are omitted, but they are for the most part indicated in the Pliocene, the last stage of the Tertiary.

* The authors cited in the foregoing compilation, besides Dr. Richardson, are Pictet's *Paleontologie*, Magazine of Natural History, Mantell's Isle of Wight, Pictorial Atlas, Natural History of Crinoida, the British Annual, Prof. Gray, Dr. Wright, D'Orbigny, Owen, etc., etc.

TABLE EXHIBITING THE DISTRIBUTION OF FOSSIL MOLLUSCA AND RADIATA IN EACH FORMATION.

GEOLOGICAL STAGES OR SUB-ERAS.	MOLLUSCA. No. species in each stage.	RADIATA. No. species in each stage.	Total of both species in each stage.
Pliocene	444	162	606
Miocene	2903	160	3063
Eocene	1478	199	1677
Nummulite	562	132	694
Upper Chalk	1208	524	1232
Lower Chalk	218	148	366
Upper Greensand	627	183	810
Gault	307	52	359
Lower Greensand and Wealden	656	124	781
Portland Beds	59	2	61
Kimmeridge Clay	184	16	200
Coral Rag	403	235	638
Oxford Clay	499	230	729
Kelloway Rock	253	25	278
Bath Oolite	407	125	532
Inferior Oolite	508	94	602
Upper Lias	273	14	287
Middle Lias	270	13	283
Lower Lias	163	12	175
Red Marls	619	114	733
Muschelkalk of Germans	104	3	107
Magnesian Limestone	82	9	91
Carboniferous or Coal	887	161	1048
Devonian, or Old Red, etc.	1054	146	1200
Upper Silurian	356	61	418
Lower Silurian	375	52	427
Total	14,947	3,000	17,947

The *Mollusca* inhabiting the seas and lakes of the modern era, are perhaps no less numerous and varied than they were in previous eras. Indeed, it will be seen by the table already presented, that they attained their greatest development in the ancient earth during the Tertiary period; while the table below, compiled from the researches of Baron Cuvier and other distinguished naturalists, will sufficiently indicate that there has been no diminution of species, whatever depreciation, if any, may have occurred in individual numbers. It will also exhibit, in a very remarkable degree, that instead of the ancient forms having passed into a higher grade of animals, they are still maintained in their original and undeviating moulds; which, indeed, nothing short of direct and specific creation could change:

TABLE EXHIBITING THE SPECIES OF LIVING MOLLUSCA.

MULTIVALVES, OR SHELLS WITH MANY VALVES.

Genera—classic terms.	Genera—common names.	No. species.
Chiton	Coat-of-mail	28
Lepas	Acorn shell	32
Pholas	Stone-piercer	12

BIVALVES, OR SHELLS WITH TWO VALVES.

Genera—classic terms.	Genera—common names.	No. species.
Mya	Truncate, trough-shell, or gaper	26
Solen	Razor-sheath, or knife-handle	23
Tellina	Tellen	94
Cardium	Cockle, or heart-shell	52
Mactra	Kneading-trough	27
Donax	Wedge-shell	19
Venus	Venus	153
Spondylus	Thorny oyster, or artichoke head	4
Chama	Clamps, or clams	25
Arca	Ark	43
Ostrea	Oyster and scallop	36
Anomia	Antique lamp	51
Mytilus	Mussel	64
Pirnea	Fin-shell, or sea wing	18

UNIVALVES, SINGLE VALVE WITH REGULAR SPIRE.

Genera—classic terms.	Genera—common names.	No. species.
Argonauta	Paper sailor	5
Nautilus	Pearly sailor	31
Conus	Cone	83
Cyprœa	Cowry	120
Bulla	Dipper, or bubble	52
Voluta	Volute, or wreath	144
Buccinum	Whelk	200
Strombus	Winged, or claw shell	53
Murex	Trumpet, or rock shell	182
Trochus	Top-shell	33
Trubo	Wreath, gig, or top-shell	151
Helix	Snail or spiral	267
Nerita	Nerit, or hoof-shell	76
Haliotis	Sea-ear, or ear-shell	19

Total genera......31   Total species............2,106

It will thus be observed that the species of Molluscan animals are quite as numerous in the present geological era as they were in any of those of previous formations. They were only surpassed by the Miocene stage of the Tertiary, which, however, contained many allied with species still living. This overwhelming fact ought to be sufficient to put the seal of condemnation upon all theories contemplating a gradual change and development of organic species, from a lower to a higher type, or from a higher to a lower type.

The third primary division of animal life comprises the *Articulata*—animals that, in the absence of a true skeleton, have their bodies surrounded by movable rings, or by coats of horny or calcareous enamel. They range higher in the scale of organization than the Molluscans, but neither are furnished with the skeleton of the *Vertebrata*. They are divided into six classes — the *Annelida, Cirrhipoda, Crustacea, Arachnida, Myriopoda*, and *Insecta*. All of these, again, comprise numerous sub-classes, families, and species.

The Annelides are worms with red blood, and have a soft, elongated, and articulated body, divided in folds and segments. Some of them form tubes to live in, either of calcareous matter exuded from their own body, or from foreign substances; to which tubes, however, they are not attached. None of this family have feet; but the greater number have *setæ*, or bundles of stiff movable hairs, which supply their place. They are generally hermaphrodite; and their food consists of insects and vegetables. Nearly all live in the water, or bury themselves in holes in mud or sand. The sea *lumbrici*, or worms, though forming a numerous and diversified family, yet require no particular notice. The *lumbrici terrestres*, or common earth-worms, so well known, are the only animals of this class which do not enter the water. They are destitute of eyes, attain to about a foot in length, and their body is divided by a hundred or more rings. They pierce the earth with ease. At least twenty species are known. It may be here remarked that the earth-worm is a great assistant and friend to the farmer and gardener. Living near the surface, their perforations aërify the soil, and allow the passage of water to the roots of vegetation. Their secretions also contribute to the chemical changes which occur in the processes of vegetation, and furnish to the young roots the very nourishment they require. Despised and trodden upon by man, the Creator seems to have intended them as a secret auxiliary in the economy of the earth, by aid of which it is clothed with flowers, fruits, and herbs. Yet many a fool, unable to comprehend the wisdom and goodness of Providence, would shrug his shoulders in affected disgust at these slimy burrowers, and wonder why they were suffered to crawl upon the earth! The wonder, indeed, is that *they* themselves should be allowed to feast and revel upon the labors of God's humble workers!

The Annelides embrace two orders, the first characterized by naked bodies, as the Nerites, and the other by a calcareous or membranous tube or sheath. To these belong the *Serpula* of the Devonian and oolitic rocks, the *Terebella*, and the *Spinorbis* and *Siliquaria* of the Tertiary. The common leech, so extensively employed in medicine, belongs to the former order. It is furnished with a three-fold and triangular jaw, with two ranges of very fine teeth, which act like a cupping-glass. The blood with which the animal will gorge itself does not go into the stomach, but into distinct vessels, hence a single meal will suffice its appetite for more than a year!

The *Cirrhipoda* are marine animals, and are usually found attached to rocks. It is inclosed in a multivalve shell, and was formerly ranked with Molluscans. There are two families, and several varieties are found in the Tertiary and oolite rocks. The species *Balanus, Acastra, Corunola*, etc., belong to one family; while *Anatifa, Pollicipes*, and *Aptychus* are characteristic of another.

The *Crustacea* comprises a large, varied, and numerous class, which is arranged into eight orders. The skeleton is in the form of an external crust, exuded from the vessels of the skin, and hardened with carbonate and phosphate of lime. At certain periods this crust is thrown off, to permit the growth of a new one. The crabs and lobsters are true types of the class; but it comprehends all animals with articulated feet, a heart for circulation, and branchiæ for respiration. Their feet comprise at least six, and their eyes three in number—the latter occurring as simple lenses, or comprising a number of compound lenses.

The order *Isopoda* is represented by the common wood-louse, which have been found fossil in amber; but a large number of marine and terrestrial genera occur in the wealden and Tertiary beds of Europe. The order *Decopoda* comprises the prawns, shrimps, and craw-fish now living; and other families are represented by the lobsters and crabs. Of the prawn family, more than a dozen genera occur in the upper lias; while of the lobster, or *Astacidæ* group, some ten or more are found in a fossil state. All these animals are still so numerous and universally diffused over the earth, that no description of them is required.

The order *Cyproida* are microscopic creatures, inclosed in bivalve shells, united by a hinge on the back. They can close the valve entirely, and protrude the feet at pleasure. There are five fossil genera known, some of which existed during the Silurian era, while certain species still survive in our seas. The remains of *Cypris* were extremely abundant in the fresh-water strata of the wealden.

"The order *Trilobites*," says Dr. Richardson, "had the carapace composed of several rings, divided into three lobes by two lateral depressions. The number of thoracic segments varies from five to twenty. The lateral lobes of the anterior segment support the eyes, which are prominent and compressed, and are often preserved in the fossils in a high state of perfection. Their feet were membranous. Many of them had the power of rolling themselves into a ball. They all belonged to the paleozoic era. They lived in numerous families and presented an immense association of individuals, but were much restricted in the number of genera and species. They are arranged into six families—the *Asaphidæ*, *Calymenidæ*, *Harperidæ*, *Olenidæ*, *Odontopleuridæ*, and *Ogygidæ*. The two first named could roll themselves into a ball, a feature which usually distinguishes their fossils from those of the others."

The Arachnida, or spiders, comprise a numerous class. Koch and Berenat have described one hundred and twenty-three species, belonging to fifty genera, of which thirteen are extinct, while none are identical with those still living. There are but few fossil specimens in the older rocks; but the presence of *Cyclopthalmus*, a genus of the scorpion, in the carboniferous stage of Bohemia, proves that the class was represented in the fauna of the primary epoch, and supplies another link to the chain of evidence that the flora of the coal resembled that now growing in tropical regions. Fossil spiders are also found in the amber of Prussia.

The web which the common spider constructs, shows it to be alike cunning, cruel, ingenious, and persevering. Their first effort is to throw out a cable, stretching from one object or abutment to another. After this they proceed to form a series of radiating spokes, which they afterwards strengthen and join together by means of transverse lines, which increase in width and length from the centre, where the animal takes up his position, and watches for his prey. As soon as a fly runs against it, his wings and sticky limbs become fastened in the fabric, and he flutters in vain to escape. In the mean time, his victim is scarcely secured, before the spider with great swiftness pounces upon it, and bears it off to his central station, where it is leisurely devoured. The exterior body of the spider is furnished with a reel, upon which his delicate yarn is spun as rapidly as it is woven. His limbs serve the purpose of a measure when constructing his net, which is always built according to the most exact geometrical proportions. The spider "knows no such word as fail." No matter how often its webs may be destroyed, he will rebuild them with renewed energy—sometimes rolling up and cleaning old material to be again employed in new enterprises.

The *Myriapoda* are represented by centipedes, which have a body composed of twenty-four feet. The *scorpion* of Europe belongs to the order, and attains a very high development in Asia, where it is called a land lobster. They conceal themselves under stones and old walls, and are distinguished for their fatal assaults upon each other. Extinct genera have been found in the same rocks with the spiders.

The class *Insecta* embraces an almost innumerable variety of genera and species, which it would require volumes adequately to describe. The distinguishing genuine features are indicated in the annexed table, to which we shall add some brief descriptions, principally collated from Bucknell's Natural History, an English work of two volumes, which has been of material service to us in the department of which it treats:

CURVIER'S TWELVE ORDERS OF INSECTS.

Classical names of orders.	General habits, features, and character.	Common names of genera and species.
Myriopoda.	Insects which have more than six feet, arranged on the body in a series of rings, without wings, but many jaws, such as	Millepeds, hundred legs, etc.
Thysanoura.	Insects with six feet, the belly furnished on the sides with false feet or appendages for leaping, without wings or jaws, such as	Padura, etc.
Parasita.	Parasites, having six feet, with suckers, but no jaws or wings, such as............	Lice, etc.
Suctoria.	Suckers with six feet, and sucker-like mouth, but neither wings nor jaws, as the	Fleas, etc.
Coleoptera.	Insects having six feet, four wings, (two upper case-formed, and two lower folded,) and jaws, as the.........................	Beetles, etc.
Orthoptera.	Insects with six feet, four straight wings and jaws, as.................................	Locusts, etc.
Hemiptera.	Insects with six feet, four wings, the two upper of unequal consistence, a sucker, but no jaws—some without any wings,	Cochineal insects, bedbugs, etc.
Neuroptera.	Insects with six feet, four equal wings, and jaws	Dragon-flies.
Hymenoptra.	Insects with six feet, four unequal wings, and jaws, ...	Bees, wasps, ants.
Lepidoptera.	Insects with six feet, four powdery wings, a sucker, but no jaws,.........................	Butterflies & moths.
Rhipiptera.	Insects with six feet, with two wings folded like a fan, a sucker, but no jaws,	Stylops.
Diptera.	Insects with six feet, two membranous wings, a sucker, but no jaws,..................	Gnats & flies.

The *Diptera*, embracing the various families of flies and gnats, are too well known to require further description than that contained in the foregoing table. They undergo a complete metamorphosis from their larval to their mature condition, and their power of reproduction is truly wonderful. The gnat deposites her eggs in water, each brood having from two to three or four hundred. The common flies are sometimes concealed in ova, while others are brought forth alive. Those found in cheese have the singular faculty of leaping to a considerable distance. The insect limestones of the lower and upper lias of Gloucestershire, and those of Oxford and the wealden beds of England, and of Solenhofen in Germany, contain beautiful specimens of the wings of flies. Many are also preserved in the Tertiary rocks, and entire specimens occur in amber.

The order of *Lepidoptera* is mainly distinguished by caterpillars, or the silk-worm, one of the most important and remarkable insects in the entire class. It feeds upon the leaves of the mulberry, but many other caterpillars feed upon the leaves of oaks, and other forest trees, and sometimes destroy the foliage of vast groves and forests in a single season. The silk-worm envelops itself in minute silken threads, of a bright yellow color, in the form of a ball or cocoon, very much resembling in form a pigeon's egg. It remains inclosed in the cocoon for fifteen days, and then eats its way out; but in cocooneries for the production of silk, this habit is anticipated, and the animal inhabiting the cocoon is destroyed by heat. The silk of the cocoon is afterward reeled, and then woven into fabrics of every possible form. Each cocoon usually produces a thread four hundred yards in length.*

Of butterflies and moths, there is an infinite variety, all undergoing similar transformations from birth to maturity. The moths are well known in consequence of their destructive effects upon cloths and furs, and which, from

---

* Efforts were made a few years ago to introduce the manufacture of silk into the United States, and it was inaugurated with an extraordinary movement in the planting of the *morus multicaulis*, the vegetable upon the leaves of which the caterpillar feeds. The experiment was overdone, and the speculation burst suddenly, like a bubble, involving many in heavy pecuniary losses. A few establishments, however, continued in operation, and that of Mr. Gill, in Wheeling, Va., was operated for many years with considerable success. That gentleman produced from his factory some of the finest silk goods in the American market; but we believe the enterprise, after a trial of ten or fifteen years, was finally abandoned.

their larvæ, become caterpillars in fifteen days. They are provided with a long proboscis, coiled in a spiral form, by which they bore into the cloths, where they deposit their eggs. The lithographic limestones of Solenhofen, and the Tertiaries of France, contain fossil specimens, both of the caterpillar and the wings of the butterfly.

Hymenoptera embrace bees, wasps, and ants in great variety of genera and species. Of these, the honey-bee is the most important and the most curious in its habits and organization. When a swarm quits an old hive, it is composed of one queen, or female bee, several hundred males, or drones, and many thousand workers, or neuters. Should there happen to be two or more queens, a murderous conflict ensues, the swarm remaining with the victor. In a new hive, they divide themselves into four parties—one of which rove the fields in search of materials, another lay the foundation of the cells, a third polish and finish what the others have begun, and a fourth bring home food for themselves and for those laboring upon the hive. Things being thus arranged, the queen bee becomes fecundated, and then examines the cells in progress, and begins to lay eggs, depositing but one in each cell, which are of three kinds—the first and smallest are intended to produce workers, the second are for males, and the others, limited to a few very large ones, are for queen bees only. Several hundred eggs are generally laid in a single day, and which become living larvæ in four or six days. These are supplied with food by the hive-workers or nurses. When the larvæ are six days old, the nurses close the cells with wax, and the entombed larvæ begin to entwine themselves with a silken sheet, and become nymphs. At the end of twelve days, they break their inclosure, and come forth perfect bees. They are now cleaned by the nurses, and then join the out-door workers. The eggs in the male cells are generally about two months later than those of the workers, and the royal cells are not even begun until the queen has deposited her eggs in the male cells. The young queens are consequently the last hatched, and pass through the same metamorphosis as the others. In the event of a hive being deprived of their queen by accident, the workers have the capacity of producing another. This is done in the following extraordinary manner: the cells are examined for the larvæ of workers which are not more than three days old. On finding such, they immediately enlarge the cells of these larvæ, feeding them with female or royal jelly, until, by dint of care and labor, a female is produced which is to replace the one lost. Besides the cells for rearing their young, others are made for the storage of honey, and which is destined for their use when none can be gathered from the flowers. On the weather becoming cold, not only are the larvæ and nymphs destroyed, but the male bees also. Having no stings, like the workers and females, they are easily massacred, or driven out of the hive, and perish by the cold. During the winter the whole hive is in a state of half-lethargy.* Fossil remains of bees, wasps, and hornets are found in the insect limestones of the lias, and generally in the rocks of the Tertiary.

The common ants also live in society, like bees and wasps, the community consisting of males, females, and workers — the workers always being without wings. Their mode of life is very similar to that of bees, but their nests are constructed in the ground, or beneath the roots of trees. It consists of a central cavity, with communicating subterranean roads or galleries. They feed on fruits, insects, or carrion, and the various classes into which their community is divided always keep distinct.

* Scripture Natural History.

In proportion to size, the ant is no doubt the strongest animal upon the earth, as well as one of the most industrious.

The *Neuroptera* are distinguished for the dragon-fly. These insects drop their spawn upon the surface of water, which, on sinking to the bottom, become larvæ. From this state they change into nymphs, and then remain aquatic for two years or more. At length they crawl out upon adjacent twigs, and as their skin dries and shrinks, wings expand from their body, when they fly away—a new creature! There are several varieties of this insect, some of which remain many years in the water before emerging into the air. In the *ephemera*, the transition from the aqueous nymph to the winged insect is instantaneous after reaching the surface.

The Ternietes or white ants of India and Africa, which are regarded as a serious evil in warm climates, have considerable resemblance to the European species, but they attain a much superior growth. They erect nests on the surface of the ground, which sometimes attain a height of ten or twelve feet. The nests are very tall, but slender, and terminate in rounded or conical peaks. They are perforated with galleries, in which the community, divided into various classes, reside—the king and queen having a central apartment to themselves. Their social government resembles that of man to a greater extent than that of the bee. The ant-hill is guarded by sentinels or soldiers, who appear to be specially created for that purpose, giving the alarm to the communities within on the approach of danger. They undergo metamorphoses similar to the other Neuroptera, but it is rather more incomplete. They surpass the bees and the beavers in mechanical art, and fully equal them in industry. In migrating, they observe all the military precision of soldiers in battalion.

The order *Hemiptera* embraces bed-bugs, garden-bugs, plant-lice, cochineals, water-bugs, etc., etc. Some of these have the extraordinary faculty of fecundating their issue to the fourth and fifth generations.

Orthoptera are principally represented by grasshoppers and locusts, of which there are many genera and species. Locusts are a great scourge to vegetation, and formed one of the plagues visited upon ancient Egypt. In some countries, however, the natives esteem them as a rare delicacy. Locusts are rare in America, but are supposed to make a visit in great number about every seventeenth year. Grasshoppers, however, which belong to the Locust family, are abundant, and are sometimes no less destructive to the crops of the farmer. They deposit their eggs in the ground in the fall of the year, and lying dormant during the winter, the warmth of the sun brings forth a wingless insect in the spring, which, in the course of twenty days, expels its outer skin, and then appears with wings. The common cricket is a member of the family.

The *Parasita* include fleas, lice, and vermin of every description.

The *Coleoptera* are made up mainly of beetles, as the May-bug, Hercules beetle of South America, the Glow-worm that, on summer evenings, lights up the way-side, the Undertaker beetle, etc., etc.—the latter so-called, because it buries the bodies of other insects and worms. These animals were represented at the close of the paleozoic period, and their fossils occur in the carboniferous rocks. Indeed, nearly all the orders thus enumerated have their fossil remains strewn in the various strata, in greater or less abundance, from that early era, to the present time; but there has been a constant variation of species in particular eras, as the genera now living are, in a great measure, different from previous geological stages, and yet present specimens that appertained to each.

Such is a brief and necessarily cursory survey of that portion of the animal kingdom, comprising its more minute creatures—creatures which, often unseen, yet fill the seas, the earth, and the air. The three great divisions of the Radiata, Mollusca, and Articulata comprise many thousand species, and many of their principal genera present more living creatures than those of the vertebrata, including man, combined!

Now, notwithstanding the low origin and humble nature of its animals, and the comparatively few specimens outside of the Radiata and Mollusca, which it affords, many geologists, indeed nearly *all* of them, give more prominence as well as *priority* of existence to the aqueous fauna of the Paleozoic period, than to its wonderful and prolific flora. In other words, the Mollusca, Radiata, and Fish of the Silurian and Devonian eras, are recognized as overshadowing the vegetation of the coal period! And it is also claimed that animal life *preceded* vegetable life, and that, therefore, the Bible is again incorrect. I have already remarked that such an arrangement is altogether unjustifiable, besides being in conflict with Revelation, and the obvious course of nature. It may be granted, as our table will show, that during the Devonian and carboniferous eras, there was a considerable development of the lower orders of animal life, including specimens of Sauroid fishes. All these flourished to some extent, as will be seen by the table, during the Silurian period, but it was utterly impossible that they could have existed in the previous metamorphic seas, when these were periodically, if not constantly disturbed and heated by the transition of the rocks. It is true the whole expanse of waters may not have been heated at the same time. The effect may have been more local than general;

but in either case its duration was constant. The heat, however, was always operating in the shallow places—always tending to elevate the bottom of lakes and marine basins; and those are the very places where animal life could alone exist—the pressure of the water, in mid ocean, forbidding its existence there. We have, then, a limit—a positive barrier of heat and boiling water, beyond which it is impossible that animal life could have existed, either on the land or in the sea. This limit, too, may be found to extend further into the Silurian rocks than is now generally conceded. In fact, the fossil specimens thus far found in the lower strata of those rocks are so few and obscure, that it would be prudent not to place too much reliance upon them. Some specimens may yet prove to be no more ancient than the great Cobham stone discovered by Mr. Pickwick. Of Radiata, we may observe that there are thus far one hundred and twelve species in all the Silurian strata, and only nine in the Magnesian limestone above; while in the intermediate Carboniferous and Devonian strata there are over three hundred species. The proportion of Mollusca is still more varied—there being nearly two thousand species in the last-mentioned periods, while those before and after have less than eight hundred. And notwithstanding that creatures half fish and half saurian flourished somewhat plentifully during portions of all these eras, it does not lessen the significance of the fact that animal existence, as a whole, was mainly represented by the Radiata and the Mollusca—the two *lowest divisions in the scale.*

But that animal life of a strictly terrestrial and of an infinitely higher order exists in the *juices of vegetation*, at the present time, is a fact which nobody will be likely to controvert. Every woodman has observed, on splitting open the solid trunks of trees, the cut-worms and bugs which inhabit their centre. The worms inhabiting the cores of apples, pears, peaches, and other fruits, doubtless, enter from the outside; but they are always hatched in the juices they feed upon. Those found in the interior of walnuts, chestnuts, and similar nuts, must originate there, and in fact constitute a portion of their oily substance, as maggots do that of cheese. Certain kinds of trees generate worms and bugs peculiar to them; and they are often so plentifully diffused as to utterly destroy the timber which they ravage. Hemlock breeds bed-bugs; while oak, chestnut and hickory abound in cut-worms, borers, and crawling ants, which operate internally and externally, and appear to be inseparable from them. Other trees attract, if they do not generate caterpillars, which, feasting upon their leaves, multiply to such an extent, that they exhaust the vitality of large numbers in a single season. This may be seen particularly in trees of dense foliage, selected to ornament private grounds. Weevils and grasshoppers originate in the crops of the farmer, and so destructive are their depredations that the product of entire plantations not unfrequently falls a sacrifice to their unappeasable appetites! In like manner, certain species of borers attack the timbering of vessels, especially the submerged portion, and it often happens that they are utterly ruined by their perforations. During the Crimean war, over one hundred and twenty vessels were sunk by order of the Russians in the harbor of Sebastopol. After the war terminated, arrangements were made for raising these vessels; but it was found that this marine *teredo* had, in the mean time, so completely perforated the solid timbering, that many of them were worthless, and unfit for use. The timbers were, in some instances, bored in such a manner as to resemble the cellular structure of a sponge. The existence of worms in timber, and in the body of fruits and nuts, can perhaps only be accounted for on the supposition that their **larvæ** or spawn, in the form of minute microscopic infusions, exists primarily in the water which the plants themselves absorb in their pores and cavities, and where they are subsequently developed.

But whether the origin of the parasitic, microscopic, and coral animals be due to or associated with the primitive vegetation or not, is a matter of no particular consequence in reference to the *priority* of the one over the other—since we have the most conclusive evidence of the existence of vegetation during the metamorphic period, when, by reason of sub-marine heat, it was utterly impossible that aquatic life *could* have existed. The coal and graphite of Rhode Island, of Scandinavia, of Cuba, and of portions of France and Germany sufficiently demonstrate the relation which the ancient forests bore to the rocks of that period. It will be vain and absurd to suggest that all these are merely *altered* Silurians or Devonians. They are *true* metamorphic rocks, occurring in their proper positions, and exhibiting all the external evidence of rocks belonging to that group. Nor do I stand alone in this opinion. Several geologists of distinction have intimated similar views, but have not expressed them with the boldness which the circumstances warrant. It is, however, time to vindicate the truth of the Mosaic record; and when a fact has significance beyond and apart from geological inquiry, it should not be tampered with in half-suppressed doubts and misgivings. It is either a fact or it is *not*—it is either true or false. If it is a fact in geology that vegetation *preceded* animal life, *both* in the *water* and the *dry land*, it deserves to be known. I say it *is* a fact; and I have thus given my reasons. I say it is also a fact—overwhelming and palpable—that vegetation distinguished *the whole Paleozoic period*, and far surpassed in *extent* and *universality* the animal creatures that flourished in its seas. These two great truths once admitted in Geology, there is no longer any serious embarrassment to its reconciliation with the simple narrative of Revelation. And it was with the conviction of the perfect harmony between them, when properly understood, that I have bestowed so much time and space to the investigation which the subject involves.

Although the intervening fourth day was characterized by violent and extensive volcanic action, by the elevation of vast systems of mountains, and by the change of sea into land, it was yet *not* a geological but more properly an *astronomical period*, as Moses himself has indicated. Geologists generally, in speculating on the revealed Cosmogony, appear to *overlook* this great fact. They treat *all* days in the light of *geological* phenomena, and thus construe plain facts into the most unwarrantable inferences and assumptions. Some writers make the geologic days as short as those of Jupiter or Saturn. The Silurian and Devonian; the Carboniferous; the Saliferous; the Oolitic; the Cretaceous; the Tertiary; the Quarternary, or Historic period; all these are sometimes erected into *Mosaic days*, notwithstanding that the rocks and fossils of some of them do not occur in certain portions of the earth, and *nowhere in regular superimposed order!* The Astronomical days, in the mean time, are unaccounted for! Now, these are the kind of writers whose skill in technical nomenclature enables them to conceal their stupidity, and to impose on the confiding world, under the name of *science*, the most shameless trash which the prolific invention of dabsters in tautology can indite.

The Secondary formation which we are now considering, comprehends a large number of layers of rock, classed into inferior systems, but under so many different names and local variations, that it is no easy matter to define their specific features. It, however, begins with the new red sandstone strata; and although some geologists, under the name of Permian, assign two of its principal groups to the Paleozoic period, from the seeming identity of their fossils, yet others, with equal propriety, group them all together, and style them collectively the new red sandstone, Triassic or Saliferous systems. The latter name is bestowed in consequence of the numerous deposits of rock-salt and brine springs found in these rocks; but they will perhaps be better understood by most persons in the United States by the appellation of *new red sandstone*, in contradistinction to the old red sandstone, over the uptilted beds of which they often occur unconformably, or in a horizontal position—thus pointing out a difference in their ages, and indicating the disturbance which overturned the old red *before* the new red was deposited. Including the Permian rocks, the new red sandstone system comprises: 1. red sandstone; 2. magnesian limestone, (including calcareous conglomerate marble, of which large deposits extend across Pennsylvania, Maryland, and New Jersey;) 3. Variegated sandstone; 4. Muschelkalk, seldom found in these measures outside of Germany; and 5. variegated marls. These rocks are generally of marine origin, and their usefulness to man, in supplying vast quantities of salt, which is absolutely indispensable to human life, and in domestic economy, cannot be over-estimated. It is not, however, an accompaniment of our American sandstones; on the contrary, all the salt mines with which we are acquainted are in older rocks. Those of Rochester, in New York, are in Silurian strata; while on the western slope of the Alleghany, along the Kiskiminetas, the Ohio, and Kanawha, in Virginia, salt is very abundant under the coal measures. In the last mentioned region, the salt-borings have in several places also tapped reservoirs of illuminating gas and liquid oil. This gas is collected on the surface, and applied to the evaporation of the salt of the brine. The new red sandstone of the Alleghany is derived mainly from the disintegration of the old red sandstone, underlying the coal. When the mountains were elevated, the salt was not transferred to the new rocks, as it probably had been at other places; but was retained where originally deposited, and where it had crystallized during the heat which elevated the coal. Under these circumstances it could not be drained off to the shallow estuaries of the new red sandstone, lying principally on the eastern slope of the mountain. There was an elevated axis which prevented this. They were therefore formed without the salt, to which they were otherwise entitled by natural drainage. The waters of the ocean, although they were in constant proximity, could not supply the deficiency, because the lakes were not shallow enough to effect the speedy evaporation. The ocean, therefore, in surging over the sandstone shores and lakes, lost little of his saltness; and comparatively few of his animal creatures—for the American sandstones are almost as poor in one as the other. But if deficient in these, they furnish inexhaustible quantities of superior and beautiful building material. Impregnated and colored with the oxyd of iron, the "brown stone" of this group has been very extensively employed, during the last twenty years, in the Atlantic cities of America. It is perfectly homogeneous in composition, and is readily carved into the most elaborate designs of architectural ornament. In England, the magnesian limestone or marble has furnished the material of which the new houses of Parliament are constructed. This stone, when found crystalline in structure, is one of the most imperishable in the earth; and it is somewhat singular that, with its abundance in the vicinity of many of our populous cities, its merits in the United States should have been comparatively overlooked. Gypsum (sulphate of lime) a species of alabaster or soft statuary marble also occurs, but sparingly, in this group.

There was, upon the whole, a very considerable diminution, both of animal and vegetable life, as compared with the periods preceding—but more especially of the vegetation. The only positive increase was in sea-weeds, (fucoides) which expanded in some regions to an enormous extent. The terrestrial vegtation was represented by the resinous pines of the coal, but under circumstances unfavorable to the formation of that mineral, though it must not be overlooked that both anthracite and plumbago are often found in small detached seams and deposits in these rocks. All the other varieties which distinguished the coal period, as Lepidodendria, Sigillaria, Stigmaria, and the Equiseta, had wholly disappeared, and mere traces of calamites, cycadea, palms, ferns, and mosses are to be found in them. Of the Radiata there was also a great decrease; but the Mollusca were tolerably well represented. The Articulata were represented sparingly by serpula and scorpions, trilobites and macrocus, but no annelides or insects proper. Of the vertebrata, fishes were somewhat numerous, as well as sauroid animals of a peculiar type.

In 1834, an account was published in Europe of some remarkable fossil footmarks in the new red sandstone, at Hessberg, in Saxony. Accounts of these impressions have been given by Drs. Hohnbaum and Sickler, Prof. Kaup, Mons. Link, and the late Baron Humboldt. The largest track was supposed to have been made by a marsupial animal, whose hind foot was eight inches long. This animal, Prof. Kaup named *Cheirotherium*, from the resemblance of its track to a human hand. Some of the tracks appear to have been made by tortoises; and M. Link, who has made out four distinct species from the tracks, suggests that some of them may have been made by gigantic Batracians—frogs, salamanders, etc.* Mr. Owen has suggested that the tracks refferred to the Cheirotherium, were made by a gigantic Batracian, or frog, whose hind feet were much larger than his fore feet. He has given an ideal sketch of the animal restored, the bones of whose head only have been discovered, and of the manner in which the tracks might have been made. Prof. Hitchcock differs with Mr. Owen, and thinks the tracks are those of a marsupial, whose hind legs are considerably longer than those in front. In 1847, Prof. Plieninger, of Stuttgart, published a description of two fossil molar teeth, referred by him to a warm-blooded quadruped, which he obtained from a bone-breccia in Wurtemberg, occurring in the upper beds of the new red sandstone, in Germany called *Keuper*. These beds are 1,000 feet thick, and comprise sandstones, gypsums, and carbonaceous slate-clay. Remains of Reptiles, called *Nothosaurus* and *Phytosaurus* have been found in it with the fragmentary bones of Owen's *Labyrinthodon*; also detached teeth of placoid fish and of rays. From the double fangs of the tooth found by Plieninger, and their unequal size, and from the number of the protuberances or cusps on the flat crowns, he inferred that it was the molar of a Mammifer, and considering it as predaceous, probably insectivorous, he named the supposed animal *Microlestes*, (a little beast of prey). Previous to this discovery of the German Professor, the most ancient of known mammalia were those of the English Stonesfield slate, a subdivision of the lower oolite. In the dolomitic conglomerates of England, remains of two distinct genera of reptiles have been found, called *Thecodontosaurus* and Palæosaurus, the teeth of which are conical, compressed, and with finely serrated edges. These saurians, (which,

---

Buckland, Bridgewater Treatise.

until the discovery of the *Archegosaurus* in the coal, were regarded as the most ancient examples of fossil reptiles,) are all distinguished by having the teeth implanted in the jaw-bone, and in distinct sockets, instead of being soldered, as in frogs, to a simple alveolar parapet. Both these families occur in the Trias of Germany. In 1844, the first *skeleton* of a true reptile was found in the coal measures of Munster-Appel, in Rhenish Bavaria, and described by H. von Meyer, under the name of *Apateon pedestris*, the animal being regarded as nearly related to the salamanders. In 1847, Prof. Von Dechen found in the coal-field of Saarbruck, the skeletons of three distinct species of *air-breathing reptiles*, which were described by Prof. Goldfuss, under the generic name of *Archegosaurus*. They were considered by Goldfuss as saurians, but by Von Meyer as allied to the *Labyrinthodon*, and therefore connected with the batracians, as well as the lizards. In 1844, the very year that Von Meyer introduced his *Apateon* or salamander, Dr. King published an account of the *foot-prints* of a large reptile discovered by him in the coal measures near Greensburg, in Westmorland county, Pennsylvania. These footprints were examined by Prof. Lyell, while on a visit to this State, in 1846. "I was at once convinced of their genuineness," says the distinguished geologist, "and declared my conviction on that point, on which doubts had been entertained both in Europe and the United States. The footmarks were first observed standing out in relief from the lower surface of slabs of sandstone, resting on thin layers of fine unctuous clay. I brought away one of these masses. It displays, together with footprints, the casts of cracks of various sizes. The origin of such cracks in clay, and casts of the same, has before been explained, and referred to the drying and shrinking of mud, and the subsequent pouring of sand into open crevices. Some of the cracks traverse the footprints, and produce distortion in them, as might have been expected, for the mud must have been soft when the animal walked over it and left the impressions; whereas, when it afterward dried up and shrank, it would be too hard to receive such indentations.* No less than twenty-three footsteps were observed by Dr. King in the same quarry before it was abandoned, the greater part of them so arranged on the surface of one stratum as to imply that they were made successively by the same animal. Everywhere there was a double row of tracks, and in each row they occur in pairs, each pair consisting of a hind and fore foot, and each being at nearly equal distances from the next pair. In each parallel row, the toes turn, the one set to the right, the other to the left. In the European *Cheirotherium* (before mentioned) both the hind and the fore feet have each five toes, and the size of the hind foot is about five times as large as the fore foot. In the American fossil, the posterior footprint is not even twice as large as the anterior, and the number of toes is unequal, being five in the hinder, and four in the anterior foot. The American *Cheirotherium* was evidently a broader animal, and belonged to a distinct genus from that of the triassic (new red sandstone) age in Europe. We may assume that the reptile which left these prints on the ancient sands of the coal measures, was an *air-breather*, because its weight would not have been sufficient under water to have made impressions so deep and distinct. The same conclusion is also borne out by the cracks of the air and the sun, so as to have dried and shrunk."

These curious discoveries of Dr. King were followed, in 1849, by the discovery of similar footprints in the *old red sandstone* near Pottsville, outside of the Schuylkill coal basin, by Isaac Lea, Esq., of Philadelphia, who read

---

* Sir Charles Lyell, Elements of Geology, p. 337.

a paper on the subject to the American Philosophical Society, in June, 1849. Mr. Lea has since published a description of the footprints in magnificent form, illustrated with beautifully-colored lithographs of the tracks, in their natural size. From this description I glean the annexed particulars:

"The position of the footprints was on the west side of the turnpike road, about a mile southeast of the town of Pottsville, and a few hundred feet below the Mount Carbon Hotel. The massive sandstone rocks here are of a beautiful red color and fine texture, evidently formed of sand and clay which has passed through much attrition. The color is due to a considerable charge of the red oxyd of iron. Minute spangles of mica are generally interspersed throughout these rocks, and assist in giving the surface of the fractures a soft and almost satin-like texture. The strata here are tilted somewhat over the perpendicular, by the upheaval of this range of mountains; but the surfaces which are exposed bear evidence of these sedimentary rocks having been deposited in a nearly horizontal position, and in a placid state of water, presenting to the animal a very slightly inclined shore, as it advanced from the waters which existed on the northern side. The impressions made at that time were upon the sands of a shore from which the waters had for a time receded, having left the shore covered with well-defined '*ripple marks*,' and a profusion of '*rain drop pits*.' The surface of the rock exposed to view was about six feet by twelve, and across the shorter diameter were distinctly and beautifully impressed a double row of tracks, consisting of six impressions, duplicated by the hind foot falling into the impression of the fore foot, but a little more in advance. The specimen taken from the mass of the rock was thirty-four by twenty-one inches. The *six double* impressions show, in the two parallel rows, formed by the left feet on the one side, and the right feet on the other, that the animal had *five toes* on the fore foot, three of which toes were apparently armed with unguical appendages. The hind feet appear to have had *four toes*. The impressions of the hind feet being made nearly on the same spot as that of the fore feet, cause some obliteration and confusion, as well as variation in size and form of footmarks. The best defined one is four and a half inches long and four broad—this is including the *double impression*. The single foot would probably measure three and a half inches long by three inches broad. The stride or step of the animal measures, from toe to toe, thirteen inches; from outside to outside the distance is eight inches. The mark of the *tail* is distinctly impressed, causing a *groove-like furrow* over the top of each ripple line, oblique to their direction, and generally five to six inches long and three-quarters of an inch wide. There are four of these tail-marks or grooves on as many ripple lines, the crests of which lines are elevated about half an inch above the intermediate depressions. The tail was evidently not a thick one, and the animal must have had a distinct and perfect step, and not a half-swimming motion, as in the crocodilians, there being no trace of the dragging of the feet. The tail must have been considerably elevated, as the alternate tail-impressions show that a vibration actually took place at every step, the four grooves not being in a direct line, but each one approaching its nearest footmark to the right or the left, alternately, and therefore never precisely on the central line between the two rows of the footmarks. These facts prove that the animal which left its imprint in this ancient sandstone stood much higher on its legs than the *Crocodilus* or the *Monitor*, and probably was not so long in proportion to the size of the feet. It is well known that the alligator leaves no foot-impression in the mud, but simply a large furrow, made by the ventral and caudal proportions. The form of the foot impressions is, however, very similar to that which is received by the mould in clay of *the Alligator Mississippiensis*, specimens of which are in the collection of the Academy of Natural Sciences, Philadelphia. If an opinion might be hazarded, as regards the probable size of the animal, based on this meagre diagnosis, I should suppose it might reach as much as seven or eight feet in length. The water, in passing over the impressions, left lines indicating its direction. Occasionally may be seen small subglobular forms, which may possibly be the ejectamenta or coprolites of some of the animals that passed over the shores of these waters."*

The footmarks of Mr. Lea were, only two years after, succeeded by a similar discovery in the upper layers of the old red sandstone of Morayshire, in Scotland. These footprints are in pairs, forming two parallel rows; and sometimes those of the fore and hind feet nearly run into each other, as in the case of Mr. Lea's. The hind foot is one inch in diameter, and larger than the fore feet in the proportion of four to three. Two years after, viz., in 1851, the fossil *skeleton of a reptile* was found in the same formation and in the same district. The bones had decomposed, but the natural position of almost all of them

---

* Isaac Lea, on the Fossil Footmarks in the Red Sandstone of Pottsville, from Trans. Am. Phil. Soc., vol x.

could be seen, and nearly perfect casts of them were taken from the hollow moulds which they left. The matrix was a fine-grained, whitish sandstone, with a cement of carbonate of lime. The skeleton exhibits the general characters of the Lacertians, blended with peculiarities that are Batracian. Hence Dr. Mantell infers that this reptile was either a fresh-water Batracian or a small terrestrial lizard. The skeleton is about four and a half inches in length, but part of the tail is concealed in the rock. Dr. Mantell has proposed for it the generic name of Telerpeton (or *afar-off reptile*), while the specific name, *Elgineuse*, commemorates the principal place near which it was obtained.

Similar footprints of Chelonian reptiles have been found in rocks supposed to be still older than the old red sandstone, on the banks of the St. Lawrence, at Beauharnois, in Upper Canada. The rocks, indeed, are supposed to belong to the lower Silurian series, but some doubt appears to exist as to their precise geological position. No doubt exists, however, as to their being lower down or older than those of the old red sandstone or Devonian system. Prof. Owen intimates that the animal which made these foot-impressions was a fresh-water tortoise, rather than a land tortoise. Supposing it to have been a tortoise, this rock is by far the oldest in which such remains or signs of that animal have been found.

I have thus presented, somewhat at length, descriptions of the footprints and fossil bones of animals which are conceived to be anomalies in nature and in geological chronology, *and in direct conflict with the Mosaic revelation.* The footprints discovered by Lea and Dr. King are supposed to have been made by a *quadruped* and an *air-breather*,—and its characteristics would consequently be terrestrial rather than aqueous, or perhaps it combined some of the features of both. The integrity of the Bible is thus assailed in various ways; *first*, in assuming that animals of a high organization *preceded vegetation; second*, that land animals existed simultaneously, if not really *before*, aquatic animals; and *thirdly*, that the ancient climate was not such as Moses describes—for he asserts most unequivocally that there was no *sun* at that time, and that "the Lord God had not caused it to *rain* upon the earth, but that there went up a mist which watered the whole surface of the ground." The geologists, on the other hand, bring forward, with their animal footprints, casts of *rain-drops*, and fissures and fractures in the sandstone, and then proceed, upon the basis thus erected, to annihilate all the law and the apostles. In connection with the footprints already described, Professor Lyell remarks: "Having alluded to the spots left by rain on the surface of carboniferous strata in the Alleghanies, on which quadrupedal footprints are seen, I may mention that similar rain-prints are conspicuous in the coal measures of Cape Breton, in Nova Scotia. In such a region, if anywhere, might we expect to detect evidence of the fall of rain on a sea-beach, so repeatedly must the conditions of the same era have oscillated between land and sea." In 1851, Mr. Richard Brown had the kindness to send me some greenish slates from Sydney, Cape Breton, on which are imprinted very delicate impressions of rain-drops, with several worm-tracks such as usually accompany rain-marks on the recent mud of the Bay of Fundy, and other modern beaches. The casts of the rain-prints project from the under side of two layers, occurring at different levels, the one a sandy shale resting on green shale, the other a sandstone presenting a similar warty or blistered surface, on which are also observable some small ridges, which stand out in relief, and afford evidence of cracks formed by the shrinkage of subjacent clay, on which rain had fallen. Many of the associated sandstones are described by Mr. Brown as ripple-marked."

"The great humidity of the climate of the coal period," continues Sir Charles, "had been previously inferred from the nature of its vegetation and the continuity of its forests for hundreds of miles; *but it is satisfactory to have at length obtained such* POSITIVE PROOFS *of showers of rain,* the drops of which resembled in their average size those which now fall from the clouds. From such data we may presume that the atmosphere of the carboniferous period corresponded in density with that now investing the globe, and that different currents of air varied then as now, in temperature, so as to give rise, by their mixture, to the condensation of aqueous vapors." The cracks in the sandstones containing the foot-marks, Sir Charles refers to the effects of the sun, as previously remarked.

Nearly every work on Geology which has lately come under my notice, is embellished with cuts of rain-drops; and those of Prof. Lyell contain some four or five specimens.

Now, these fossil footprints, rain-drops, and sun-cracks in the old red sandstone, directly impeach the veracity of divine revelation, and it will be no sufficient answer to the array and combination of evidence brought *against* its integrity by such men as Sir Charles Lyell, Prof. Hitchcock, Hugh Miller, Sir R. Murchison, Dr. Mantell, Richardson, and many others, to suggest (as many writers professing to defend it, have suggested) that it is *not* a record of scientific fact. The Bible may be scientific, or it may not be; but it professes to tell the *truth,* and that is the reason we believe in it. If geologists *prove* it to be *false,* our confidence in it must necessarily be *weakened;* for believing, as we do, in its *holy inspiration,* we regard it *utterly incapable* of mistake or inaccuracy, no matter whether the facts embodied in its statements be based on the truths of science or *not*. Truth is truth. If the geologists are correct, then the statements of Moses *cannot* be; if the Bible be true, then the geologists are all deceived. Instead, therefore, of offering plausible excuses and evasions, I propse to meet these and similar contradictions of the Bible fairly, squarely, and boldly; and if I do not turn the insidious daggers thus pointed at the Christian community, into the bowels of their authors, it will be because my confidence in the Bible has been mistaken, and that my faculties of practical inquiry and observation have deceived me.

Now, what *are* these footprints described by Lea, Lyell, and King? Of the latter I cannot speak with positive knowledge, although I have been on the spot; but as to those of Mr. Lea, I have passed the very rocks which have yielded his specimens, almost every day for the last thirteen years. I am as familiar with those rocks as I am with the faces of the people who live in their vicinity. Before we can entertain the idea of the tracks having been made by a land, or an air-breathing animal—a quadruped capable of locomotion, it must first be shown that, during the Devonian period, the tracks occupied the *position of dry land*. If they really constituted dry land during that period, all the theories proposed by the class of geologists to which Mr. Lea appears to belong, are at once overturned; for, occurring as they do, half a mile from the nearest workable seam of coal, how was it possible for the immense layers of sandstone, coarse conglomerate, and arenaceous schist to have been deposited? Are we to suppose that, *after* the animal had imprinted its tracks, there was an instantaneous submergence of the land for more than eight hundred miles? or only a partial submergence of fifty or a hundred miles in length? That the whole extent of the sandstone rocks, underlying the coal, would have to be thus submerged, if the animal were an inhabitant or frequenter of the land, is perfectly overwhelming, because in no other way could the *intermediate sedimentary rocks have been deposited.* If, then, the soft mud upon which these tracks were imprinted, had been suddenly *lowered into the water*, or the water as suddenly brought over them by a tidal flood, is it not miraculous that the movement of the waves—the oscillations of vast bodies of excited waters, should not have obliterated all traces of the impressions? Why, the ordinary ebbing and flowing of a high tide, apart from any extraordinary convulsive movement of the land, would alone suffice to wash out any footprints made on the soft mud of the shore!

Mr. Lea observes that the surface of slabs of sandstone immediately adjacent to those which contained the footmarks, reveal ripple-marks, as left by the water. He infers from this that the rippled sandstones constituted an ancient *shore*, upon which the fossil reptiles moved. This idea is still further supported by finding other slabs, including those with the footmarks containing casts of *rain-drops;* while nearly all the slabs show *sun-cracks*, as if the shore had been exposed to the sun and to showers of rain.

Now, in the first place, I do not regard the so-called ripple-marks as such in fact; and in the next place, if they *are* ripple-marks, it can be shown that they were necessarily formed *under* the water, and that, if they afterward became a temporary *shore*, they would have been effaced by the withdrawal of the water. Ripple-marks are furrows; but those at Mount Carbon are *not.* They are *indentations*, running in *parallel rows*, and have perhaps as valid claim to be considered *foot-marks* as the obscure specimen obtained by Mr. Lea. These indentations are on a very hard sandstone—a homogeneous stratum six feet thick. Seams of it are unusually arenaceous. Above this stratum there are fissures filled in with soft mud; and it is from *this* that the foot-marks were obtained. Ripple-marks are invariably formed in the eddies or between the projecting rocks of rivers and lakes. They are the result of obstructions to the channel of the stream, which, holding comminuted mud and sand in suspension, precipitate it to the bottom wherever the movement of the waters is temporarily arrested. Water has precisely the same effect upon the soft mud and fine sand thus deposited, that the blasts of winter have upon drifted snow. Every gale that sweeps over the earth, scoops out the snow, and piles it up in irregular wave-like furrows along the fences and the way-side. Water is thus, in many respects, similar to the atmosphere we breathe—that is, it is an aqueous body, containing currents and eddies, like the air; all of which are caused by the irregularities of the surface over which they flow. But when, after scooping out these furrows in the bottom, the water recedes and leaves the land exposed, its withdrawal, its surging motion, gradually levels down the soft materials, and effaces all traces of the previous ripples. Therefore I say the ripple-marks, if they be such, were produced by and *under* the water, and, instead of affording evidence of *dry land*, prove exactly the reverse; for an hour's exposure of such soft mud to the sun and wind would have dried it up, and obliterated every trace of the ripples. But, it must be borne in mind, that these ripple-marks and footprints were obtained in a *deep valley*, where the strata are *perpendicular*, and surrounded on every side by mountains *towering eight hundred feet* above it. If there had been a *shore* at any time, it would consequently have been *eight hundred feet from the footprints* on the summits of the mountains! And besides, the strata, instead of being laminated in perfectly straight lines, with regard to each other, would have been *wedge-shaped, or sloping;* but we have no such evidence. On the contrary, the lines of lamination and stratification are *perfectly parallel*, betraying not the slightest indication of a previous *shore*. So much, then, for the ripple-marks; as to the rain-drops and sun-cracks, I will consider them presently. In the mean time, we shall assume that no dry land appeared here at any time, until after the deposition of the coal; and if this proposition be correct, it is perfectly clear that the tracks found by Mr. Lea are *not* those of a *terrestrial air-breathing quadruped.*

But, supposing that the tracks were really made by an animal—(a supposition which, however, can only be entertained on the basis of the most extraordinary credulity), it is still impossible that it could have been an *inhabitant* of the land. It would have required *food*—unless, indeed, it was so essentially an *air-breather* that it could live on that element alone. If it was an inhabitant of the land, upon "what meats did it feed?" Where was the land vegetation to have appeased its hunger? There is not a *trace* of such vegetation, nor a trace of any mollusc, worm, or fish whatever. *No human being has ever found any thing in that entire formation which could have sustained animal life!* Even fleas or worms could not have found nourishment on the land, had they existed. A few poor little specimens of algæ are found in the rocks, but they all grew and were buried *in* the water. **There were** no land plants, whatever. If, therefore, **any** *animal* existed, it belonged to the water; and if it made any foot-marks, they must have been made *under* the water. Had Mr. Lea combined with his telluric researches— (which, by the way, have been of a highly valuable and interesting character in many other departments of science, especially that portion relating to the practical geology of the coal measures)—had he evinced a taste for the amusement of angling, during his summer sojourn at **Mount**

Carbon, I venture to say that his splendid illustrations of the footprints of the *Sauropus primœvis* would have been accompanied by full length portraits of the living descendants of that distinguished Devonian myth. Had he taken a rod and line, and wandered along the shaly and shady nooks of Tumbling Run, a stream which empties into the Schuylkill directly opposite the spot where he obtained the foot-tracks, he would have met two gigantic dams, erected by the Schuylkill Navigation Company, to hold back supplies of the aqueous fluid, to make up any deficiencies which may occur, upon the line of that work, during seasons of drought. These dams are from sixty to eighty feet in height, and the water thus held back makes two very broad, deep, and beautiful artificial lakes. The water is very clear and pure, and derived from numerous springs that bubble up their cooling draughts in the narrow valley. If Mr. Lea had dropped a line in those waters, baited with a crawling, slimy *lumbrici terrestres*, (of the class *Annelida*, and division Articulata!) ten chances to one but that he would have drawn up a creature entirely capable of making footmarks such as he described, only on a scale somewhat diminished. He would, in short, have hauled out a *living* Sauropus, swimming in the same pond with trout and eels.

Assuming the privileges of an original discoverer, I will here describe how I came to find this remarkable animal. I was rambling along the shores of the Tumbling Run, taking my usual morning walk (sometimes extending ten or fifteen miles around Pottsville), when I perceived that the lakes were being drained of their water, the summer drought compelling the Navigation Company to withdraw the supplies which these great reservoirs afford. The escape flume was therefore dry, but between the irregular stony bottom were several little shallow pools of water, in which, as I approached, there was a great floundering of young trout, as if they were afraid that my motives were not honorable. To convince them of their error, I drew up my sleeves and gathered them out of their little retreats, which the thirsty summer sun would have licked up in a few hours more, and carefully put them back in the lake. While engaged in this benevolent amusement, my attention was called to the individual in question. He was an inhabitant of the water, and shared in the accident which, but for my timely arrival, would have consigned him with his finny associates, to *terra firma*. He was walking around on the bottom of his hydrogenous basin, as if in search of a shelving stone or a quiet sequestered nook, in which to retire from the more active scenes of aqueous life. His *pedestrian* qualities excited my regard. I determined that he too should be saved! But first, I desired an exhibition of his functions. A twig, introduced in gentle proximity to his caudal appendage, had the effect of persuading him to accelerate his movements, and I found that he was equally at home as a swimmer and a pedestrian. Finally, I converted my handkerchief into a net, and with this safely secured him. I then left for my hotel, and after a walk of fifteen minutes, reached my *sanctum sanctorum*, where, in a glass jar, I intended him for the post of honor among my relics of the Devonian and carboniferous rocks. To my astonishment, however, I found that *atmospheric air* did not agree with him; the admixture of oxygen with nitrogen did not meet his case so well as oxygen combined with hydrogen. In fact, the poor thing was dying; and now I reproached myself for having put too much confidence in *science*. Having been taught by the descriptions of the Geologists to regard creatures like him as *air-breathers*, I took it for granted that a little nitrogen mixed with oxygen and the vapors of hydrogen, could do him no harm, especially during a mere experimental trial of *fifteen minutes*. But I was deceived. The *Sauropus Tumbling Runensinens modernis* died! *Resquiescat in pace*, and may he not become a fossil to haunt and perplex the scientific world hereafter!

He was a beautiful animal;—from five to six inches in length, of a dark green color, with bright golden or vermillion specks, like those of the trout. The structure of the head was somewhat similar to that of the toad, and may therefore be denominated batracian. It very much resembled the ideal *Labyrinthodon pachygnathus* of Prof. Owen, except that, unlike that supposititious animal, it enjoyed whatever luxury might appertain to a somewhat lengthened caudal appendage. Its fore-feet were furnished with four toes, and its hind feet with *five*. The hind toes, like those of the *Labyrinthodon* (supposed to represent the *Cheirotherium*,) are about twice as large as those in front. It bore no resemblance to salamanders, water efts, crocodiles, alligators, gechas, chamelions, tortoises, or serpents. It appears to have been a *species of lizard*, and answers all the requirements of that supposed to have made the footmarks. Its tail was long and sword-like, and answered a useful purpose in swimming; while in walking, it swayed from right to left, in the manner described by Mr. Lea. In some respects, it resembled the *Amblyrynchus cristatus* of South America, which is said to be the only *marine* lizard now known. "This marine saurian," says Mr. Darwin, "is extremely common in all the islands throughout the Archipelago. It lives exclusively on the rocky sea-beaches, and I never saw one even ten yards in shore. The usual length is about a yard, but there are some even four feet long. It is of a dirty black color, sluggish in its movements on the land; but when in the water, it swims with perfect ease and quickness by a serpentine movement of its body and flattened tail, the legs during this time being motionless, and closely collapsed on its sides." The teeth of the Tumbling Run *Sauropus modernis* are like those of fish—the jaws being furnished with plates, on which are warty protuberances, such as characterize catfish and eels.

Prof. Agassiz, at the meeting of the American Scientific Association, in 1851, remarked of Mr. Lea's *Sauropus* footprints, "that he did not believe they were made by an *air-breathing animal;* he thought they *might* have been made by fish of the ancient type, but he did not believe that any air-breathing animal had been found even as low down as the *New* Red Sandstone!" Here is the testimony of one of the greatest Naturalists of the age, very plainly and plumply putting an "extinguisher" on all the deductions and assumptions of Geologists as to the character of the animals which imprinted the footmarks. I will not say that the *Sauropus modernis* actually made the footprints of Mr. Lea, because I do not believe they were made by any animal whatever, either of the air, the land, or the water; but I *do* assert that the footprints it makes exactly answer the description of those of the supposed ancient animal! *Reductio ad absurdum!*

Prof. H. D. Rogers, at the same session of the scientific body above referred to depreciated the footprints of Mr. Lea, by denying that they *occurred* in the old red sandstone. He alleges that the footprints were *inside* of the coal measures, and denied that there was any identity between the old red sandstone of America and that of Europe! All this, however, seems to have been evolved in a spirit of professional jealousy. He did not like to see Mr. Lea reaping glory from a discovery which *he,* himself, as the official geologist of the State, should have made; therefore, he suggested that the footprints did *not* occur in the Old Red Sandstone at all. "These footprints in the *red shale* formation at Mount Carbon," observes the envious professor, "are of an age essentially later than that attributed to them; they occur in a geological horizon only a few hundred feet below the conglomerate which marks the beginning of the productive coal seams, in which series similar footprints, attributed to batrachian reptiles, have previously been met with in western Pennsylvania. Instead, therefore, of constituting a record of antique reptilian life, earlier than any hitherto discovered by at least a whole chapter in the geological book, they carry back its age only by a single leaf." Mr. Rogers, as if to retrieve laurels that should have adorned his illustrious brow alone, proceeded to hunt up footprints for himself. He searched around the old sandstones and conglomerates, and in his great Report of the State Survey (great for its enormous dimensions, as the celebrated Daniel Lambert was for his weight), he remarks that "about five hundred feet *lower down* in the formation, or further south, in the same locality, the Geological Survey brought to light another species of footprints of much smaller dimensions; and soon afterward two other varieties, etc., etc. He gives a very lame description of these footprints; and adds that "considerations of economy have compelled him to omit engraving them!" This is very remarkable. Many thousands of dollars were appropriated by the State to pay for publishing his work—a work executed in *Scotland*, and bearing a Scottish imprint; yet an engraving, the execution of which would certainly not have cost over ten or fifteen dollars, could not be afforded! The engraving of "sun-cracks" on the *same* page, immediately below the sentence first quoted, must have cost at least twenty-five or thirty dollars, and is of no earthly significance whatever; yet the footprints, if they be such in fact, cannot but be regarded as of *extraordinary interest*, but are altogether omitted from "considerations of economy." If Mr. Rogers were sincere—if he, himself, really *believed* in the footprints which he says he discovered in a geological zone still *lower than* that of Mr. Lea, a simple engraving on wood, exhibiting the tracks, would have sufficed. Mr. Lea, with a liberality which does him infinite credit, has expended large sums to lay before the world, the discoveries he has made, some of which Mr. Rogers should have introduced into his Report, since they appertain to Pennsylvania; but to shirk the whole question, upon "considerations of economy," is a reflection upon the liberality of the State which has appropriated something like a hundred thousand dollars to aid his explorations; besides being personally uncourteous to gentlemen like Mr. Lea, who sustain *their* paleontological researches with the most munificent and lavish expenditure.

But the truth is, that all these foot-prints are too obscure and unreliable to deserve any consideration for one moment. All that we *know* of the ante-medieval periods of Geology, are unfavorable to the existence of land animals, or animals of any kind except those occupying the lowest scale of organized life. All the speculations of geologists, based upon such frail data as these footprints afford, are precisely of a character with the great Cobham stone discovered by Mr. Pickwick. It will be remembered by those who have read the history of that remarkable discovery, what a sensation it produced,—how Mr. Pickwick, the immortal discoverer, delivered a lecture before a general meeting of the Club, in which he entered into a variety of ingenious and erudite speculations on the meaning of the inscription. "It appears," says Mr. *Boz*, (the worthy editor of the posthumous papers of the Pickwick Club,) "it appears that a skillful artist executed a faithful delineation of the curiosity, which was engraven on stone and presented to the Royal Antiquarian Society, and other learned bodies—that heart-burnings and jealousies without number were created by rival controversies which were penned upon the subject—and that Mr. Pickwick himself wrote a pamphlet containing ninety-six pages of very small print, and twenty-seven different readings of the inscription. That three old gentlemen cut off their eldest sons with a shilling apiece for presuming to doubt the antiquity of the fragment—and that one enthusiastic individual cut himself off prematurely, in despair at being unable to fathom its meaning! That Mr. Pickwick was elected an honorary member of seventeen native and foreign societies, for making the discovery; that none of the seventeen could make any thing of it, but that all the seventeen agreed it was very extraordinary! Mr. Blotton, indeed—and the name will be doomed to the undying contempt of those who cultivate the Mysterious and the Sublime—Blotton, we say, with the doubt and caviling peculiar to vulgar minds, presumed to state a view of the case, as degrading as ridiculous. Mr. Blotton, with a mean desire to tarnish the lustre of the immortal name of Pickwick, actually undertook a journey to Cobham in person, and on his return observed that he had seen the man from whom the *stone* was purchased,—that the man presumed the stone to be ancient, but solemnly denied the antiquity of the *inscription*, inasmuch as he represented it to have been carved by himself, in an idle mood," etc., etc. Hereupon followed a new controversy. Blotton, it will be remembered, wrote a pamphlet, giving his interpretation of the inscription, copies of which he addressed to the seventeen learned societies. The virtuous indignation of the seventeen learned societies being aroused, "several fresh pamphlets appeared; the foreign learned societies corresponded with the native learned societies,—the native learned societies translated the pamphlets of the foreign learned societies into English,—the foreign learned societies translated the pamphlets of the native learned societies into all sorts of languages; and thus commenced that celebrated Scientific Discussion, so well known to all men as the *Pickwick Controversy!*"

But, before we leave the Footprint Controversy (which has already surpassed the great Pickwick Controversy), we must say a few words in relation to the rain-drops and the sun-cracks. These are said to occur side by side with the footmarks, but are frequently found isolated. Now, *in the face of all the learned societies and professors, I say no such thing as rain-drops exist in the Devonian rocks!* The whole thing is a mistake—an absurd and ridiculous illusion.

The so-called rain-drops were produced on the slabs *after* the elevation of the mountains in which the strata are imbedded. When the laminated and stratified rocks were upheaved, and set on their edges, like shingles, they gradually contracted on cooling, while the disturbances occasioned by their uplifting, fractured them internally and externally. As the seams of rock parted, the surface water of the mountains, charged with fine mud and sand, percolated through them, and gradually *filled up the crevices and cracks*. In some cases, the percolation of the water, charged with sediment, was slow, and it would trickle down the smooth sides of the slabs in drops, one after the other. The earthy material thus held in suspension would be deposited, while the water itself was evaporated. In time, the surfaces of rocks thus ramified with cleaved laminæ, would become *coated with little warts or protuberances*, of the same clayey nature as the rocks themselves. In this manner stalactites are gradually formed in caverns. Wherever the rocks are calcareous, the warts and stalactites are calcareous; wherever they are aluminous, the "rain-drops" are aluminous. These so-called rain-drops are being formed every day. I observe them everywhere in my rambles among the rocks. Icicles in winter afford another illustration. Drop after drop accumulates on the cornices of houses, or upon the limbs of trees. If the water contained soft, plastic mud in solution, every drop would leave behind a small amount of sediment, which, by the slow succession of drops, would finally accumulate into protuberances varying in size from a pin's head to an ordinary rifle ball. The cracks and fissures attributed to the sun were produced partly by the original disturbance which changed the position of the strata; partly by the immediate action of the elevating heat; and afterward by the contraction of the strata in the process of cooling. These cracks were afterward filled up by surface drainage, and when this happened to introduce different earthy materials from that composing the rocks, the cracks themselves will exhibit the fact. Many of them are filled in with quartz crystals, and these are being constantly elaborated by supplies of water circulating through their labyrinthine crevices and laminæ. It is hardly possible to break a piece of laminated red shale or sandstone, without finding in it the traces of surface drainage.

But a moment's reflection would satisfy any man of common sense and observation, that it would be utterly *impossible* for rain-drops to accumulate and be preserved on the rocks in the manner suggested by Lyell and his cotemporary geologists. If they fell upon a *sandy* or *clayey* shore, they could make no impression on the ground, for the reason that the *succession* of drops is too rapid, and occur in too great a profusion. The impression of one would be effaced by another; while, after the rain had ceased, the surface of the mud would be so saturated with the rain, so perfectly liquefied, that no impression could by any possibility remain. Besides, if the sun had been *hot enough to crack the ground*, it would have *absorbed the rain-drops* immediately. The whole idea is the most absurd that I have ever encountered in the entire records of geology. Although, as I remarked before, all Geologists exhibit engravings of these rain-drops, no one has ventured to surmise *how* they were fossilized. It was sufficient for one man to say that they *were* rain-drops, without assigning any *reason* for it, and then all the others blindly adopt his opinions. The prefix of "Sir," or "Doctor," or "Professor" to a man's name, gives confidence to those who have no independence or capacity to think or examine for themselves; and hence the common sense of the age is borne down with the most consummate trash, under the guise of science, that could possibly be conceived.

I have in my collection many specimens of these so-called rain-drops, as obtained from different formations. I have found them in the body of the coal veins, and incrusting the sides of their stratified benches or walls. These often occur in *flattened drops*, because the crack was too *narrow* to admit of the full expansion of the round or spherical form of the drops, while the earthy matter left behind is correspondingly flattened, as a shot would be if compressed between iron plates. Those in the coal, as might readily be inferred, are composed of sulphur and iron, and present a bronze-like color. In the slates by the side of the coal veins, the rain-drops are irregular in size, varying from a pin's head to buck-shot. Like those of the coal, they also consist of sulphur and iron. Having been deposited by water which held these ingredients in solution, in trickling through the fissures of the coal, the sulphur has eaten holes in the slates, as might readily be conceived. These holes were originally filled with the sulphuret of iron (pyrites), but some of the balls have since fallen out. Now, that these balls and cavities should have been produced by showers of *rain*, would indicate a meteorological phenomenon, even greater than the footprints—for it shows that, if they were thus formed, there was a time when the earth *rained down sulphuric acid!* The rain-drops in limestone regions, would also indicate that the atmosphere at another, or perhaps the same time, and in the same vicinity, rained down showers of *carbonic acid.* Yet Sir Charles Lyell, and the author of the Vestiges of Creation, both remark that such rain-drops even show the *"direction in which the showers came!"*

It is absolutely sickening to me to dwell longer on this branch of the subject! I blush for the credulity and *stupidity* of a world that can swallow such absurdities, when their sole object and unavoidable tendency is to *undermine, and bring into contempt,* the holy word of the great Jehovah! But, alas! *Ce monde est plein de fous!* Must it ever be so—must it *always* be deceived by its philosophy—its "science falsely so-called"? Are we *never* to have an age of *common sense,* when reason and experience can be heard without a resort to the subtilties of classic verbiage and labyrinthine technicality? Disrobe Science of the unmeaning tautology that surrounds it, and what is it but the expressed idea of the laws of the Creator, ready to teach in plain English, as well as in bad Greek and Latin? Many of its recognized votaries and expounders are the mere peddlers in words—phraseology; inventors of classification; discoverers of Cobham relics—footprints—old teeth and bones! Let men come forward who *observe,* and let us hear what they have learned, *not* merely in books, but in the fields and among the rocks—in the mountains and in the valleys. But, for heaven's sake, let us hear no more of fossil footmarks, until the rocks favor us with the fossil *bones* of the animals that made 'em, so that we may at least learn—

> "Whether the snake that made the track,
> Was going south, or coming back!"

We happen to have higher and older authority than the

geologists to believe, as we have previously premised, that during the Devonian and coal periods, even if the land *had* been exposed as indicated, there was no sun to crack and parch the crust, nor yet showers of rain to leave behind the petrifaction of their fluid-drops. And in reference to *aqueous* animals, Moses says nothing to contradict the belief in their existence ; but on the contrary, leads us to infer that they *did* exist. For after the production of the coal, and the permanent establishment of the solar influence, he commands the "*waters* to bring forth *abundantly* the moving creature that hath life, and to *fill the seas,*" thereby leaving us free to infer that they *may* have previously existed, but in somewhat sparing numbers.

Besides the Cheirotherium footsteps in the new red sandstone, these strata were also distinguished by a family of monsters called *Ichthyosaurus*, half-fish and half-reptile, which, although living in water, breathed the air, and subsisted mainly on the smaller animals of the seas they inhabited. The *Plesiosaurus* was equally gigantic, and perhaps even more remarkable—having had a neck twice as long as its body, the head of a lizard, the teeth of a crocodile, the extremities and paddles of a whale, the ribs of a chameleon, and the trunk of a quadruped. The *Megalosaurus* was a gigantic lizard, which occasionally frequented the land. The *Pterodactyle* was also a huge lizard, but, by means of wings, was enabled, to some extent, to fly through the air, and thus to dart down upon its prey. Tortoises, toads, and crocodiles, of great dimensions, prevailed very extensively in this era ; and it may be said of *all* the monsters that lived during this time, that while they each combined some of the features of fish with those of tortoises, reptiles, whales, and birds, they yet partook of none of the distinctive characteristics of any. They were indeed, neither fish, flesh, nor fowl ; but a singular combination of all. They thus constitute a class of themselves, different alike from the preceding, and those of the subsequent eras. The whole of these strata having been deposited under water, (and generally under the sea,) and there being no remains of land animals, properly so understood, we are entitled to infer with Prof. Agassiz, that none had yet appeared on the earth; unless, indeed, we regard *birds* as such. But all the tortoises, lizards, toads, and other reptiles that flourished at this time, were properly aqueous, although all of them may have occasionally wandered on shore, or been abandoned there through their own negligence, by means of sudden withdrawals of the tides, or similar causes. Under these circumstances, it is probable that they made the best use of their pedal extremities which the novelty of their situation required; and instead of swimming, they learned to crawl or walk. In this way we may account for some of the foot-tracks discovered in the new red sandstone, because the *fossil skeletons* of the animals that seem to have made them, are found in the same rocks. In the new red sandstone of Connecticut, the foot-tracks of birds occur in great abundance, along with these of toads, tortoises, and lizards. That they are really the tracks of birds, seems to be generally admitted. Unlike other tracks in the lower measures, they occur very extensively, and have been found in extraordinary abundance in no less than sixteen different places. The animals that made them, would appear to have belonged to the order *Grallatores*, or waders, and some of them were of gigantic size—even exceeding the racing ostrich of the desert, or the extinct denornis and apteris of New Zealand, whose fossil bones show that they stood from seven to twelve feet in height. These gigantic waders wandered along the shallow estuaries in search of food, or some of them may have skimmed the air, from one island to another. The land, which was then constantly rising, was no doubt always strewn with the spoils of the sea; and this would not fail to tempt the fowls to the shore, while its mud was yet soft, but rapidly drying under the effects of the sun that, since the fourth day, had *now* emerged in new-made brilliance—hence the impressions of their feet on arenaceous mud that has since become indurated.

It may seem strange that God should have introduced marine animals and birds at the same time, and reserved mammalia for a subsequent era. But when we reflect that, by means of birds, he caused the seeds of vegetation to be scattered over the earth, as the dry land continued to emerge, the reason is rendered sufficiently obvious. The vegetation of the coal period was not *adapted* to animals. No animal is known to feed on the foliage of pine-trees, impregnated as it is with resinous and sticky juices. But, after that period, *new species of trees and vegetables were introduced*, and birds were then created to scatter the seeds over the earth, so that when the higher order of animals should appear, they might find the means of sustenance.

Prof. Hitchcock has devoted much attention to the investigation of the footmarks in the new red sandstone of Connecticut; but his enthusiasm seems to have carried him a little beyond the line of the *practical*. He has divided his footmarks into some fifty species, and upon the merest assumptions of their order and genera, has proceeded to assign *names* to the individuals who, unable to appear in person, have obligingly left behind their footmarks "on the sands of time." Moved by what we cannot but regard as a *silly-mania* for mere names, the professor very properly styles his first order *Sillimanium;* the next, (with perhaps more truth than the name itself might imply), he calls *Lyell*ianus; in a similar *mood*, he styles another otozoum *Moodii;* while other individuals or species are named Danarus, Baileyanus, Emmonsianus,

Adamsanus, Deweyanus, etc. Thus, all these distinguished gentlemen are complimented with *anuses*. It has long been suspected that professional Geologists have a Mutual Admiration Society, one of the main objects of which would seem to be the invention of nomenclature for all sorts of unknown animals, by means of which their *own* names may be transmitted to admiring posterity. The propriety of this scheme may be justly questioned; for the expediency of naming animals wholly unknown, and which may in fact never have existed, except in imagination, cannot be perceived. It was a prudent admonition of a plain farmer to his son, never to attempt the numerical solution of Rasores in gestation until they could be resolved *ab ovo!* or, rather in plainer English, "Never count chickens until they are hatched!" Our geologists, however, being a little more scientific, follow the instructions of the Dutch burgomeister in the play, when in pursuit of a criminal: "First," said he to his policemen, "first, *imprison* him; secondly, *arrest* him; and thirdly, "be sure to *catch* him!" But, granting the occasional expediency of temporary names to distinguish unknown and extinct animals, the propriety of borrowing those of distinguished individuals is certainly not without objection. We can readily suppose (as a friend has suggested), a meritorious member of the Mutual Admiration Association, with the Celtic appellation of *O'Alinmy*. Now, if Professor Hitchcock (who is such an adept in names) wished to compliment Professor O'Alinmy, by bestowing *his* name upon one of these bird-tracks, the species would be known as *O'All-in-my-i*, and this, it is very obvious, would have a disparaging effect upon the poor birds, in the estimation of the vulgar mind!

The class of birds is well defined. Warm-blooded, they all breathe the air not only by lungs, but also by means of auxiliary air-sacs, as well in some measure by their very bones, the internal cavities of which are proportionally larger than those of other animals, and are usually filled with air instead of marrow. The class comprises seven principal order: *Raptores*, or birds of prey; *Incessores*, or perchers; *Scansores*, or climbers; *Rasores*, or scratchers; *Cursores*, or runners; *Grallatores*, or waders; and *Natatores*, or swimmers. The fossil remains of birds are very scarce in the older formations, but occur in very great abundance in the Tertiary formation.

The fishes comprise a very numerous and diversified class of the verbetrate animals. They are divided into four leading orders, as the Placoid, Ganoid, Ctenoid, and Cycloid, each of which branches out into a large number of families, genera, and species. The ancient fishes, as we shall presently show, were different in their organization from those now living—having been allied, in some measure, to the subsequent class of reptiles.

The Reptiles, like the fishes, also comprise four orders, styled respectively the *Chelonians*, or tortoises; the *Saurians*, or lizards; the *Ophidians*, or serpents, and the *Batrachians*, or toads and Salamanders. Nearly all these animals (the footmarks to the contrary notwithstanding), appear to have been introduced, *for the first time*, in the strata of the Secondary formation—the formation which we are now discussing. They appeared, as we have shown, in the new red sandstone, but flourished more abundantly in the Oolite. And the fact is not without significance; for as the arterial blood of all reptiles is invariably mixed with a proportion of venous blood, the temperature of their bodies is nearly that of the surrounding atmospheric medium. Their organic functions are thus much influenced by atmospheric changes, and when the temperature falls to 40° or 50°, it almost invariably terminates their lives. Hence this fact has an important bearing on the climaterial properties of the ancient earth—nearly all the theories of which contemplate a *warm, humid,* and nearly a universal temperature.

It has been shown, too, that the bone cells of reptiles, birds, and fish, have forms and dimensions peculiar to each; and that while changes in their general and specific forms have constantly occurred, from one era to another, *none can be detected in the organization of the bones themselves*, which are peculiar to each class. The bone-cells of Mammalia, according to Mr. Quicket's investigations, average about one-two-thousandths of an inch, and if we adopt this as a standard of comparison, it is found that the bone-cells of birds will fall below, and those of reptiles will far exceed it; while those of fishes are so entirely different from mammals, birds, and reptiles, both in shape and size, that they cannot be mistaken for either. By the aid of the microscope, we are thus enabled to show that the physiological laws relating to the structure and growth of bone have ever been the same from the first creation of vertebrate animals in the far remote period, when saurian fishes were introduced into the seas of the Silurian eras, down to the present hour; that the colossal *Iguanodon* was provided with bone-cells formed after the same type as the tiny lizard that crosses our path; that the bones of the gigantic *Dinornis* exhibit no difference in structure from those of its representive, the *Apteryx;* that the bones of the *Mastodon* and *Megatherium*—those terrestrial giants of the pre-Adamite earth—are modeled after the type which we see in our domestic quadrupeds, and in man himself.

Immediately above the new red sandstone there are several layers of clay, limestone, marl, and shale, of an aggregate thickness varying from six hundred to one thousand feet, which has been named in England the *Lias* group. It forms the base of the oolite rocks, and sometimes gradually passes into them. The strata are intermixed, as well as separate; but all of them are prolific in the remains and shells of marine animals, as well as of the huge reptiles already referred to. Many of the crustaceans are special to this group; while an important change occurred in the structure of the fish. Those of the previous eras were somewhat allied to the saurian animals in their anatomical structure, but they now began to assume the distinctive features which characterize those of the present age. The change of anatomy and physiological character, while it constantly proceeds in the animal and vegetable creations of the earth, as we ascend from one geological period to another, was not regulated by a perceptible transition from species to species; but was invariably *direct and specific*. The creatures, therefore, that flourished in one period were peculiar to *that* period; and, if they appeared afterward, still preserved their ancient structure. The crustaceans of the Silurians and Devonians occur to a greater or less extent *in all the subsequent formations:* but they remain *unchanged* to the present moment. The change, whenever it occurs, is in the introduction of *new species*, and the extinction of former ones, or their *temporary absence* in intermediate strata. The idea of *progressive development*, or gradual transition from a low to a higher order of animal and vegetable life, as proposed by the author of the *Vestiges of Creation*, and seemingly sanctioned by Mr. Darwin, in his recent work on the *Origin of Species*, is, therefore, erroneous and untenable. There is no such thing in the whole history of the creation. On the contrary, every period is, more or less, a *specialty*, and is sometimes wholly obliterated by those great convulsions which, at different times, visited every portion of the globe. It was this *alternate* repose and convulsion that *shifted* animal life from period to period, and thus produced the *new forms which we notice in them.* Had the creatures of the earth, during these convulsions, been placed *in one spot*, they would have been destroyed over and over again; but it was happily so arranged that their destruction in *one* locality would afterward be compensated by an *influx* from another. It is this *alternating change* in, and during the great *creative days*, that has led the author of the *Vestiges* astray. He can find no animal at all resembling man until the sixth day. He can find no true land animal, of any kind, until that day. He can find no bird or any of the huge monsters that distinguish the fifth day, in any of those preceding. Whatever development there is in nature, is the development of creative law alone, as pointed out by Moses—the development of progressive *days* or geological *formations:*—*not* the gradual development which traces a worm into a serpent, thence into a quadruped, and finally into man. Yet such is the disposition to receive and respect the most absurd and wildest philosophical theories, when presented under the fascinating exterior of scientific verbiage, that this work attained unprecedented popularity, and left the door wide open for visionary speculation and specious infidelity.

After the Lias, we have a group which is very extensively developed in England, and some other portions of Europe, and there called the Jurassic or Oolite. The name of Oolite has been bestowed in consequence of the resemblance which one of its principal limestone layers presents to the roe of a fish, or to clusters of small eggs. The group, however, comprises ten or twelve seams of different kinds of rock and earth, as sand, clay, marble, slate, and oolites of various kinds. These layers have been divided, with the usual hair-splitting nicety of Geologists, into upper, middle, and lower divisions. The whole is of marine origin, and abounds in fossils. The sea still produced its rank weeds in great abundance; but a change was taking place in the vegetation of the land. Almost all the species which distinguished the coal had disappeared, and there was a sudden and extraordinary development of species allied to existing willows, poplars, sycamores, and elms, with but comparatively few ferns and resinous pines. In France, so rich are these rocks in corals and shell animals, that the celebrated D'Orbigny collected over four thousand species belonging to the families of Radiata and Mollusca. In view of their extraordinary variety, it would be useless to encumber our pages with particulars. The articulated animals, however, were comparatively scarce, unless we except those of insects, scorpions, and spiders. These were more numerous than ever before; though it may be well to observe that some writers, and among them the author of the *Vestiges of Creation*, speak of them as appearing in *this era for the first time*. And the author of the *Vestiges* does not stop with the simple announcement of the fact, but, as usual, proceeds to comment and to speculate upon it. Every little incident is turned to practical account, and made to swell the current leading to the theory of progressive animal development. With this view he observes: "It is remarkable that the remains of insects are found most plentifully near the remains of *pterodactyles*, to which undoubtedly they served as prey." According to strict principles of logic, the very converse of this proposition ought to hold good. If insects served as *food* for the flying Saurian, they ought to be scarce wherever that animal is found; but their occurring *plentifully* is *prima facie* evidence that his depredations could not have been serious. But it is unfortunate for the development hypothesis in this instance, that some of these minute animals existed previously, and long before the pterodactyles made their appearance!

But to resume: The vertebral animals were also but

sparsely represented in the Oolite, except, as in the Lias, by the prevailing saurians. These, however, increased immensely, both in number and form, and constitute the leading feature of the fauna of this period, if not, indeed, of the entire Secondary Formation. The change before noticed in the structure of the fish, had now become universal, or nearly so. Formerly, the vertebral column of all fish extended into the upper lobe of the tail, as it does in the shark; but from the Lias onward, this column terminated into a bilobe, or into two branches of the fin (or before it reached the fin), and so continues in nearly all the species now existing. There were other minor differences peculiar to species; but the change here noticed was universal, with but a few exceptions.

Geologists have indulged in a great deal of speculation in reference to certain detached bones found in the rocks belonging to this group. *First*, a large and curious bone from the main Oolite was produced, which the celebrated Cuvier pronounced to be the ulna of a whale; but which Mr. Owen, an anatomist of hardly less pretension, referred to a species of saurian animal. *Second*, in the Stonesfield slate of the same group of rocks, large numbers of the *lower jaws* of mammiferous quadrupeds have been found, belonging to three species and two genera, for which the names of *Amphitherium* and *Phascolotherium* have been adopted. Cuvier pronounced one of these specimens to belong to a small ferine mammal, the jaw of which resembled that of our opossum, but differing from existing genera by having a greater number of molar teeth. The first specimens found had ten such teeth in a row; but others, found some years after, had sixteen in a row, of which twelve were molars. The question was raised, whether this fossil belonged to a mammifer, a reptile, or a fish; and although a great diversity of opinion existed among comparative anatomists, the geologists appear to have settled the dispute among themselves, and established the bones as belonging to *land animals* of the *marsupial order of mammalia*. The specimens of *Phascolotherium*, with scarcely less exemption from doubt and obscurity, are also elevated to the rank of pouched animals. While the geologists reason or speculate themselves toward this conclusion, they throw the ulna of the whale, previously discovered, overboard, or degrade it to the level of the reptilian species.

The changes of opinion, in reference to these fossils, have been made in the face of previously-expressed views, but evidently without sufficient ground to support them. The bones are as mysterious as they were before; comparative anatomists are still disagreed. Yet, upon the mere *supposition* of a geologist that they *may* have belonged to a humble species of the marsupial order, other geologists, who never saw the specimens, rank them as such; and thus, in time, by common consent, they are established, until a new discovery, or the scrutiny of an original investigator elicits their true nature.

Notwithstanding the supposed marsupial and placental character of these specimens, in speculating upon the absence of land quadrupeds during this and the Wealden eras, Prof. Lyell says, "that the absence of the bones of whales, seals, dolphins, and other *aquatic* mammalia, whether in the chalk or in the oolite, is certainly very remarkable." Now, considering that the aquatic animals here mentioned, are of a higher order than the marsupials of the fossils (admitting that they are such, but they may just as likely belong to reptiles or fish), and that they did not, and some of them could not, frequent the shallow lakes, estuaries, and lagoons, inhabited by the saurians, chelonians, batrachians, and similar creatures, there would seem to be nothing remarkable in the premises. On the contrary, it *would* be remarkable to find the *skeletons of whales* side by side with reptiles, lizards, and fish. The whole order of Cetacean mammalia, with perhaps a few exceptions, were inhabitants of *deep water*, rendered necessary by their enormous proportions; and while this ought to account for the *absence* of their bones in the shallow deposits of the Oolite seas, it ought also to explain their absence until the appearance of the subsequent Tertiary rocks, upon which their skeletons were deposited, as the *then* floor of the sea, or the posssible floor of its great currents, into which the ocean was divided then, as it is now; so that in the case of whales, their carcasses would naturally be borne away into the deepest channels, and *not* be floated into the shallow basins along the land coasts. If, instead of whales, they had been trilobites, or fish, or lizards, their fossils would have been strewn amongst these rocks; but being nothing less than *whales* and considerably larger than any other animals which the world has ever yet produced, we must look for their remains in the *deep sea-currents* which they inhabited, and these we shall have an opportunity of inspecting when we reach the *Tertiary periods*, or the sixth day.

Above the Oolite rocks, in England, occur an interesting series, called the Wealden. The word is derived from woods, or *wealds*, and has reference to the extensive forests in the southeast of England, where the rocks were first observed. They consist of limestone, sandstone, conglomerate, and clay, abounding in the remains of fresh-water and land animals and vegetation. As it is believed to be the only *fresh-water* deposit occurring in the Secondary Formation, it is extremely interesting as exhibiting the remains of creatures and vegetable life that flourished on the "dry land" during that period. These deposits occur in England, Scotland, and some other portions of Europe, but nowhere in the United States. The **main deposit consists of a stratum of clay, from 150 to**

200 feet thick, having various shades of blue and brown streaks, with subordinate benches of limestone and sand; and containing fresh-water shells, and bones of reptiles and fish. After this, another stratum occurs, from 400 to 500 feet thick, consisting of gray, white, ferruginous, and fawn-colored sands and sandstones, with fragments of lignite, and carbonized vegetation; but no animals or shells. Below are layers of friable sandstones of various shades; and then compact, bluish-gray grit, in lenticular masses, the surface often covered with mammillary concretions, and the lower beds frequently passing into conglomerates, with quartz pebbles, containing ferns and stems of trees, bones of saurian reptiles, birds, turtles, fishes, and shells of the genera Unio, Cyclas, Cyrena, and Paludina—all fresh-water. Further down is clay of a bluish-gray color, alternating with sand and shale, and containing bones and shells more sparingly, with fragments of ferns and stems of vegetables. Another layer of white and yellow sand and sandstone exhibits only ferns and pieces of lignite, or imperfect brown coal. A still lower stratum of sand, alternating with clay and shale, affords only ferns and lignite. Finally, a layer of shelly limestone, alternating with sandstone, shale, and marl, exhibits a few shells of the genera Cyclas and Cyrena, with specimens of lignite and carbonized wood. Among these strata (which extend 200 miles in one direction, and 220 in another,) remains of forests *in situ* have been found in several localities. The trees had invariably been severed near the ground, leaving the summits of the stumps jagged, as if violently detached by a hurricane. Although fragments of lignite occur in the strata, no information is given by the English geologists as to the conversion or non-conversion of these weald forests *into coal*. Vegetation, indeed, would appear to have been almost as prolific as that of the coal period; and although

it consisted almost exclusively of *ferns and palms*, yet these are the varieties which, according to the geologists, furnished the great bulk of the coal of the Devonian basins; while the manner of deposition at the mouths of great rivers and estuaries of the sea *was precisely similar.* Here is the same vegetation and the same circumstances in operation which they refer to the carboniferous era;— *but where are the layers of coal?* Shale, and clay, and remains of vegetation, alternating with sandstones, limestones, and conglomerate;—*but where is the coal?* The solution of the problem, is this: after the oolite, every *trace of the resinous vegetation disappeared*—not a sign of it is to be seen in the subsequent Cretaceous era. There were a few specimens only of Cycadeæ and monocotyledonous palms, more of ferns and aquatic fucoides; but absolutely not a trace of Conifera, Sigillaria, Lepidodendrons, and Calamites. The proposition hitherto advanced, therefore, that notwithstanding the extraordinary abundance of the ferns during the carboniferous era, they *contributed nothing whatever to the coal*, is fully exemplified here, where scarcely any thing else than ferns prevailed.

In some of the Weald deposits, there is an alternation of marine and fresh-water strata. This alternation is continued three or four times, and is precisely similar to that of certain coal-fields. Geologists explain it in a similar manner, by supposing the gradual sinking of the strata under the sea, and their subsequent elevation into a basin to receive the detritus of the land. This subsidence and elevation is supposed to have been gradual, and the work of long periods of time; yet it must be admitted that, like the theories proposed for the deposition of the coal, a great deal of *special pleading* has to be employed to make all the points of the proposition harmonize.

The fishes of the Wealden are specifically different from

those of the succeeding strata, and belong principally to the genera Pycuodus, Lepidotus, and Hybodus. Several species of crocodiles have been identified, as well as large numbers of turtles. The most remarkable animal, however, was the Iguanodon—a monster reptile, occupying the same relation to its species, as to size, that the elephant does to the mammalia. Although detached bones and teeth only have been found, these serve to identify it with species existing in St. Domingo, and which are herbivorous and terrestrial. There were several other enormous and peculiar reptiles in this epoch, including some which flourished previously. Of these, the pterodactyle still survived; but notwithstanding its supposed depredations upon insects, these latter exhibit an extraordinary increase, and many new genera were introduced, evidently for the first time.

We now reach the last group, in ascending order, in the great Secondary formation, namely, the *Cretaceous*. Geologists apply this name to it in consequence of its immense strata of chalk, although in the United States these layers are wholly absent. It is, however, supposed that the chalk is here represented by cotemporary layers of ferruginous sand; but as the identity of fossils is not clear, they may yet call for some special distinction as to time or the operating circumstances of their deposition.

The strata of this extensive and widely-diffused system are literally made up from the spoils of the ancient seas— embracing coral reefs, and sponges, the shells of belemnites, ammonites, baculites, turrilites, and the bones of fishes, reptiles, and other animal remains, with fragments of plants and sea-weeds. The vegetation alone was inferior, and consisted almost exclusively of marine plants —those long, slender grasses which still grow along the shallow coasts, and in the stagnant rivers and lakes of the present era. There was hardly a trace of land vegetation, except a few specimens of palms and cycadeæ. Even the ferns, whose beautiful and diversified leaves were scattered in every rock from the foundation of the earth, had now temporarily disappeared, or were but obscurely represented by one or two species.

The microscope establishes the interesting fact of the origin of the chalk strata, wherever they occur in Europe, from the shells and secretions of marine animals. During the Oolite, and previously, the minute Radiata, with the co-operation of the higher shelled animals and crustaceans, built many islands in the midst of the ocean, some of them extending hundreds of miles, in every direction. Their instincts, as we have before observed, lead them to combine together in the form of trees, or incrusting or hanging from rocks. To do this, they are supplied with a sticky paste which they secrete in large quantities, and thereby elaborate carbonate of lime from the water which they inhabit. When the ocean became dotted with these coral reefs, new animals made their appearance, who browsed upon them like cattle in luxuriant meadows. This is still done by certain fish in the coral islands of the Pacific, and in opening them their intestines are found invariably to be filled with a calcareous excrement and with milky juices. Besides the fish, there were also peculiar worms that bored into the porous coral, and elaborated a similar juicy excrement. The disturbing operations of these animals, aided by the friction of the waves, finally wore down the coralline islands, and impregnated the waters with fine calcareous mud. The seas, all over Europe, where these strata are found, were white and milky with the thin detritus thus held in suspension by the water; and the natural result was its transportation to the adjacent shallow coasts, where it would be precipitated in the form of sticky chalk, and upon the subsequent elevation of the strata, the whole would solidify under the ordinary drainage of interior springs and rivers, as well as by means of rain and snow, in the manner of stalagmites. The same process, or the first stage of it, is still in operation in all the coralline islands of the Pacific; and it has been found, on comparison, and under the microscope, that there is no difference between the calcareous mud around them and that of the ancient chalk. The accumulation of layer after layer, in this manner, was in many respects similar to the deposition of the coal strata. Many of the beds of limestone of previous formations have originated in precisely the same manner; but having afterward been subjected to heat and pressure, the stone has been changed, and the impressions of the polyparia obliterated. Were these beds of chalk exposed to slow heat, under pressure, they would crystallize, and be converted into the finest granular marble.

The fossils of the chalk are specifically distinct from those of the subsequent Tertiary formation. They are the finely sculptured spoils of an ocean which rivaled the Atlantic and the Pacific in extent, and which it must have required vast ages to accumulate—strewn, as they now are, over large portions of Great Britain, France, Russia, Germany, Sweden, and Denmark, as well as in Asia and Africa.

In some of the layers of chalk, in Europe, there is a large admixture of flint, which often indicates the plane of stratification of the beds. These silicious nodules are the remains of animal sponges, sea urchins, and other minute creatures, whose spicular secretions were silicious instead of calcareous. Among the shells preserved in the chalk, are the Tenebratula, which has appeared in almost every previous era, but began to dwindle away upon its approach to this. The Belemnite made its appearance in the Oolite, and is remarkable for having been supplied with an ink-bag, by which, upon occasions of danger, it was enabled to muddle the water around it, and thus protect itself from attack. Among the great variety of the fossil shells, some of them were very beautiful and highly sculptured, as those of Turrelites, Ostrea, Scaphites, Hamites, etc. The Saurian animals, which flourished so extensively in previous eras, began rapidly to diminish; while even fishes were comparatively scarce. Among the former, however, was a monster called the *Moscesaurus*, which measured twenty-five feet in length, and the anatomical structure of which is presumed to have been intermediate between that of the Iguanodon and the Monitor. Crocodiles and turtles existed, but without material change from the preceding era. Bones of birds have also been found, in various places, in these strata.

Included in the Cretaceous group, but below the chalk, is a deposit which always accompanies it, called the *green sand*. It embraces layers sometimes eight or nine hundred feet thick, and consists mainly of sharp silicious sand, intermixed, or sometimes interstratified with marly calcareous sand, and laminæ of mica. The lower benches alternate with ferruginous sands, and beds of clay and sand, with seams of hard limestone and chert. Although the deposit is generally stratified, it is yet thoroughly intermixed, as if the sea had been charged with the drainage of extensive rivers in addition to the silicious and calcareous remains of its own waters.

The Cretaceous group of the United States is very extensively developed along the Atlantic coast in Rhode Island, New Jersey, and Delaware; while the whole vast region of country in the southwest, which we had frequent occasion to refer to while discussing the coal period, as having then been under the dominion of the ocean, was redeemed during this period. The region we allude to would, of itself, form no inconsiderable ocean; including, as it does, a large portion of Alabama, Mississippi, Tennessee, Arkansas, Texas, Missouri, Kansas, and Nebraska. All this vast space belongs to the Cretaceous era, and was only redeemed from the sea toward the dawn of the sixth day. A large portion of the Mississippi and the Upper Missouri traverse the Cretaceous strata; and it is a singular fact, that such is the porous nature of the soil adjacent to those mighty rivers, that they absorb the great bulk of their surplus of water during times of extensive overflow, and return it during seasons of drought. This fact is frequently illustrated at a distance of more than thirty miles from the Mississippi—the wells dug in the earth alternately yielding and withholding supplies of water with the varying volume of the river itself! If, by any convulsion of nature, the lower valley of the Mississippi should sink but a few feet below its present level, the ocean would again invade it, and it would become, what it was before, a mere adjunct of the Gulf of Mexico.

The Cretaceous strata of New Jersey consist principally of green sand and green marl, with seams of coralline limestone overlying, containing fossils which Prof. Lyell thinks agree, upon the whole, with those of the Upper European beds. He collected some sixty shells, of which five were identical with the European species, while others were cretaceous in their generic forms. Fifteen out of the sixty shells were regarded by Prof. Forbes as good geographical representatives of well-known cretaceous fossils of Europe.

The Secondary Formation, comprehending the New Red Sandstone, the Lias, Oolite, Wealden, and the Cretaceous systems of rocks, occupies very extensive areas in North America, or rather that portion of it belonging to the United States, for these rocks nowhere occur to any extent outside of our political confederation. This may account, in some measure, for the superior fertility and variety of our soil; nevertheless, the fact is a singular one, when we consider that, large as is our domain, it does not comprise one-sixth part of that vast and almost uninhabited portion lying to the north, and embracing the Canadas and Russian and British America. But the reason may be explained by remembering what we have before stated, viz., that nearly the whole of the North American continent was elevated during the Metamorphic and the Carboniferous eras, and that it was in all probability the first, as it was incomparably the largest, tract of country ever redeemed from the primitive seas, *at any one period*. In South America, the Secondary rocks scarcely occur at all—there being but two or three isolated little districts in Peru and Venezuela in which they are represented. In proportion to the surface, the greatest development of these rocks is in Continental Europe. A large portion of that country lying between the Mediterranean on the south, the Caspian Sea and Persian Gulf on the east, the Baltic on the northwest, and the Ural mountains, in Russia, on the east, was occupied by these seas during the period in question—there having been, here and there, little islands of the primitive rocks scattered irregularly through the great field of waters. We accordingly find the Secondary rocks in the ascendant in Ireland, England, France, Germany, Portugal, Spain, Austria, Turkey, Syria, Russia, and that portion of Persia along the eastern shore of the Mediterranean, and north of the Euphrates. The Secondary strata are very sparingly exhibited in Asia and Africa, except that in the latter they form a zone from the Red Sea west through Tripoli and Algiers to the Atlantic coast—having the great desert of Sahara on the south, and the Mediterranean sea on the north. Small districts have been ascertained to exist in China and Japan; but little is as yet known of the geological structure of either of these vast regions.

Now, in the beginning of the great Secondary Formation (or the Fifth Day), God commanded the "waters to bring *forth abundantly* the moving creature that hath life, and fowl that may fly above the earth, in the open firmament of heaven." While we are left to infer the previous existence of marine life, as before suggested, the waters are *now* commanded to bring forth *abundantly*;—and God blessed the creatures thus produced, and ordered them still to increase and multiply, and "*to fill the waters in the seas.*" Have we not just seen an absolute and literal fulfillment of this order? Was not sea after sea gradually *filled up* with the countless millions of corals, molluscans, crustaceans, fish, and enormous reptiles? The whales which, in the cold seas of the north, attain a length of ninety feet, in the torrid climate that then prevailed, when every other creature expanded into giant-like proportions, became absolutely too monstrous to frequent the shallow seas thus gradually being redeemed from the ocean, and it is only in the deep waters of the ensuing Tertiary that we can reasonably hope to meet their fossilized remains. But with the seas thus redeemed and laid open to our inspection, we see the extraordinary and wonderful development of marine life; and no one can fail to perceive in their fossil remains the complete and overwhelming realization of the divine injunction, nor to acknowledge, with profound humility, the power and majesty of the great Creator. But not only of the seas: the atmosphere having now, for the first time, become clear, buoyant, and open, was fitted to support the winged messengers of the air; and accordingly we early find the imprints of their feet, and then, at every successive stage, their fossilized bones—thus showing that *they*, too, promptly responded to the creative order. During the coal period, such a command would evidently have been premature; but Moses, comprehending the true circumstances of each period, has introduced the order of creation precisely as it occurred.

Thus, in response to the divine word, the seas were rippled and foamy with the living creatures that filled them; the air was blackened with flocks of birds, of varied plumage; while colossal fowls, to which the eagle or the racing ostrich of the desert are as pigmies, screeched discordantly along the margin of the ocean inlets, or waded shallow lakes in search of food; enormous whales, and fish, and minute corals, filled the seas; crawling reptiles, and scaly crocodiles, and lizards, crowded the muddy marshes, and wallowed in stagnant pools; aquatic fowls and pterodactyles skimmed the lakes and rivers; while soaring birds spread their pinions in the breeze! So completely and so perfectly was this command realized, that we may say with entire truth, "the dust we now tread upon was once alive."

> And God said, Let the waters generate
> Reptile with spawn abundant, living soul;
> And let fowl fly above the earth, with wings
> Displayed on the open firmament of heaven
> And God created the great whales, and each
> Soul living, each that crept, which plenteously
> The waters generated by their kinds,
> And every bird of wing after his kind;
> And saw that it was good, and blessed them, saying,
> Be fruitful, multiply, and in the seas,
> And lakes, and running streams, the waters fill;
> And let the fowl be multiplied on the earth.
> Forthwith the sounds and seas, each creek and bay,
> With fry innumerable swarm, and shoals
> Of fish, that with their fins and shining scales,
> Glide under the green wave, in sculls that oft
> Bank the mid sea; part single, or with mate,
> Graze the sea-weed, their pasture, and through groves
> Of coral stray, or sporting with quick glance,
> Show to the sun their waved coats dropt with gold;

Or in their pearly shells at ease attend
Moist nutriment, or under rocks their food
In jointed armor watch; on smooth the seal
And bended dolphins play; part huge of bulk,
Wallowing unwieldy, enormous in their gait,
Tempest the ocean; there Leviathan,
Hugest of living creatures, on the deep
Stretched like a promontory sleeps, or swims
And seems a moving land, and at his gills
Draws in, and at his trunk spouts out, a sea.
Meanwhile the tepid caves, and fens, and shores,
Their brood as numerous hatch from the egg, that soon
Bursting with kindly rupture forth discharged
Their callow young; but feathered soon and fledge,
They summed their pens, and soaring the air sublime
With clang despised the ground, under a cloud
In prospect: there the eagle and the stork
On cliffs and cedar tops their eyries build:
Part loosely wing the region, part more wise
In common ranged in figure wedge their way,
Intelligent of seasons, and set forth
Their airy caravan, high over seas
Flying, and over lands, with mutual wing
Easing their flight; so steers the prudent crane
Her annual voyage, borne on winds; the air
Floats, as they pass, fanned with unnumbered plumes.
From branch to branch the smaller birds with song
Solaced the woods, and spread their painted wings
Till even; nor then the solemn nightingale
Ceased warbling, but all night tuned her soft lays,
Others on silver lakes and rivers bathed
Their downy breast; the swan, with archèd neck
Between her white wings mantling proudly, rows
Her state with oary feet; yet soft they quit
The dank, and rising on swift pennons, tower
The mid aerial sky. Others on ground
Walked firm; the crested cock, whose clarion sounds
The silent hours, and the other, whose gay train
Adorns him, colored with the florid hue
Of rainbows and starry eyes. The waters thus
With fish replenished, and the air with fowl,
Evening and morn solemnized the *Fifth Day.—Milton.*

THE SIXTH DAY—GEOLOGICAL.

24 And God said, Let the earth bring forth the living creature after his kind, cattle, and creeping thing, and beast of the earth after his kind: and it was so. 25 And God made the beast of the earth after his kind, and cattle after their kind, and every thing that creepeth upon the earth after his kind: and God saw that it was good. 26 And God said, Let us make man in our image, after our likeness; and let them have dominion over the fish of the sea, and over the fowl of the air, and over the cattle, and over all the earth, and over every creeping thing that creepeth upon the earth. 27 So God created man in his own image, in the image of God created he him; male and female created he them. 28 And God blessed them, and God said unto them, Be fruitful, and multiply, and replenish the earth, and subdue it: and have dominion over the fish of the sea, and over the fowl of the air, and over every living thing that moveth upon the earth. 29 And God said, Behold, I have given you every herb bearing seed, which is upon the face of all the earth, and every tree, in the which is the fruit of a tree yielding seed: to you it shall be for meat. 30 And to every beast of the earth, and to every fowl of the air, and to every thing that creepeth upon the earth, wherein there is life, I have given every green herb for meat: and it was so. 31 And God saw every thing he had made, and, behold, it was very good. And the evening and the morning were the sixth day.

THE extensive formation now known as the Tertiary, not a great many years ago, was included with that of the Alluvium, or modern formation. The investigations of Cuvier and Brongniart, of the strata immediately surrounding the city of Paris, in 1810, established the fact of the existence of a series of strata, of marine, river, lake, volcanic and land origin, which, on account of the extent, variety, and character of their fossils, were entitled to the rank of a separate and independent formation. Since that time, similar strata have been examined elsewhere in England, France, Germany, and numerous portions of Continental Europe, the fossils of which, while they exhibit a gradual transition toward those of existing species, in other respects point to an era at once independent and isolated from the past or the present. It appeared from the researches of Deshayes, in France, on the fossil shells of this interesting formation, that it naturally arranges itself into three leading groups or divisions, as determined by their approaches to the character of existing species. Thus, the fossils of the oldest group show an average of about four per cent., as compared with the species now living on the earth; those of the middle group exhibit about eighteen to twenty per cent., while those of the third, or upper group, have fifty per cent. Prof. Lyell, availing himself of this Paleontological discovery, called the older strata *Eocene*, the middle *Miocene*, and the upper *Pliocene;* but, as if this were not sufficiently comprehensive, he afterward erected several sub-groups, as the upper, middle, and lower Eocene, and the newer and older Pliocene. He also introduced various other local and general names for particular strata; so that, although the whole formation is comparatively new in geological discovery, it is already characterized by as many technical names and local sub-divisions, as that of any other. The names proposed by Lyell, however, in this case, are sufficiently simple, and they have consequently been generally adopted; they indicate the gradual transition of one stage into another, until finally arriving at the post-Pliocene, diluvium, or boulder strata, we find ourselves surrounded by the physical *debris* of the present—by that genial Sabbath of Nature, during which the work of Creation *ceased*, and the great Author rested from his creative labor. The transition from stage to stage is very gradual, and sometimes scarcely perceptible; for beginning with an identity of but three or four per cent. of the molluscan fossils, the *last sub-group* of the Tertiary contains no less than ninety-four per cent. of species now inhabiting the adjacent seas and lakes. A description of the Tertiary, however, in consequence of the technical names being *new*, and owing to the numerous subordinate groups into which Lyell and others separate it, is not only difficult, but somewhat unsatisfactory. Mr. Lyell devotes fifty-five pages of his *Elements* to its consideration, of which nearly one half are actually occupied in explanations of a purely technical character;—that is, they have little bearing directly upon the elucidation of the subject as a *whole*. The first part of the formation was exclusively marine—a continuation, under different features, of the Cretaceous; the middle part was also principally aqueous, but exhibits a strong tendency in its fossils to the fauna and flora of the present, accompanied by the introduction, for the first time, of land animals of prevailing genera and species. The latter part was distinguised by intense and wide-spread volcanic action; the change of land into sea, and of sea into land; by the elevation of mountain systems, the prevalence of universal cataclysms, the diffusion of erratic boulders, icebergs, and moraines, and finally terminated by gradually merging into the geological laws now in force.

The Nummulite rocks of the Swiss Alps, and of the Pyrenees, are the oldest members of the Tertiary group in Europe. Indeed, some geologists are disposed to rank them with the preceding Cretaceous system; but although they are elevated nine or ten thousand feet above the level of the sea, the fossils they contain are very similar to those of the Paris and London basins. They consist of limestone, slate, marl, and sandstone of various colors and textures, and nearly all abounding in the **character-**

ristic fossils of the Nummulite. From their resemblance to metallic coins, these fossils were formerly called by the Germans *devil's money*, and formed the theme of many a wonderful legend and romance. The strata often attain a thickness of several thousand feet, and besides their wide distribution in Europe, in isolated basins, they occur extensively in Asia, Africa, and America. In certain portions of the Alps, where they have been disturbed and exposed to heat, they have been changed into crystalline marble, quartz rock, and mica-schist.

The Tertiary formation, as developed in different parts of England, exhibits considerable variation in lithological character. It is generally at least two thousand feet thick, without counting absent strata, some of which, but generally the upper, are always wanting. It is seldom, therefore, that the *complete* formation is found in one region; if it were, the aggregate thickness would be immense. In England, it is well represented in Hampshire and the Isle of Wight, and the order of superposition (without regard to Lyell's Miocene and Eocenes) may be stated as consisting of, 1. sands and plastic clays, the latter being extensively used in England, Germany, and France, in the manufacture of pottery ware; 2. the London clay, so-called because the cities of *London and Paris* are both erected on such a stratum, and which is also used to some extent in manufactures, especially brick; 3. fresh-water debris, comprising river sands, calcareous marl and mud, and river shells and land plants; 4. clay and marl with *marine shells*, but of different species from those of the lower London clay; 5. upper fresh-water deposits, including white and green marls, and calcareous limestones, which form an almost solid aggregation of fresh-water shells, principally of Paludina concinna, Lymnea pyramidalis, and Planorbis euomphalus. The Paludina have a spiral termination, but are scarcely larger than a grain of wheat; nevertheless they often constitute the great bulk of the solid contents of the rocks in which they are imbedded. The fossils of the London clay consist largely of Cassidaria carinati, Pleurotoma prisca, Trochus agglutinans, and Turritella edita, (also a spiral, but very long and slender.) Fragments of trees are found which, although converted into hard stone, still exhibit the perforations of the Teredinæ, a boring mollusc, nearly related to the Teredo, the pest of the Indian seas. In the Paris basin, besides the usual alternation of marine and freshwater deposits, there occur very extensive beds of gypsum, more familiarly known as plaster of Paris. The finer portions, being alabaster, are employed in the arts of sculpture, especially for small ornaments; while the ordinary gypsum, containing a few per cent. of carbonate, in addition to the prevailing sulphate of lime, makes an excellent plaster or stucco for houses. It is likewise valuable, in certain cases, as a fertilizer.

The vegetation of the Tertiary shows a remarkable increase over the preceding Cretaceous—in fact over all the groups combined of the Secondary Formation. It is a matter of no astonishment, therefore, that this period should have produced beds of coal—especially as there was a great increase of *coniferous trees*, perhaps nearly equal to that of the carboniferous era. To this fact (which is one deserving particular notice, since not a *trace* of these trees is to be found in the era immediately preceding) may be attributed many of those isolated deposits of lignite, mineral bitumen, pitch, and impure vegetable gums, oils, and resins, so plentifully distributed in basins over the earth. These same trees, with those of the Cycadeæ, had also a prolific development during the Oolitic period, and the result was the deposition of such coal basins as that near Richmond, in Virginia, which, in consequence of its *greater age*, is more completely mineralized than those of the Tertiary period. Whenever these trees occur, there is coal; but whenever they do *not* occur, there is none; and the coincidence is full of significance in connection with the theory of the origin of coal which I have proposed in this work. But besides the coniferæ, there was an equal, if not a greater development of other dicotyledonous trees, and of species allied to those now existing in the forests of the earth. Among these may be mentioned the families of poplars, willows, elms, sycamores, maples, birches, magnolias, oaks, etc. The monocotyledons also were developed to a greater extent than ever before, and are represented at the present time by more than one thousand species, principally of palms, naiades, and tropical trees. The ferns and mosses were also more numerous, as well as marine plants and weeds; but all the other varieties which distinguished the carboniferous era, and which are usually supposed to have furnished the great bulk of its coal, have no fossil representations whatever.

The Infusoria and Polyparia, which flourished in nearly every preceding era, still continued, on a scale of even increased magnitude, their extended operations in the bottoms of seas. Vast beds of limestone and silicious concretions have been elevated from the ocean, which are literally derived from their delicate secretions and the remains of their minute skeletons. Insects, which appeared irregularly, and with the varying development of vegetation, again spring into existence, after their *total absence* during the Cretaceous period. They began gradually to increase from the dawn of the Tertiary, and at its close were more numerous than ever before, and scarcely less so than at the present moment. Spiders, scorpions, serpulæ, and other articulated creatures, also expanded very materially over previous periods; but there was no perceptible increase of marine molluscs, except in

Nummulites, Miliotites, and other microscopic shell animals, and in the variety of land and fresh-water testacea. These latter were evidently as numerous during the Tertiary as they are now, and, as has already been suggested, were superseded by the present species. The most extraordinary change in the marine fauna of the Tertiary was in *Fishes;* and in this, as in every other movement affecting the animal species, the infidel theory of progressive development, or of gradual infusion of a lower species into a higher order of creation, is at once utterly disproved and annihilated. Nearly every leading comparative anatomist has adopted a system of classification of his own; and, in dealing with the ancient fishes especially, there is necessarily abundant room for diversity of opinion. The animal kingdom is divided into four leading parts—the principal one being the Vertebrata, which is itself divided into four classes, as fishes, reptiles, birds, and mammals. Each class, in turn, is arranged into orders; the orders into tribes; the tribes into families; the families into sub-families or genera, and the genus into species. There is in each division, a central *nucleus*, or *typical characteristic of structure*, around which all the individuals can be arranged; and the only difficulty that has arisen has been as to *which* division—species, genera, family, order, or class—the individual specimens properly belong. Cuvier, and most of the older ichthyologists, classified the extinct and living classes according to their *internal skeletons*, their brain, nervous, generative, and circulatory organs. But as in the fossil specimens it was often impossible to determine their true character by these tests alone, owing to their imperfect preservation, Prof. Agassiz adopted a classification founded mainly on the structure of their *scales* and that of their *external skeleton*, which are generally not only well preserved, but very often the only portions not entirely obliterated. In this way many specimens have been identified, which defied classification according to the other systems. Agassiz, therefore, divided the fish, both living and fossil, into four orders,—the *Placoid*, the *Ganoid*, the *Ctenoid*, and the *Cycloid*. The fish of all the preceding geological eras, up to that of the Cretaceous, belonged to the two orders first mentioned. The Placoids were distinguished by having their body covered with broad horny and enameled plates, with bristling protuberances like the teeth of a file, instead of smooth imbricated scales; and for presenting, in lieu of a well-defined *osseous* skeleton, a soft cartilaginous structure. Of the six families which the order comprised, at least two have typical representatives in our seas. These are the sharks, the rays, the dog-fish, and some other cartilaginous species—all very numerous in some regions, and the first occupying to the creatures of the deep, the same position as that of the vulture to those of the earth and air. The shark has been known to follow in the track of vessels for hundreds of miles in quest of spoil; and to attend, with eager watchfulness, during naval engagements, nothing daunted by the deadly conflict of arms, ready to seize the human victims that might fall a prey to their voracious jaws. Some of the species of rays, armed with stings, (as the *raia pastinaca*,) or with an electric apparatus (as the *torpedo narke*), are also enabled to prey upon the weaker and smaller fishes which they encounter. The Placoids and Ganoids appeared for the first time during the upper Silurian era, but flourished more extensively in the carboniferous than any other period. They sent out lateral families, genera, and species at different times, but always as *distinct subordinate groups*. While there has been a constant *variation* in every *progressive formation* (the old families running out and the new ones coming in) the general features that characterized the *order as a whole* were invariably distinct and separate. And while some of their typical representatives are as distinct and numerous *now* as those of the previous eras, they remain *true* and *fixed* to their ancient instincts, and preserve, in some measure, the anatomy and family features peculiar to the ancient types. Through all the changes of time and of circumstances, they are as essentially *placoidian* to-day as *other* placoidians were in the far remote antiquity. In the great mint of Creation, generic and specific moulds have been *changed*, as all else was changed, from era to era; but no one but a philosophic fool would venture to infer from thence that the *solid gold and silver of the stamped coins has been basely amalgamated.* Nature has never been convicted of counterfeiting; on the contrary, the integrity of her stamp has ever been maintained, from the highest to the lowest of her creatures, and in the face of the most wicked and outrageous abuses. To charge her with infusing lead into silver, or silver into gold, or of bringing down god-like man by gradually merging into him the fishes of the sea and the crawling reptiles of the land, is merely equivalent to charging upon the Almighty Creator an *ultimate design of merging and fusing all his splendid variety of created life into one channel, or into one species!* Such an idea of fusion or of development, so far from having been contemplated in the scheme of Creation, is absolutely beneath contempt. And yet the late Hugh Miller wrote two very learned and elaborate books, mainly to confute it—as if a bare-faced libel on the Almighty *needed* any refutation whatever. But in his attempt to put down the theory of progressive development, he suggested another, scarcely less unfortunate or absurd—that of *gradual degradation.* Much respect has been accorded to Miller because of his zeal in behalf of religion; but many of his propositions, though maintained with a great deal of pseudo-scientific acumen, and with extraordinary beauty and force of diction, are utterly untenable, unphilosophic, and *insidiously inimical* to sound truth. His pages are full of special pleadings; and often proceeding upon obscure and doubtful premises—upon an isolated shell, or tooth, or stem, or algæ—he boldly launches out into the most gorgeous and dashing inductions, cutting hither and thither, and putting whole armies of Lamarkians, and Okens, and "Vestiges" to flight, like another Don Quixotte vanquishing a drove of swine. But he was a good and a great man; and his practical explorations of the old Red Sandstone are among the most valuable contributions to modern geology. He was, however, more of a poet and a word-painter than a philosopher; and loved to dally with pretty ideal visions, and high-sounding words, rather than to pursue stubborn facts. He was more a Theologist than a Geologist; and while he deserves credit for unsurpassed zeal in his endeavors to harmonize the two, he has yet placed obstacles between them, which are utterly insurmountable upon the basis of scientific and philosophic inquiry.

The Ganoid order of fishes embraced thirteen different families, most of them numerous, and several belonging to the Sauroid type. Like the Placoids, this order was most numerous during the Carboniferous period. Their body, like that of alligators, was generally inclosed in a coat-of-mail—the scales being hard and horny, and of a rhomboidal shape. The skeleton was more bony than that of the Placoids, but less than those of Cycloids. In this particular their position was *intermediate* between the two, but in other respects they were perfectly distinct from both. The sturgeon, however, which belongs to the Ganoid order, has a cartilaginous structure; but differs from the shark in the arrangement of its gills, which resemble those of the true fish. Its mouth is small, and destitute of teeth, and, like the Cycloids, its head is cuirassed, and supplied with bony bucklers. While this fish is found everywhere in the ocean, it also frequents rivers, where it deposits its spawn. During the Oolitic period, as we mentioned while speaking of that era, the Sauroid fish had an extraordinary development—but those varieties with heterocircal tails were utterly extinguished, and no less than fourteen genera, with *homocircal* tails, made their appearance. Of these genera, two representatives remain—the *Polypterus* of the Nile, and the *Lepidasteus*, or pike, of our American rivers and lakes. Of the families of *Syngnathidæ* and *Diodontidæ*, several types exist, but their fossils are rare in the Tertiary. With a few exceptions, and they of a somewhat doubtful character, the true Saurian fishes all expired during the Cretaceous period—a period which effected a more complete and general revolution of animated nature than any previous era. With the simultaneous and almost universal extinction of the Ganoid order, and of the Saurian type, those of the Ctenoid and Cycloid made their appearance. But they did not come *gradually*, as if developed from the extinct races, their immediate predecessors; *but they came in myriads*, and in numberless distinct families, genera, and species; and *as they came, so they remain!* While the Cycloids and Ctenoids are both distinguished from the ancient fishes by their osseous or bony structure, as well as some other prominent features, ichthyologists have devised various schemes of classification to separate them from each other, the most popular of which, based on the number and character of their fins, also constitutes a feature in the system of Agassiz. It is hardly worth while here, or proper, in view of the space it would require, to enter upon a description of their genera and species, since *nine-tenths of all the fish now living* belong to these two great orders—the other one-tenth being feebly represented by those which are typical of the preceding eras. Both of them comprise a great many families—the Cycloid being well represented by those of the Pike, the Carp, and the Salmon, while the Ctenoids are equally represented by the Perch, the Mackerel, the Mullet, and the Gudgeon. As far as specific character is concerned, there is no fish now living that really presents a true counterpart of any that distinguished the ocean previous to the Cretaceous formation. The change then introduced was so thorough and so sudden, that the whole ichthyologic field was reorganized on a new basis, to *meet the requirements of the approaching era of man.* This was not the work of a slow, precarious and gradual development; but it was *instantaneous and direct.* It was not the effect of an empirical law; but the work of the great Creator, flowing spontaneously from the exercise of his almighty Volition. Bearing this great *fact* in mind, and remembering also, in connection with it, the *true osseous skeleton* of the orders thus created, with which the merest urchin of the hook and line is familiar, where is the force of Miller's theory of *gradual degradation,* on the one hand, or of the Lamarkian (or Aristotelian) theory of *gradual development* on the other? If the ancient fishes combined in their form and organization some of the features of birds, reptiles, and turtles, it only proves, if it proves any thing at all, that they were not *true fishes.* Because they inhabited the seas, it does not follow that we must regard them as true fishes. Ducks and geese frequent the water, and whales and dolphins *live* in it;—yet they are not *fish.* Fishes were made for man; and God did not *create* them until the dawning of his era. Moses, indeed, in his retrospective vision of the fifth day, says nothing whatever of *fish.* He speaks of the waters *"bringing forth moving creatures and great whales,"* and these, we repeat cannot be understood as implying *true fish*. Moses and all his people were as familiar with the existing type of fish as we are at this moment; and the word *fish* is used again and again in the Bible. That he, therefore, did not contemplate *true fish*, properly so understood, in the phenomena of the great Secondary Geological Day, is very plain, for the simple reason that he did not *name them;* while, that he *did* contemplate them in the *Sixth Day*, is equally plain, for the reason that he *does name them!* Now, all this is significant. The "moving creatures" of the Fifth Day could not properly be called fish, nor reptile, nor turtle, nor bird, nor beast. Even comparative anatomists, and scientific men of all departments of zoology and physiology, generally exceedingly prolific in nomenclature, have been much perplexed in adopting names by which to *distinguish* the peculiar and varied creatures which characterized that vast formation. And after all, the term applied by Moses is perhaps more expressive than those of professional naturalists. "*Moving creatures*" contemplates a variety—a heterogeneous assemblage, otherwise unclassified, rather than a *distinct* class or order; and hence the propriety and significance of the term. Had Moses used the word *swimming* creatures, or *crawling*, or *flying* creatures—or had he used the word *fishes* direct—the impolicy of the expression would here be palpably manifest—for a whole army of scientific sticklers, sitting in their high places like eagle-eyed vultures on frowning precipices, would long since have pounced upon the unfortunate word, and voraciously assailed the integrity of his inspiration and his facts. While its non-introduction, under the peculiar circumstances, is extremely fortunate, its omission can neither be ascribed to accident nor ignorance. It was design—deliberate and premeditated. For after the creation of the true and existing orders of fish, on the dawn of the Sixth Day *he ex-*

pressly refers to them as constituting a portion of the vast domain of man. The term "moving creature," as applied to aquatic fauna, is no longer used;—but "*the fish of the sea*" are now *distinctly* and *unequivocally referred to*, well knowing, as he did, that the *previous species* had been extinguished, and that under no circumstances could *they* have contributed to the sustenance or the varied requirements of the human family. The theories, therefore, of progressive development and of retrogression and degradation, are alike absurd and visionary; and they are only the more palpably so, for having been elaborated with consummate skill and a great display of pseudo-learning.

We have previously alluded to the supposed absence of the fossil remains of whales and dolphins in the rocks of the Secondary Formation. Isolated and detached bones have indeed been found, which have been referred to these animals, and their existence, during that period, is not doubted by geologists; but, on the contrary, astonishment is expressed at the absence of their fossil remains. I suggested that this might be explained on the basis of their enormous dimensions, requiring them to select the *deepest channels of the primitive oceans,* or at least forbidding their sojourn in the shallow estuaries which were then *being converted into dry land.* The Greenland whale is an animal from sixty to eighty feet in length. Less than a century ago, when the fishing grounds in the regions of the North and South poles were first frequented, some were found over one hundred feet in length, and comprising a bulk greater than that of *one hundred elephants.* The ordinary whale, as it is now captured, is still, by far, the largest animal that ever inhabited the earth; and we have the very best evidence to infer that, before the species were prematurely cut off by whalers, they attained a size commensurate with their age, and far greater than any now living in our modern seas. We have a right to infer this, not only from our practical experience, but from the language of Moses. He does not speak simply of whales, but of *great whales*—such as were found inhabiting the Polar Seas when whale-fishing was in its infancy. The order of Cetacean Mammals, to which they belong, comprises several other species, besides Dolphins and Porpoises—all of which are much smaller, and generally more local in their habitats, than the great Arctic whales. The term *great whales*, therefore, has a meaning, as every word which the good old prophet uttered has a distinct application. If he meant to include the *whole order* of whales, large and small, the use of the word "great" would have been improper; or if he meant to distinguish even the ordinary arctic or antarctic whale, such a word would have been out of place. But there is a literal and distinct meaning involved in it; and we are to understand from it, what our own experience has taught us, that the *ancient whales* were far more enormous than any which exist in modern times; that they were necessarily inhabitants of the deep sea, and that in consequence of their extraordinary bulk, their carcasses could not be associated in the *same shallow basins and estuaries with the remains of smaller animals and testaceans:* and that, consequently, we must look for their fossil remains in the bottoms of those deeper and more remote seas, some of which may have *become dry land only during the Tertiary period*. Isolated bones of these monsters of the deep have been found, as I before stated, in the formation during which they were created; but for the reasons here suggested, their more complete skeletons or osseous remains occur in by far the greatest abundance in the Tertiary measures which immediately succeeded.

From the known fact that the Jews were unacquainted

even with the northern or southern whales, certain Zoologists have suggested that Moses must have referred to some other animal of the Sauroid or Ophidian orders. This is a somewhat anomalous suggestion to emanate from a gentleman, (W. J. Bicknell, of England), who subscribes himself author of a "Scripture Natural History," and as a "Licentiate of Theology." If the term "moving creature" did not sufficiently comprehend the whole division of Chelonia, Sauria, Ophidia, Batrachia and Fishes, we might venture to infer a misapplication of the words "great whales"—especially if no such animals had existed. But so long as the divine authenticity of the Mosaic narrative is maintained on the basis of physical progress, discovery, and law, we see no sufficient reason for interposing needless apologies. When the inspired Cosmogonist shall have been *convicted of lying and deception*, it will be time enough for his friends to volunteer explanations for mitigation of sentence; but in the mean time, the Bible can afford to stand *as it is*, and dispense with all such well-meant but rather equivocal apologies and explanations.

If the Paleozoic Formation was remarkable for its vegetation beyond all other features, and the great Secondary for its wonderful and prolific marine fauna, then the Tertiary is even more extraordinary than either in the development of the higher classes of *land* animals. This far exceeds in extent, variety, and perfection, any thing which we have yet met with in the stratification of the earth, and forms the crowning glory of the magnificent and stupendous whole. And yet, while we have seen the footprints of supposed *Cheirotheria*, as far back as the New Red Sandstone, (if not, indeed, in the Coal measures, and the *Old Red Sandstone?*) and the scattered teeth and jaw-bones of the *Thylacotherium*, and other supposed Marsupial mammals, in the Oolite,—not to mention the fossil remains of other land animals higher in the scale of physical organization—is it not at least curious, (and in connection with progressive development,) is it not extraordinary and incredible that, during the Cretaceous era, *none of these, nor hardly a trace of any terrestrial species of mammalia whatever manifested themselves!* If true terrestrial mammals occurred at all, during any portion of the great Secondary Formation, it was in few, doubtful, and isolated species; while in the Cretaceous sub-era, which terminated the Fifth Geological Day, *no traces of any kind have been found!* But on the opening of the *very next day,* when Moses lifts the vail of the Tertiary, after the good Creator had commanded "the earth to bring forth the living creature, cattle, and creeping thing, and beast of the earth," what is the scene that expands before the view? It is one, the living magnitude of which bewilders the senses, and impoverishes the most versatile resources of pen and pencil to describe. All at once, as if the Creator had exerted his miraculous power to their highest tension, the whole surface of the earth is covered with strange, wonderfully-made, many of them gigantic, "living creatures!" Myriads and myriads of cattle—vast herds of roving monsters—birds of varied plumage—beasts of every conceivable form, proportion, instinct, habit, and species, crowd the teeming vales and the sloping hills, wallow along the shallow lakes, bask in the vernant sun, or browse leisurely amid the green forests! The past history of the earth, remarkable and prolific as portions of it had been, affords no parallel to this. Nor was the ocean by any means deficient in life. If possible it was more populous than ever before; but, like the earth itself, with new and strange creatures. Where now, let me ask again, is the theory of development? And what becomes of the still weaker theory of *degradation?* Moses himself unmistakably indicates a development of formations. He carries us by stages from a lower to a higher order of Creation; but *not* by a law of gradual operation. It is the development of successive *creative acts*, and nothing short of the most immediate, direct, and miraculous creative action can account for the sudden introduction of the terrestrial fauna of the Tertiary upon the surface of the earth. The whole animated Creation was changed—changed, as it were, *instantaneously*, as one act of the drama succeeds another.

> " The first *five* acts already passed,
> A sixth will close the drama with the day—
> Time's noblest offering is the last!"

The great class of Mammalia, for which the latter part of the Tertiary or Sixth day was pre-eminently distinguished, includes not only the higher quadrupedal animals that sustain their offspring during infancy by the secretions of milk in the mammæ, but also comprises the *human species.* The class forms two divisions: the first known as the *Diadelphian*, and the second as the *Monadelphian.* The Diadelphian division is small, and includes but a few living families, all of which are natives of America and Australia. They are distinguished for being furnished with pouches, by means of which they sustain their young for a time, after bringing them forth in an immature condition. In some of the species, however, as the Mexican Opossum, the pouches are not developed; and the young opossums seek protection of the mother by winding their long prehensile tails around hers, and grasping the fur of her back in their mouths. In this way, she can avoid her enemies with half a dozen of her progeny on her back. The Australian Kangaroo is the most perfect Marsupial or pouched animal now living. After the birth of her young, which she brings forth singly, and after a gestation of only thirty-nine days, the mother places it in her pouch, which is situated in the folds of the skin below the abdomen. Here it receives the lacteal nourishment to sustain it until able to shift for itself, which it will undertake to do after a pouch-life of nine or ten months. It will leave the pouch occasionally, at intervals, during this period, but return to it on the least indication of danger, where it may often be seen peeping out, as if considering how far it might be compatible with prudence to venture forth. The Kangaroo readily becomes domesticated, and they formerly occurred in large numbers in certain portions of Australia. The animal is also remarkable for having its kind limbs twice as large and long as those in front, for which reason its gait consists of successive leaps, instead of a walk. The American Opossum, so far as its pouch is concerned, is similar to the Kangaroo, but much smaller. Most persons who read these pages are doubtless familiar with this animal, as it inhabits nearly all the Middle States. In Australia there are three or four species of opossum belonging to the Marsupial order—including, also, squirrels and rats. It is supposed that the extinct Thylocotherium, Phascolotherium, and some other obscure fragmentary remains previously alluded to, belonged to this order.

The Monadelphian division comprehends the great bulk of the higher animal species, including man. It is distinguished from the Diadelphian, among numerous other striking features, by the full organic structure of its species at the time of their birth, requiring only, for a time, the subsequent care and nourishment of the mother. It comprehends no less than ten different orders, based principally on their organs of touch, of mastication, and locomotion. These orders are themselves so numerous, that they are subdivided into many families and groups. Of the first order, we have the *Cetacea*, comprising three families, and including the different varieties of whales, dolphins, and porpoises. The remains of living and extinct genera of this order are plentifully diffused in the Tertiary, but mainly in the more recent strata. Among these are the remains of a gigantic animal, supposed by some to have been a cetacean, by others a pachyderm, called the *Dinotherium.* It is supposed to have had a proboscis, like the elephant, with two enormous teeth attached to the lower jaw, but curving downward. In other respects, it was fashioned somewhat like an elephant; but is presumed to have been more aquatic than terrestrial in its instincts. Some writers, speculating upon its habits, have described it as swimming along the shores of lakes, and attaching itself by means of its curved tusks to trees, in which position it would employ its long trunk to feed upon the tender shoots and foliage. Its skull has been found in two or three instances, usually in the Miocene strata of the Tertiary. Another Cetacean, styled the *Zeugloden,* but formerly regarded as a reptile, has been found in the Eocene of Alabama.

The order of *Ruminantia* may be recognized by the structure of their feet, which terminate in two toes, inclosed by a bony hoof. The name, however, is bestowed in consequence of the fact that all the species belonging to the order *chew the cud,* and have a singular organization of the stomach. Some zoologists assign but three families to it—those of camels, stags, and oxen. Others include sheep and goats, the true position of which seems to be somewhat obscure and doubtful. Including them with this order, it not only becomes the most numerous as to *individual* numbers, but is by far the most useful and important in domestic economy of any other in the entire range of the animal kingdom. The two-hunched Camel, which has been styled the *ship of the desert,* and the swift-pacing Dromedary of the Orient, are both intimately associated with the earliest history, and the progressive struggles and labors of man. To the Arab they are alike invaluable and indispensable—furnishing him with food and milk, while their hair is woven into tents and wearing apparel, their hides converted into leather, and their dung into fuel. Strong hopes are indulged that recent efforts of our Government to acclimate and render available the services of the caravan-camel in the southwestern plains of the United States, will prove eminently successful. Three or four years have now elapsed since their introduction, and though some of the animals have died, they have thus far generally exceeded the expectations formed of them. The Dromedary differs from the Camel in having but one hunch on its back; and in being a much swifter traveler. It is said to be able to travel with ease one hundred miles per day. Besides possessing four compartments of the stomach, from one of which (forming a sort of temporary repository, which is peculiar to all Ruminantia,) the food is returned in the cud, to be again chewed before undergoing the usual process of digestion, the Camel is also furnished with a tank, by means of which it carries supplies of water for its own private use. It is thus able to resist the pangs of thirst to an extraordinary extent, and to traverse regions of arid plains where ordinary animals would perish in a day. The wise provision of Providence is thus exemplified in a most remarkable manner—adapting his creatures in every instance, as it would seem, to the peculiar circumstances which surround them. Extinct genera of the *Camelidæ* are rare. Fossil remains of them have been found in the glaciers of Siberia, while specimens of existing genera occur abundantly in the Eocene strata of France, Asia, and America. The *Cervidæ*, or Stag family, is very numerous, and, as indicated by the name, comprehends all known varieties of the deer—as the Moose or Elk, the

Reindeer, the Stag or Red Deer, the Axis Deer, the Fallow Deer, the Chevrotain, the Roebuck, the Antelope, and the Nyl-ghau and Cameleopard, or Giraffe. Few of these animals have been domesticated, and they are chiefly valuable for their flesh and the amusement they afford the hunter. The Laplander, however, turns the Reindeer to practical account—not only employing it in sledging over the snow and ice, but mainly subsisting on its flesh, and deriving clothing from its skins. With the exception of the Cameleopard and the Chevrotain, which are natives of Southern Africa, and the Axis deer of India, all the rest are found in different latitudes in America—most of them, ideed, as the Elk, the Stag, the Fallow Deer, the Roebuck, and species of the Antelope, being peculiar to the United States. There is a strong family resemblance in all these animals—three of which are spotted, as the Giraffe, the Axis, and the Fallow. The Giraffe is remarkable for its height, long neck, and slender build; but notwithstanding its seeming awkwardness, it is said to be a swift runner, and has to rely on this faculty as a means of defense or escape from its enemies. Its bones have been found in the sub-apennine Tertiary of France; while all the later Tertiary strata, and especially all caverns, in Europe, Asia, Africa, and America, absolutely teem with the dissevered skeletons of nearly every member of the order now living. They formed then, as they do now, a common prey to predaceous animals; hence the abundance of their gnawed bones in nearly all caverns and strata of recent formation. Extinct genera are rare, though not wanting. Of these, the *Sivotherium* is supposed to have been intermediate between ruminants and pachyderms. The head contained horns, and, some persons conjecture, a proboscis also, like that of the elephant. Remains of it have been found in the later Miocene of the Himalaya mountains in India. The *Capra* comprise the common domestic goat, the Ibex and the Chamois of the Swiss Alps and the mountainous regions of Savoy, Piedmont, and Germany, and the goats of Angora and of Syria. The goat is found in nearly all climates, hot and cold; is remarkably sure-footed in traversing mountain steeps and precipices; generally yields nutricious and wholesome milk when herded and domesticated, and its skin is valuable for various purposes in manufactures and domestic economy. The sheep (*ovis aries*) comprise various species, but have been materially changed by domestication. Contrary, however, to all development hypotheses, they still continue to be *sheep*—having no more bones in their skeleton now than had the flocks of Abraham and Isaac. Great Britain is the principal field of sheep husbandry in the world; but certain States of our Union have lately been making very rapid advances, especially Texas. The western part of Pennsylvania, and the adjacent districts of Ohio and Virginia, have long been known for their extensive production of sheep, with special reference to the quality of the fleece. In England more attention has been given to the quality of the flesh than the wool; hence John Bull is as famous for his mutton chops as for his roast beef. Speaking of roast beef, reminds us of the *Bovidæ*, or Ox family. Little, however, need be said of them. The word oxen originally implied black cattle, of both sexes, and in this sense may be applied to *all* the members of this extensive and valuable family. It comprises the domestic ox of Europe and America, the ox of Syria, the American Bison, the Buffalo, the Indian Zebu, and the Musk ox of Hudson's Bay. The domestic ox formed a leading element of individual and national wealth in the most remote ages of antiquity—among the Jews their skins sometimes formed a medium of exchange, in the absence of money. The wealth of the old patriarchs consisted altogether of cattle, even lands being held as of secondary value. Emasculation was forbidden by the Mosaic law, in view of the valuable services of the ox as a beast of burden, and in the general agricultural system of those primitive days. The property possessed by cows of affording milk long after their young are withdrawn, is a feature in physiology which is said to pertain to no other animals. And in view of their universal adoption into the domestic economy of man, it suggests the inference of their creation *expressly for his benefit*. The domestic oxen have been much improved by cross-breeding, but whether their condition, as a whole, has been strengthened or elevated, is a question which admits of doubt. At any rate, it need not be considered here. The Syrian ox differs from that known to us by presenting a considerable rise or protuberance over the fore-shoulders. The Zebu of the East Indies is an animal not materially different from the other species. They are natives of Asia, but are also found scattered throughout Africa. The Musk ox has a large and powerful body supported on short legs, and is usually found in herds of thirty or forty in the cold regions of North America. The Buffalo, although a native of Africa, is very numerous in America, and, like the Bison, forms immense herds on our western prairies. It is estimated that fifty thousand of these animals have been met with in single herds in those vast regions stretching out west of the Mississippi. Their skins are valuable, as every one who has wrapped himself up in the folds of a buffalo-robe, during the blasts of a pitiless Nor'easter, can gratefully attest. Specimens of the remains of the extinct genus *Leptitherium* have been found in the caverns and the upper Tertiary deposits of Brazil; while those allied to living specimens are quite abundant all over Asia, Africa, Europe, and America. Indeed, during the latter stages of the Tertiary, these animals were almost, if not quite, as numerous as they are now; and roved over the country in immense herds, like the Buffalo of the West.

The order of *Pachydermata* is so named because all its species are distinguished from other animals by the thickness of their skins or hides. During the Tertiary period, it was the most wonderful class of animals, in every respect, which ever appeared on the earth. Although much of its ancient character is lost, it still furnishes some of the largest and most powerful animals now living, as also some of the noblest and most esteemed. It presents but three families, comprehending the horse, the rhinoceros, and the elephant. The horse not only includes every known variety of that animal, of which there are many, remarkable either for their size, their strength, speed, docility, or other qualities; but also includes the ass, wild and domestic, the mule, and the African Zebra. The degree of intelligence which this animal has attained during his familiar intercourse with man, is no less astonishing than gratifying. When we see the thousands of wild horses of the plains, and reflect that the noble animal we bestride has been reclaimed from such a condition, and elevated to a degree of intelligence under which his services are invaluable, it shows the subtle power and majesty of *mind*, and evinces the perfection of that dominion over the beasts of the earth, with which man was endowed by his Creator. And the only progressive development which we could ever discover in nature, is that wrought upon the inferior animals by the influence of man; and for this he has the full sanction and express authority of the common parent of all. And yet, in the face of such influence—in the face of all efforts, whether dictated by sordid gain, or that vaulting ambition which impels some men to usurp not only temporal and civil, but the highest laws of nature and morality, man has been unable to produce a genuine race of mongrels, having the *inherent powers of reproduction.* It is true the mule may be regarded as a sort of hybrid between the horse and the ass, but both parents are alike members of the *Equus* family—they are themselves *horses*. Yet even the mule is obtained in violation of all the instincts of the mother; and nature has affixed the stamp of barrenness upon the offspring, as if to condemn and to arrest further innovations upon creative *order*. Nature, however, seems to have sanctioned the production of this animal for a special purpose, since it is endowed with qualities possessed by neither of its progenitors. Such is its sure-footedness that it can ascend and descend mountain passes and declivities, where unaided man himself would hesitate to venture. In the Andes and the Alps, the mule does not descend by steps, but drawing its feet close together, and falling back on its haunches, it slides down the steepest mountain slopes, bearing her rider or freight in safety to the base. No other animal could do this, or could ever be induced to try it; for it requires not only extraordinary agility and suppleness of the limbs, but a *cool daring of instinct*, which seems to be an innate peculiarity of the mule. But while mongrels, prodigies, and monsters may be produced,—if for no other end than to warn man against the enormity of the abuse,—nature, in no instance, *permits their perpetuation*. The natural vitality of the true species will either *absorb them*, or their own deformity entails the *inability of reproduction*. This fact, which is alike fatal to the idea of development or of degradation, will account for some of those apparent anomalies in the ancient zoology, which have puzzled modern anatomists. *Nature produced monsters then*, as she does now; *but they were always confined to single individuals.* Fossil remains of the horse, the ass, and the zebra, have been found in the upper Tertiary veins, in all parts of the world; but I believe they always indicate an identity with the existing species, and thus point to the fact, not without interest, that, notwithstanding all the local changes which the animal has undergone, the family, *as a whole*, remains precisely where nature originally placed it in the scale of created life.

Belonging to the pachyderms proper, there is a large group both of extinct and living species. Among the latter, —to which some of the former were closely related,—are the hippopotamus, the wild boar, the hog, the peccary, the tapir, and the rhinoceros. The hippotamus is one of the most compact and powerful animals ever created, and is well described by Job, who, indeed, appears to have been a naturalist of the highest ability. "He moveth his tail like a cedar; the sinews of his thighs are wrapped together. His bones are as strong pieces of brass: his bones are like bars of iron. He is the chief of the ways of God; he who made him hath furnished him with his sword. Surely the mountains bring him forth food, where all the beasts of the field play. He lieth under the shady trees, in the covert of the reed, and fens. The shady trees cover him with their shadow; the willows of the brook compass him about. Behold, should the river swell, he hasteneth not; he is fearless should even the Jordan come up to his mouth. Who can take him openly, or draw a cord through his nose?" The hippopotamus, as indicated by the sacred writer, is an amphibious animal, but not carnivorous. It feeds exclusively on herbage, and such is the enormous capacity of its stomach, that it can hold six or seven bushels of half-chewed vegetable matter. A hippopotamus on exhibition in the Zoological Gardens of Regent's Park, London, has brought forth young on two or three occasions, and we observe, by a paragraph in the newspapers, that she invariably destroys them shortly after birth. Whether this is generally peculiar to the animal in its unrestrained condition is of course hardly probable; if it be, then nature has undoubtedly made some provision for the escape and support of the young, or otherwise the species would long since have been extinct. Remains of this animal have been found in the Tertiary and alluvium of Europe, Asia, and North America. The rhinoceros, in its general appearance, is not much unlike the hippopotamus. The former is distinguished by a sharp horn issuing from its nose, with which it can gore and lacerate any animal bold enough to attack it, and by a long tail terminating in a switch. The other animal has quite a short and stumpy tail, and no nasal tusk. The hide of the rhinoceros is so hard that it often resists a rifle-ball, while the claws of the tiger and the lion make but little impression upon it, except in vulnerable spots. It is said to be fond of wallowing in muddy pools and flats in the vicinity of large rivers. Its present habitat is the East Indies, especially in the valley of the Ganges. That of the hippopotamus is principally in the south of Africa. The rhinoceros is believed to include ten or more species, of which at least four have been identified in the upper Tertiary of Europe. The tapir of South America represents a genus of animals which appear to have been numerous at some places, during the upper Tertiary period. It presents some resemblance to the hog, the hippopotamus, and the elephant—though its present size hardly exceeds that of the animal first named. Its skin is remarkably hard and tough, and has but a thin coating of bristling hair. Its tail is stumpy, its legs short and thick, terminating in hoofed toes, and its upper jaw is furnished with a fleshy prolongation or incipient proboscis, similar to that of the elephant, but not one-fourth the comparative length. The animal is amiable in its habits, and is easily domesticated. Its flesh is eaten by the Indians of South America, and is said to be as palatable as that of most other herbivorous animals. The genus *Laphiodon* of the

Miocene, of which twelve species have been found, and the *Paleotherium* of the Eocene, were both very similar to the existing tapir. The size of the ancient animals, however, varied considerably—some having been scarcely larger than a hare, while others attained the ordinary proportions of a horse. Other varieties of the extinct tapir combined the structural peculiarities of the horse, the hog, the rhinoceros, the hippopotamus, and the camel.

There were five or six extinct genera, each of which was distinguished by some peculiarity of structure, which it is not necessary here to describe. The wild Boar, which belongs to the *Pachyderm* order, is generally considered to be the parent of the domestic swine. From the fact that his upper snout is furnished with tusks, we think this proposition rather questionable. The domestic swine are, no doubt, derived from progenitors originally wild and ferocious; but they were very probably a different genus from the tusked animal now existing. The Mexican peccary is very similar to swine, but instead of two, one of its species (the *Babiroussa*) has four tusks; and yet there is no instance of a hybrid race among them. The peccary is gregarious, exceedingly ferocious, and lives principally on roots and vegetables. The remains of animals allied to the existing genera of swine have been found, but somewhat meagrely, in the upper strata of the Tertiary. The domestic variety is recent, and was no doubt created for the office of scavenger, for which his habits and instincts incomparably qualify him. If, indeed, he is not remarkable for the fastidiousness of his appetite and general deportment, it may be said to his palliation that man, with a great deal more *pretension*, is scarcely more so; since he does not hesitate to devour the very animals whose valuable services as a common scavenger he affects to despise! *Tempus edax rerum*, and so does man!

The family of *Elephants* forms the remaining group of the order of *Pachydermata*. The living species are found in Africa and Asia, but the size of those of the latter is considerably greater than the former. They seldom exceed ten feet in height, or four thousand pounds in weight. They are regarded as an unusually intelligent animal, but it is doubtful whether they are capable of higher mental culture than dogs or horses. They have been extensively employed in wars, in caravans, and in the pompous pageants and processions of the East, and have thus had the full benefit of a long and close intercourse with man, by which means the intelligence of the species has been developed. The proboscis of the elephant is prehensile, and is made up of no less than forty thousand aggregated muscles, curiously interlaced. It possesses amazing strength and flexibility, and with the enormous ivory tusks, forms a leading feature in the appearance and structure of the animal. During the Tertiary period, the family of Elephants included several varieties now extinct. Among these were the *Mastodons*, of which there were several genera. They were considerably longer and somewhat higher than the living elephants, and appear to have been very numerous all over the world. Their bones have been exhumed in great abundance in the valleys of the Ohio and the Mississippi, and in the States of New York and Virginia. The ivory obtained from the tusks of extinct elephants has, from the most remote ages, formed no inconsiderable item of commerce and manufactures. Each tusk usually attains a length of eight feet; and the supplies of the world, for the last century, have been principally derived from Russia, where the antiseptic qualities of the frigid climate have no doubt tended to their preservation. Tusks from the living species still constitute a principal trade among the natives of Africa.

The *Edentata* is an order which presents a transition from those previously described, in the form of the hoof, which is in them superseded by claws or toes armed with nails for scratching and digging. There are three or four families—two of which are represented by the living ant-eaters and the armadillos; the others, long since extinct, by the gigantic Megatherium. The Sloths, which include two genera not materially different, were for a long time a great puzzle to zoologists, who could not reconcile the structure of their limbs with the usual ambulatory movements of such animals. It was finally ascertained, however, that in their native forests, they were in the habit of traveling from tree to tree by means of the connecting limbs, and, the better to facilitate their movements, always selected *windy days*, when the branches would be brought in closer connection; but unlike squirrels, they traveled with their bodies suspended *under the limbs*, and their feet upward, like flies adhering to a ceiling. It was ascertained that they also slept in this novel position, and, in short, reversed the usual order of nature in all their movements. The animal called the Ant-eater is generally about four feet long, exclusive of tail, and is provided with a bird-like snout, serving as a sheath for a long pointed tongue, which it thrusts into an ant-hill, and withdraws coated with the living prey. It is found with the Sloth in Africa, but its chief country is South America. There are three extinct genera of this family, all of which are found in the pampas and caverns of South America, with the remains of those of existing genera. The second family comprehends the manis and the armadillos—all of which are inclosed in scaly coats-of-mail. The manis is also an ant-eater, but unlike the others (which are clothed with hair), it has a long lizard-like body, pointed snout, and short legs with long pointed claws. Its length is usually about eight feet, of which the tail comprises one-half. True to its serpentine instincts, it rolls itself up on the approach of danger, and, by means of its hard osseous scales, intimidates the boldest denizens of the wilderness. This is also peculiar to the armadillos, whose horny covering is arranged into flexible sections for that purpose. They are all burrowers, feeding upon roots and vegetables, and fly to their holes on the apprehension of danger. The armadillos, unlike the manis, have the general form of swine, though in one variety the body and tail are considerably elongated. The extinct genera of this family were very numerous during the Tertiary, and one of them, the *Glyptodon*, attained extraordinary dimensions, and had some resemblance to the *Megatherium*. This animal was nearly as large as an elephant, and its skull was similar to that of the Sloth. Its mouth possessed gigantic power, and, like the other monsters which distinguished the Tertiary, was well adapted for crushing the roots upon which it fed. Its feet were large and armed with tremendous claws, and the "entire frame was an apparatus of colossal mechanism, adapted exactly to the work it had to do; strong and ponderous in proportion as this work was heavy, and calculated to be the vehicle of life and enjoyment to a gigantic race of quadrupeds, which, though they have ceased to be counted among the living inhabitants of our planet, have, in their fossil bones, left behind them imperishable monuments of the consummate skill with which they were constructed. Each limb and fragment of a limb formed co-ordinate parts of a well-adjusted whole; and, through all their deviations from the form and proportion of the limbs of other quadrupeds, affording fresh proofs of the infinitely varied and inexhaustible contrivances of creative wisdom."* One

---

* Dr. Buckland, Bridgewater Treatise, p. 164.

of the most extraordinary animals of the Edentata order is the hedge-hog, of which several varieties are found in Australia. Like all the other ant-eaters, it has a long slender beak, with a prehensile tongue; short feet equipped with claws, and a body shaped like that of the domestic pig; but instead of scales or hair, it is covered with a substance resembling feathers, which serve to turn the rain like those of fowls. Another variety of the animal has the bill of the duck, and is called the *Ornithorhynchus* in consequence. This, however, is coated with hair. They all burrow in the ground, and swim and dive in the shallow rivers for food, precisely like ducks, for which purpose they also have webbed feet. The common European hedgehog has also a long pointed snout, but its food consists mainly of insects, worms, and fruits. Its armor is something like that of the porcupine, and, rolling itself into a round ball, its bristling points keep its numerous enemies at bay. It is a harmless and inoffensive animal, and though in many respects resembling the ant-eaters, it belongs to another order of animals.

The order of *Rodentia*, which is the fifth in the Mammalian series, comprises all animals distinguished for their gnawing propensities, and of which squirrels, mice, and rats are familiar and somewhat obtrusive examples. The squirrels embrace many genera, distinguished principally by color or some other minor features—as the red, the gray, the black, the ground, the flying, and the long-tailed varieties. The rats also comprise many different groups, as those popularly known as the house, the field, the ship, the water, and the musk rats. The mouse is equally varied in type, and even more numerous. Besides these, the marmot, the beaver, the porcupine, the hare, the rabbit, the guinea pig, the cavy and the chinchilla of South America, all belong to the gnawing fraternity. The dental apparatus of these animals is well calculated to subserve their natural proclivities and habits—the teeth growing as rapidly as they are worn down, and being formed in such a way that they answer the same purpose as a file and chisel. The beavers are thus able to cut and rasp down considerable trees, while the rats and mice gnaw their way to the secret recesses of spoils and plunder. All these animals, but more especially the rats and mice, have an uncompromising enemy in man, as well as in the feline and canine races; yet such is their wonderful prolificacy, that they still maintain their ground, and this, too, in the face of their own intestine wars and cannibal appetites. Like the persecuted Jew, they are found all over the earth, on the sea and on the land; in the palaces of the rich and the lowly hovels of the poor; in the well-filled granaries of the farmer and in the filthy sewers and subterranean recesses of the city. Essentially cosmopolitan, like the descendants of poor Cain, they are "fugitives and vagabonds" wherever they may go, and have no friends on whom they can rely in the hour of peril, even among their own species or around their own domestic hearth. In China, indeed, their social position is somewhat superior to that occupied in republican America or under the constitutional monarchy of England; yet, even *there*, they are only recognized at the festive board, where they are invariably involved in the *stews* and *broils* of the bill of fare !

The fur of the squirrel, and of some of the rats, the beaver, and the hare, was formerly of some value, but they have all depreciated of late years with the progress of textile manufactures. Nearly all these creatures abound in America, and it would be useless to occupy space with a description. The fossil genera in the Tertiary are numerous, embracing no less than five or six different types, while the remains of those still living are widely diffused in caverns and the modern alluvial deposits.

The *Cheiroptera* embrace the different genera of bats, of which there are four or five, among them the ternate or vampire of the East Indies. The bat has its claws or fingers elongated to stretch out a thin membrane, like a fan, by means of which it flies. It is covered with a delicate fur, like the mouse, and ventures forth at night, generally preferring for its nocturnal haunts old, dilapidated, and abandoned buildings. The vampire of India is a blood-sucker, and is so insidious in its attacks upon persons asleep, that instances have occurred of their bleeding to death in bed. The wound is generally inflicted on the foot, and it is so small that it hardly exceeds the puncture of a pin; yet from this the animal will suck enormous quantities of blood, which will continue to flow long after it has gorged itself to repletion. All the fossil species of this order are identical with those now living, and they are found in nearly all the later strata of the Tertiary, and the caverns of the modern era. Their fossils are mainly confined to Europe and South America, where the living species still flourish extensively.

The order of *Amphibia* is made up of seals, embracing the marine and river seal, the sea-bear, the sea-lion, and the sea-horse or walrus. Their extremities are so modified that they all swim with ease and dexterity. Their principal habitat is in the cold latitudes of the North— especially Greenland, Nova Zembla, and Hudson's Bay. The common seal is a long animal, gradually tapering from the middle to the head and tail, and provided with two legs, which answer the purpose of paddles in swimming. In some genera the head is thick and massive, and in others tapering from the neck, like that of the leopard. The sea-lion has its legs in front, with an enormous breast and neck; the sea-horse or walrus has a much larger body, and, in addition, is supplied with two long and sharp tusks, which he uses in climbing and in grappling sea-weeds. The sea-horse may be seen in large numbers sitting on floating ice or along the margins of northern seas, keeping up a low bellowing. The sea-lion is not materially different in its habits, but, contrary to the general impression, they are said to be courageous and determined, and not readily repulsed if attacked. The old animals roar tremendously, while the younger ones and the calves bellow like sheep. Fossil specimens of this order have been found in Miocene beds of England and France.

The eighth order is composed of the *Carnivora* proper, and comprehends the great bulk of mammalian animals. The *felis*, or cat group, embraces lions, tigers, panthers, jaguars, leopards, pumas, lynxes, and the wild and domestic cat. The dogs, or *canines*, include the mastiff, the bull-dog, the gray-hound, the blood-hound, the terrier, the Newfoundland, the Mackenzie river, the Esquimaux, and the St. Bernard dogs; besides the wolf, the fox, and the jackall. The *ursus*, or bear type, include all the varieties of that animal, as well as the raccoon, the badger, and the ratel. The civet, the ichneumon, the weasel, the ferret, the pole-cat, the sable, the otter, etc., comprise other and independent varieties.

From his powerful build, his extraordinary agility, and his unflinching courage, zoologists have assigned the lion the first rank in the brute creation. Some travelers, however, who have seen the lion and the tiger in their native jungles, assign greater courage and equal strength and agility to the latter; and it is questionable whether the high position usually awarded the former, might not be successfully contested. The lion seldom weighs over five hundred pounds, is from four to six feet high in front, and seven to nine in length. His jaws, which are enormous, operate like the cutting edges of shears—the lower jaw moving in an upward and downward direction, but not in a lateral or horizontal one. In this, as in many other respects, all the felines are readily distinguished from the ruminants, who grind their cuds by a sort of *circular process*. The padded paws of the felines, added to their muscular development, also give them an exceedingly soft and graceful movement, while the whole carriage of the lion is lofty in the highest degree. The African lion is by far the largest and the most spirited, and holds undisputed sway over all the feline species, if not over all the brutes of the earth. The lioness is usually smaller than the male, is destitute of his flowing main, and somewhat different in structure. She produces two or more at a birth, and watches her young with much care and jealous solicitude, in which her lord heartily co-operates. The tiger, although his general appearance is different, resembles the lion in many particulars of his structure, in size, habits, and inherent propensities. His coat is yellow, but regularly banded with stripes of black, which extend over the head, legs, and tail. His height varies from three to four feet, and his length from six to nine. He is more slender than the lion, and less bold ; but has all the gracefulness of a kitten. The tigress is also a careful mother, and will encounter any danger in behalf of her young. The tiger is found around the deserts of Asia, and in the East Indian Islands, but his chief habitat in modern times is Hindostan and China. The Royal Bengal tiger is the model of his species, and one of the most magnificent animals in the world. Singapore, an island twenty-five miles long, in the eastern archipelago, means the *place of lions;* but for a long time the term has been inappropriate, since the lions have been extinguished, *and their place supplied by tigers.* This would support the inference already suggested, that the tiger is probably the strongest and most terrible in combat of the two. The tigers live in the tall grass and jungles of the island, but a large number swim over from the peninsula, from which it is separated about a mile. Here they watch the movements of the Malay and Chinese laborers, as they go to the fields to work, and suddenly springing upon them, a single blow on the back of the neck, which they break, lays them dead upon the ground. It is stated, on high authority, that from three to four hundred persons are killed annually in Singapore by these ferocious monsters. Though large rewards are offered by the government for their destruction, besides which the skin is worth fifty dollars, but little progress has been made in their extermination. The leopard, it is generally thought, is of the same genera as the jaguar and the panther—there being, in fact, but little difference except in the spots of the skin. They are natives of Africa, but are found in equal abundance in India. In size they are inferior to the tiger, the panther, or the jaguar. The leopard has a white and yellow skin, dotted with irregular dark spots; the jaguar's is similar, but the spots are much larger, while that of the panther is a dark yellow, with compound or rosette spots —that is, dark rings with black dots in the interior. The panther is found generally throughout Asia and Africa, and certain genera exist in America. The jaguar, however, is known only in South America, and abounds principally in Brazil and Paraguay. Nearly all the animals thus far mentioned, except the lion and tiger, are expert climbers, and often pursue their prey to the tops of trees. Humboldt, in his Aspects of Nature, says the yell of the jaguar may be heard in the forests of Brazil, mingled with the shrill long screeches of the monkeys, as they leap in terror from tree to tree. Horses, cattle, sheep, monkeys, fish, turtles, snakes, birds—all these are the indiscriminate and varied victims of the animal. The chetah of India, is a leopard which approaches some of the qualities of the cat, though it is greatly larger. It has been trained for hunting the wild antelope, but has not been thoroughly tamed or domesticated. The puma is a native of North and South America, and has sometimes been called the American lion; but, as compared with the African congener, he reflects little credit upon his native heath. In South America he is hunted with dogs, and then lassoed; or if the dogs pursue him to his refuge in the branches of a tree, he is dispatched with a rifle-ball. He is a thin gaunt animal, of a dirty yellowish color, but has the powerful limbs and paws of the lion and tiger. The ounce is a rare animal, one or two specimens of which have been obtained in India, but whether it exists to any extent in other parts of Asia or of Africa, has not been ascertained. It is supposed to form a distinct species of the feline group. It is spotted like the leopard, but its hair is long and somewhat shaggy. The ocelot is another native of South America, and has a skin remarkable for its brilliance and beauty. The prevailing yellow of the leopard, in this animal, is tinged with a tawny hue, and covered with long reddish-black spots, which extend to all its limbs. Like all the others, it is an expert climber, and lies concealed among the foliage of trees, ready to pounce upon monkeys, or other animals of inferior strength. It is difficult to tame, and has the reputation of being the most restless and ferocious of the feline race, though it will avoid dogs until hard pressed. The lynx is celebrated for its clear and extended vision, and for its wide geographical distribution. It was, and is still found in nearly all parts of Europe, Africa, Asia, and America, but with ever-varying local characteristics, especially in the quality of its fur—all of which may be attributed to the effects of climate, and to other physical circumstances. The skins of the lynx comprise a leading item in the fur trade of Russia, America, and the British provinces. It used to be very numerous around our northern lakes; and the Hudson's Bay company, not many years ago, obtained from six to nine thousand skins annually. But the lynx has now nearly disappeared in that quarter, and the fur trade has dwindled to a mere cipher, compared with what it was fifty years ago. The ancient haunts of the wild beast, and the no less savage Indian, are now becoming the abodes of civilized man; and the beautiful transparent lakes which separate our states from the British possessions, are furrowed by the steam and the sailing vessel, instead of the frail fur-laden barges of the red man, seeking the far-off stations of the traders. The fur of the lynx is long, soft, and silky, and presents a delicate brownish gray aspect. It was always held in high repute in the market, but varied in value from three to thirty dollars each. The animal, in its natural condition, is as playful as the lamb; nevertheless it feasts upon the weaker animals, and from its position among the branches of trees, often pounces upon deer and sheep, whose blood affords a rare delicacy. The wild cat is supposed by some naturalists to be identical with the domestic animal; but this has been denied. The structure, it is true, presents no perceptible difference; but the color of the wild cat is generally steadfast with the local species, while that of the domestic animal is constantly varying, even with the same brood. Besides, the wild cat is thought to be irreclaimably ferocious; and there is no known instance of its having been tamed and domesticated. The wild animal sometimes attains an enormous size—the domestic animal varies but little. There is a peculiarity in the formation of the eye of the cat, and to some extent in all the feline family, by which they are enabled to see to better advantage at night than during the day—hence their predaceous instincts are strictly nocturnal. Although cats are not named in the Bible, they were domesticated among the Egyptians, and mummies of them have been found in Thebes, and sculptured figures on the monuments of the Pharaohs.

The *Canine* family includes our good friend *Ponto*, whose wonderful sagacity is referred to by Mr. Alfred Jingle, in the Pickwick Papers. If there is one feature more prominent than another to distinguish him from other animals, it is his extraordinary friendship, fidelity, and devotion to man. No matter who or what you are, or where you go, your faithful dog, like the devoted spouse of Wilkins Micawber, "will never desert you." The lonely hermit in the wilderness; the traveler by the wayside; the miser counting his greasy gold; the shop-keeper amid his wares; the children romping on the hills, or the family mansion sheltering its sleeping inmates—all are jealously guarded by this faithful watchman, quick to give the alarm, and often bold to pursue and seize the stealthy invader. The dog is the only real and disinterested friend which man can rely upon through all the changing phases of worldly fortune. He is the first to welcome you home, and the last and most reluctant in his *congé*. He shares with you all the excitements of the hunt; points out the fluttering game, runs down the bounding deer, pursues the fox to his hole, and the climber to his tree; and, at the risk of his own life, will attack the bear, the tiger, or the lion, to save yours! Of the different species, it may be remarked in brief that the *Mastiff* is, par excellence, the house-dog. He is large and stately, with immense ears, a long tail, and possesses unusual intelligence, and a demeanor dignified, quiet, and unobtrusive. The Bull-dog has a short stumpy tail, a thick head, or snub nose, short ears, and a surly demeanor. His courage, however, is wonderful, and he will give battle with his last expiring breath. An old law of England once required that no bull should be slaughtered until he had first been baited. This act was passed by Parliament expresisy to encourage the amusement of bull-baiting; and it was owing to the univerality of this barbarous sport that the peculiar character of the bull-dog was developed. In seizing cattle, the bull-dog generally attacks the lip, the tongue, the eye, or some other vulnerable part of the face, where he often hangs until released by the complete mutilation or dismemberment of the organ. This dog, unlike most of the species, attacks his enemy without barking; and, when occasion seems to justify it, he will not hesitate to attack a man. Although bull-baiting has long since lost its ancient respectability in England, the stamina of the ox and the fierce expression of the dog have been daguerreotyped on the national character, if not sometimes upon individual feature! The *sobriquet* of *John Bull* is consequently not only significant, but it is *dog*-matically appropriate. Of hounds, there are three rather distinct characters—the blood, the gray, and the fox hounds. The blood-hound is the largest, and has considerable resemblance to the mastiff. His ears and tail are long, his forehead broad, nostrils wide and long, and face narrow. He is distinguished for the accuracy and perfection of his scent, in consequence of which he has often been employed in war to hunt down the enemy, especially negroes, and midnight robbers. He has been termed the policeman of the canine family; but his fondness for blood often renders him also an executioner. He is even now employed in portions of the South, and very extensively in Cuba. The gray-hound is an exceedingly light and fragile animal, and is the swiftest dog of the chase. Its powers of scent are very inferior, however, and it is only valuable for its extraordinary speed in pursuit of game. The fox-hound is more remarkable for its powers of endurance in the chase than for swiftness or accuracy of scent. In form it is not materially different from the blood-hound, except that it is generally lighter and smaller. The terrier has some resemblance to the bull-dog, with which he has often been crossed; but its instincts are more particularly directed against rats. Another species, however, is a good field-hunter, and likes the pursuit of larger game, as the fox, the squirrel, and the opossum. The Lurcher may justly be termed the natural thief of the race. He has a sneaking look, and a shaggy, ill-conditioned coat—in his personal appearance rather a Jacques Strops than a seedy Robert Macaire. He has a fine scent, which serves at night the same purpose as the dark lantern of the burglar. He pounces upon his victims without any premonitory barks or growls, and sneaks away as stealthily as he came. His robberies became so notorious in England, in connection with professional poachers, that the breed was finally proscribed by law, and has lately been nearly extinguished. Specimens of more or less purity, however, may be found in the United States, and in almost every village or wayside cottage. The common cur unfortunately has but too large an infusion of his thieving propensities, with the additional one of barking furiously at every passing object. This latter quality was never borrowed—it is inherent in the very nature of the cur—of all curs, whether canine or human. We now turn to three varieties of the dog, whose noble bearing, fine proportions, industrious habits, general intelligence, and benevolence of character, it is pleasing to contemplate. These are all inhabitants of cold, dreary, and inhospitable climes, as if adversity subdued the baser instincts, and nourished only the nobler and more generous impulses of the canine race! The Esquimaux dog has the ears of a fox, with the head of a wolf. His tail is large and bushy, and curls into two coils over his back. His body is protected from the cold blasts and snows of his native region by long and fine hair. To the natives he is perfectly indispensable—in fact, without his services they could not long survive. Harnessed in teams to sledges, they travel at the rate of sixty miles per day over snow and ice. The teams are preceded by a leader, who follows the instructions of the driver. Furnished with a keen scent, they will pursue their way through the most tremendous storms of snow, and endure the most intense privations and fatigues. Besides transporting burdens on sledges, these dogs are skillful hunters, and invariably assist the Esquimaux in capturing the seal, the reindeer, and the bear. They have not the docility of our domestic dog, because their masters are themselves deficient in the elements essential to impart it; but they are in all respects intelligent, useful, and reliable; and, in the position which discerning Nature has assigned them, are infinitely more valuable, in a comparative sense, than the horse is to us. The Newfoundland dog is often six feet in length, from his nose to the tip of the tail, which latter is seldom over two feet. His weight and general proportions correspond, from which it will be seen that he is the giant of the canine family. He is a splendid-looking animal, and combines with unsurpassed strength the gentleness of a lamb. If the blood-hound is the policeman, the lurcher the thief, the bull-dog the butcher, the Esquimaux dog the traveler, the mastiff the watchman, and the gray-hound the sportsman, then we can recognize in the Newfoundland dog the *old-fashioned country gentleman*. Compared with the others, his deportment is dignified and courtly; but unlike his modern congeners of the human family—(the conventional gentlemen whose exquisite polish, like the daguerreotype, always reveals the parvenue)—he is eminently practical, and devotes the faculties with which nature has endowed him to the use and benefit of man. In the snowy regions of Newfoundland he is an important auxiliary to the woodman, and, harnessed in sledges, draws the wood from the forest to the landing without the aid of a superintending driver. He also carries messages and packages; and being an excellent swimmer and diver, for which he is provided with webbed feet, his services are often equally efficient in the water as on land; and he has rescued many an individual from a watery grave. Unlike other dogs, his aquatic habits extend so far that he can make a meal of fish, whether cooked or raw. Byron had a Newfoundland dog, called *Boatswain*, to which he was greatly attached, and upon the death of the faithful animal, he erected a little monument to his memory at Newstead Abbey. The following epitaph was perhaps only a grateful tribute to the real merits of the poor animal:

> "When some proud son of man returns to earth,
> Unknown to glory but upheld by birth,
> The sculptor's art exhausts the pomp of woe,
> And storied urns record who rests below;
> When all is done, upon the tomb is seen,
> Not what he was, but what he should have been;
> But the poor dog, in life the firmest friend,
> The first to welcome, the foremost to defend,
> Whose honest heart is still his master's own,
> Who labors, fights, lives, breathes for him alone,
> Unhonored falls, unnoticed all his worth,
> Denied in heaven the soul he held on earth:
> While man, vain insect! hopes to be forgiven,
> And claims himself a sole exclusive heaven.
> Oh, man! thou feeble tenant of an hour,
> Debased by slavery, or corrupt by power,
> Who knows thee well must quit thee with disgust,
> Degraded mass of animated dust!
> Thy love is lust, thy friendship all a cheat,
> Thy smiles hypocrisy, thy words deceit!
> By nature vile, ennobled but by name,
> Each kindred brute might bid thee blush for shame.
> Ye! who perchance behold this simple urn,
> Pass on—it honors none you wish to mourn:
> To mark a friend's remains these stones arise;
> I never knew but one—and here he lies."

Near one of the most savage passes of the Alps, on the great Mount St. Bernard, is located the convent of that name, inhabited by a society of monks who devote their lives to the entertainment and rescue of travelers who may be overtaken by snow storms or avalanches of frozen sleet, ice, and earth. Such are the dangers of this frightful mountain gulf, that many travelers are constantly and unavoidably lost, being suddenly overtaken and buried in snow-drifts, or killed by the descending avalanches, which sometimes carry large masses of rock, earth, and trees with them over the frowning precipices. The good monks have reared a race of dogs to assist them in their benevolent and angelic labors ; and such is the degree of intelligence and efficiency which they have acquired, that they have been the instruments of rescuing numbers of unfortunate travelers, and conducting them safely to the cheerful hearth of the convent. Every dog is furnished with a bottle of spirits suspended around his neck, or a cloak or blanket, intended for the benefit of the exhausted wayfarer. The scent of the dogs is so keen that they can readily trace a man in the snow, although he may be buried many feet beneath the surface. When an unfortunate is thus found, the dogs raise a vociferous barking, which is the signal for the monks to come to their aid. In many instances bodies are found after life is extinct, when they are conveyed to the convent and preserved for the future identification of their friends or relations. On one occasion, a mother was crossing the mountain with her son, a lad of some seven or eight years. An avalanche overtook her, and buried her in the snow. When the dogs arrived, she had already perished ; but the boy had sufficient strength and intelligence to mount the back of one of the animals, and was thus borne in safety to the convent. This dog for some years after, carried a medal commemorating the event, but finally perished himself while engaged in his benevolent work. Such incidents as these raise the animal above the level of his species, and challenge our highest admiration and gratitude. The St. Bernard dog is a large and fine-looking animal, with long hair, and a bushy curling tail. His face expresses intelligence and docility in the highest degree. There are several other varieties of the dog, as the spaniels, the mongrels, the poodles, etc.; but, after contemplating the character of the St. Bernard, it would be a degrading reflection on that noble animal to speak of any others, and especially such as we have indicated. While there is an extraordinary variety in the size, color, habits, and instincts of the race, it is a no less singular fact that their physiological and osseous structure is invariably the same. The little lap-dog, with his red ribbon and bell, and his *passé* nurse and mistress, has exactly the same number of bones as the giant of Newfoundland. The conformation of all is the same; the deviations are merely local, to meet the requirements of the position to which the Creator has consigned them.

The Canine group, however, is not confined to dogs. It also includes the wolf, the fox, and the hyena. The wolf has a keen scent, and, like the rest of the canine race, is a good hunter. The deer of the forest, and the sheep and oxen of the field, often fall a prey to his voracious jaws. Under circumstances of intense hunger, he has not hesitated to attack man; but the instances of this are few. A species of wolf is still very numerous and predaceous in the plains of the South and West, and they are very unpleasant neighbors to the frontier settlers and the "squatter sovereigns." In some instances, the cabins of the planters are surrounded by deep trenches, covered with oscillating platforms, well supplied with bait. The wolves, as they gather upon this platform, are precipitated into the pit by their own weight, and before morning, like the Kilkenny cats, it is found that they have generally dispatched each other by their own fraternal quarrels. The wolf of New South Wales neither barks nor growls, but erects the hair of its skin, when surprised, like the quills of the "fretful porcupine." Its prevailing color is brown, interspersed with black stripes from the fore shoulders to the tail. The fox embraces many varieties, but the principal genera are the red fox of Europe and America, the white Arctic fox, and the jackal of Asia and Africa. The habits and cunning of this animal are well known. Although an unpleasant neighbor to the farmer, he has afforded more amusement in the chase than any other animal whatever. In England, indeed, his species would long since have been extinct, had not the sporting gentry given liberal bribes to the farmers for their toleration. He is a great burrower; and after making a nocturnal visit to the poultry roost, he distributes his spoil in his several subterranean mansions. The fox has a thick fur, a long sweeping tail, and a cat-like face. He spends his days in sleep, and his nights in depredations—hares, rabbits, squirrels, poultry, and birds are his ordinary food; but sometimes he ventures upon lambs, and, if hard-pressed, will descend to shell-fish and crustaceans. He is a crafty villain, and has little in his character to deserve sympathy and commiseration, or to shield him from the desperate fury of the hounds.

The Arctic fox, like the Polar bear, changes his coat from an ash color to that of a white on the approach of the severe winters of his native region, and the hairs then become very soft, woolly, and long. In his character, he is not materially different from the red variety already noticed. He is found only in the colder latitudes of Europe, Asia, and North America. Jackals inhabit Asia and Africa, and usually travel in large packs, when they sometimes venture to attack their superiors in strength.

The habits and appearance of these animals are very similar—they are all burrowers, all great sleepers during the day, and prowling depredators at night. The hyena has two leading genera—the striped animal of Africa and portions of Asia, and the spotted one of the Cape of Good Hope. They are distinguished for their savage strength, their rapacious appetites, their long rough hair, and their general ungainly and ferocious appearance. They live mostly in caverns, and those of the Tertiary period are literally strewn with the bones of the animals upon which they feasted.

The family of bears includes several genera of that animal, besides those of raccoons and badgers. Their teeth are similar in number and arrangement to those of the dog, and although the bear is carnivorous to some extent, it seems to prefer vegetable food, especially fruit and tender herbs. They reside during the winter in caves and mountainous recesses, where they rear their young. It is an unoffending animal until molested, or when acting in defense of her young, for which the mother has an extraordinary affection, and will lay down her life in their behalf. It can climb trees with facility, and is equally at home in the water—the white or polar being, in fact, amphibious. The brown or black bear has a long-haired skin, and thick muscular legs, with padded feet and sharp projecting claws. Its walk is very soft and stealthy, and it betrays no inconsiderable cunning in its midnight prowls—often approaching the cabins of woodmen in the forest, and helping itself to whatever spoil may be left exposed. A touching incident is related in Lord Mulgrave's Voyage for the Discovery of a Northwest Passage. A female bear and her cubs were approaching the ship, when the sailors fired, killing the cubs and wounding the mother. Regardless of her own sufferings and danger she scorned to withdraw and leave her young behind. She would not understand that they were dead; she placed food before them, and by every endearing motion endeavored to raise them up with her paws; she withdrew and looked back, as if expecting them to follow; but perceiving that they lay motionless, she returned, and with inexpressible fondness walked around them, pawing them, licking their wounds, and moaning the while. At last, as if receiving the unwilling conviction that they were dead indeed, she turned toward the ship, and uttered a fierce and bitter growl against the murderers, to which they replied by another volley of shot that laid her beside her young.*

Bear-baiting was one of the classical amusements of our English ancestors, two or three centuries ago. It has been rarely tolerated in modern times; but bear-dancing is not an unusual feature in the entertainments of the ring, garnished with the grotesque movements and the stereotyped sallies of *Mr. Merryman.* The Polar or white bear differs from the ordinary species, mainly in color, and in having his feet clothed with hair, for the double purpose of giving him warmth and of enabling him to maintain his footing on the smooth ice and sleet of his native glaciers. The smaller animals of the land and the sea are his ordinary food; but being an excellent swimmer and diver, he is a successful adventurer in seals and large fish. The former he approaches by a series of adroit dives, and when he gets sufficiently near, a single bound and hug serves to secure the victim, whether it be leisurely reposing on an iceberg, or floundering in the water.

The raccoon is very abundant in the United States, and some of the genera, which consist of four or five, are no doubt natives either of North or South America. The common animal has a soft tread, a thin tapering snout,

---

* Scripture Natural History.

broad head, white face, and a long switching tail, having rings of black. Its general color is brownish grey Its upper lip, *a la* genus homo, is furnished with a thick black mustache. The raccoon is a great sleeper during the day, and rolls itself into a ball to avoid the glare of the sun, which appears to be inimical to its delicate vision. It is extremely cunning, and very tenacious of life—as much so, perhaps, as the opossum. Some twenty years ago, during the excitement of politics, the 'coon accidently became the recognized and victorious ensign of the whig party. And to politicians generally, its character is not without emblematic significance. During the brief and temporary predominance of the "Know-nothings," it was generally thought that the party of Clay and Webster was dead; but the truth is, (as I think indications now disclosing themselves sufficiently demonstrate,) the old *'coon was merely asleep*, and he yet lives to be borne in triumph in the conservative processions of the people And why not? His character is certainly as fair as that of the predaceous vulture emblazoned on our national escutcheon,—a bird distinguished for little else than its prowling and tyrannical habits. Dr. Franklin regretted that he (the bald eagle) " had been chosen as the representative of our country. He is a bird of bad moral character; he does not get his living honestly. You may have seen him perched on some dead tree, where, too lazy to fish for himself, he watches the labors of the *fishing hawk*, and when that diligent bird has at length taken a fish, and is bearing it to his nest for the support of his mate and young ones, the bald eagle pursues him, and takes it from him. With all this injustice, he is never in good condition, and like those among men who live by sharping and robbing, he is generally poor, and often very lousy. Besides, he is a rank coward; the little king-bird, not bigger than a sparrow, attacks him boldly, and *drives him out of the district.* 'He is therefore," says the Doctor, "not a proper emblem for the brave and honest Cincinnati of America, who have driven all the *king-birds* from our country!" We have said that the 'coon is cunning; and it is of a sort that may be denominated *Yankee* cunning. For example: being an amateur in crustaceans, he is in the habit of suspending his tail, in a very insinuating manner, over the haunts of crabs and lobsters, who, naturally mistaking it for a delectable morsel, clutch it, whereupon they are suddenly jerked out of the water by the astute fisherman, and a sumptuous repast is derived from their tender joints! There are two varieties of the raccoon in South America, known as the *coati-mondi*. In color they are brown and red, and are nearly as large as the fox. They often do much mischief to the plantations, but their food mainly consists of insects and reptiles. They are also expert climbers, and like the squirrel, descend trees head foremost. The *suricate* is also a native of South America, and is easily domesticated, as are, in fact, nearly all the animals belonging to this group. The badger combines in his nature some of the features of the hog, the bear, and the genet. It is a burrower, and possesses unusual strength in its fore legs, the feet terminating with sharp claws. It emits a strong and unpleasant odor, resulting from a secretion in its extremities. Its food consists principally of fruits, insects, frogs, eggs, and those birds which build on the ground or on low bushes. Their burrows are usually lined with straw and other vegetable material to render them warm and comfortable; and though emitting a disagreeable smell, they are yet extremely tidy in their subterraneous abodes. They are widely disseminated in Europe, Asia, and North America. The glutton is an animal intermediate between the badger and the coati-mondi; but as yet, little is known of it. It is a native of South America, and in captivity evinces a playful and sportive disposition. The ratel is found in Africa and India, especially in the department of Bengal. Like the fox and the raccoon, it spends the day in sleep, and prowls upon birds, rats, and other animals at night. It is a skillful burrower, and the graves of the dead are not exempt from this propensity, unless protected by a covering of thorns or stones. It has a partiality for honey, and under the guidance of the forest cuckoo, it searches out the stores of the hive, which are usually erected on the ground, or suspended from low bushes. Upon finding a hive, the ratel burrows near it, and approaches directly beneath it by a subterraneous passage. He then emerges to the surface, and boldly besieges the magazine of honey. In captivity it is said to be very frolicsome, and resorts to various antics to attract notice. The upper portion of the body, from the head to the tail, is covered with gray, while the sides and under portions are black. Its tail is moderately long, but its whole appearance is peculiar.

The civet, the ichneumon of Egypt, and the genet are all distinguished for having an apparatus to secrete a substance which imparts a a strong scent, more or less musky. The secretion, in the civet, is deposited in pouches in its posterior glands, which may be extracted from the animal, by means of a spoon, at the average of a drachm per week. The musk is very powerful, and substances scented with it will retain it for a long time. Every portion of the flesh of the animal is thoroughly impregnated with it. The scent is somewhat similar to pomatum, and is generally highly esteemed. The civet is about three feet in length, and its habits are similar to those of the fox, which it resembles in the face, but has a longer snout. Its color is brownish-gray, traversed with numerous bands of black. The civet is a native of Africa, but has been domesticated in Europe for the production of musk. The *zibet* of the island of Java is a near relation of the civet, and is distinguished for its burrowing and thieving propensities—being, in this respect, an exact counterpart of the fox. The ichneumon is a native of Egypt, but is found in all parts of Asia and Africa. Although scarcely larger than a cat, it is the deadly foe of young crocodiles, snakes, and other noxious animals of the torrid zone. Its mode of attack is to dart, with extraordinary agility, upon the head, and seize them in the most vital part. A single shake usually serves to break the necks or backs of reptiles, after which they all become powerless in its grasp. It is also provided with excretory functions, but unlike the civet, its secretions have a most offensive odor. In appearance and habits it is not unlike the weazel—its feet being armed with claws, its nose long and tapering, and its eyes small and flashing. The genet is beautifully spotted, like the leopards, and does not differ materially from the ichneumon except that it is larger, and instead of emitting an offensive odor, it is rather pleasant, being that of a delicate musk. It abounds in Turkey and Spain; and is characterized not only for the beauty of its skin, which is valuable, but also for its cleanliness, mildness of disposition, and skill as a hunter. The weasel stands at the head of a group of animals which are also distinguished for imparting, while living, a most offensive odor, but most of which are nevertheless valuable for their skins. These are the weasel, the ferret, the stoat, the martin, the pole-cat, the skunk, and the sable. These animals all present a long, round, and serpentine body and neck, and though swift runners and swimmers, they have the shuffling movements of snakes rather than quadrupeds. The weasel is small, not as large as a common cat, has eyes remarkable for their quick perception, and a color varying from a brown to a dirty white. While it is found in all parts of Europe, it exists in great abundance in the United

States. It feeds principally on rats, mice, moles, and birds. The ferret is larger than the weasel, and is a native of Africa. It was formerly employed to hunt rats and rabbits, which it would pursue into their holes in the earth—hence the name. The stoat much resembles the weasel, except in color, which is more generally white, particularly in high northern latitudes. It has sometimes been called a white weasel. The martins comprise several varieties, all of which are very numerous in the United States and Canada, and the northern parts of Europe. It is larger and longer than the weasel or the ferret, and is furnished with a long and switching tail. Upward of fifty thousand skins of this animal have been obtained, in the northern and western regions of the United States and the British possessions, in one year; and though it is allied to the class of *mustela vulgaris*, the odor from the skin is delightfully musky and pleasant. The pole-cat is well known for its offensive smell, and its nocturnal depredations—poultry, rabbits, pigeons, etc., being its principal victims. It is a great climber, and can make its way over the smoothest walls. It is also a very courageous warrior, and defends itself with considerable skill from the attacks of dogs, who often beat a retreat, probably in consequence of its disgusting smell. He has been termed the prince of marauders, a title which his cunning, skill, agility, and peculiar qualities seem fully to merit. Notwithstanding his odor, his skin, when dressed with the hair, retains nothing offensive. The skunk is an American variety of the pole-cat. The upper part of the body is white, while the lower is black. It has long shaggy hair and tail, short legs, and a long pointed snout. It seldom ventures out in winter, except in the Southern States. Its food consists mainly of rats, mice, and toads. When attacked, it emits a fetid discharge, which is offensive in the highest degree. The sable is an **inhabitant** of Russia and Siberia, and its fur is by far the most valuable of any of the family. It is long, fine, and silky, and generally of a very bright color. The sable changes the color of its hair with the change of the season, the dark thin hairs being gradually superseded by white ones, which form its usual winter coat. The skins of the sable imported annually from Russia vary from one to two hundred thousand, and they often command enormous prices. The otter, although it does not belong to this group, has a similarly long and slender build, but is more aquatic in its predilections. Its fin-like legs, feet webbed and oar-shaped, and its long rudder-like tail, enable it to make the swiftest and most astonishing evolutions in its native rivers and lakes. Its fur is short, but extremely fine, and large numbers of its skins have been exported from the American fur grounds.

Such is a brief glance at some of the leading features which characterize the numerous, varied, and remarkable animals comprising the order of Monadelphians. We could do little more than mention the families; nor was an extended description essential to our purpose. We wish to convey an idea of the extent and variety of the higher fauna of the earth on the close of the Sixth Day; and our object will have been sufficiently attained in the enumeration of the principal known families and groups. As to their fossil remains, they are everywhere scattered throughout the higher Tertiary, and it is almost needless and superfluous to dwell on the particulars. The fossil remains of extinct and living genera of bears occur in great abundance in France, Germany England, America, and very likely all over Asia and Africa. Badgers, dogs, hyenas, and a large number of the *feline* type, both of extinct and living genera, have also been found in caverns, and sometimes as low down as the Pliocene Tertiary, in all parts of the globe. And their remains invariably approach those of the living species in proportion to their proximity to the present geological epoch. But, before we enter upon a general review of this interesting formation, or make further comments upon its animal and vegetable creations, two other orders yet demand our attention.

The ninth order in the mammalian division is the *Quadrumana*, which comprises apes, baboons, and monkeys. All the representatives of this order are remarkable for having four hands, terminating in five toes or fingers, and for the general resemblance which certain of them present, when in an erect posture, to the human species. This is particularly the case with the apes, of which the Oran-outang is by far the largest and most perfect specimen. The anatomical structure of this animal is said to be similar to that of man; and, as compared with some of the lower types, it must be admitted that the partition between the two races *seems* to be very thin and transparent. We saw a specimen of the Oran-outang twenty years ago, and had daily opportunities of witnessing its habits and movements. It was brought from Africa by a fellow-townsman, Dr. S. M. E. Goheen, who retained it about his premises for several months, chained to a kennel. The pictures presented of the animal, in works on Natural History, are usually much exaggerated, especially when represented in an erect position, *which is wholly unnatural to it.* It is very rarely, and only when attempting to climb, that the Oran-outang stands on its hind feet; and it requires much skill and patience to train them to the effort. The face, too, is invariably flattened; in the original specimens it presents no regularity of feature whatever, as compared with the human countenance; but, on the contrary, is disgustingly deformed and inexpressive. In some of the pictures, however, which we have seen, it *even surpasses many specimens of the human face*—being furnished with a regular beard, and cheeks as smooth and flesh-like as if fresh from the hands of the barber. Moreover, it is often represented as exceedingly plump, round, and muscular in its physical conformation; whereas the very reverse is the case with the living animal, the body and all the limbs being exceedingly attenuated, notwithstanding that it possesses much of the agility and elasticity of movement which characterize other quadrupeds of the same size and mode of life. While we cannot pretend to determine, from absolute knowledge of the osseous structure of the Oran-outang, the actual extent of its similarity to that of the human frame, we know that the animal *as a whole* is altogether different; that, as compared with the human form, it presents a series of malformations from its toes to its head; that many of its organs are irregular and misplaced; and that while it is *suggestive* of a resemblance to man, the ideal image is entirely dissipated by the reality which a personal acquaintance with it affords. All its instincts are low, and in general intelligence it is by no means the *equal* of the dog, the fox, the elephant, or the horse. The Oran-outang is a native of Africa, but has also been found in India. It is a very rare and solitary animal, and, like all the others in their natural state, shuns mankind as much as possible. The Chimpanzee and the Barbary ape are other varieties of the ape tribe, but they differ little from the former, except that their physical features have still less correspondence with those of man. All these animals are clothed with long irregular hair, and the fingers or toes of their hind feet are of the same length and equally as prehensile as those in front. In the Barbary ape, the hind feet are considerably longer than those in front, and are very thin and sinewy. The face of the Chimpanzee is broad, with protruding jaws and an immense mouth; that of the Oran-outang is somewhat smaller, but, upon the whole, less regular. The lips of all these animals are thin and flabby, and devoid of that fullness and incarnation which gives expression to the human face. This arises, in part, from the fact that all the specimens of the *Quadrumana* have pouched cheeks to a greater or less extent. The monkeys, especially, conceal large stores of provisions in their jaws, and the apes do so to a limited extent—hence the enormous protrusion of this portion of their face. The Barbary ape, unlike the others, is furnished with immense ears, and with the single exception of its protruding jaws, the head and face are very similar to the bull-dog. And although, in its external appearance, it bears no similarity whatever to man—no more, in fact, than to a dog—yet we are assured that its osteology is the same, or, rather, that it has the same *number of bones*, joined in a manner very similar, but deviating in length, in local form, and in general conformation. All the ape tribe appear to be natives of Africa, though specimens of the Oran-outang have been found in India, and in some of the islands of the Indian Ocean. The baboon has but few varieties, and are all distinguished from the apes by their tails, their long hair, and their more repulsive features. The ribbed-nose baboon has a series of ribs arching over his cheeks on each side of his nose, which latter organ, like that of the dog, extends on a line with the mouth, and gives these extremities of the face a canine appearance. The whole structure of his head is frightful, and is rendered more so by the varied colors of the nose and the ribs—the one being a bright red, and the other a light blue. The pig-tailed baboon is a smaller animal than the other, and is provided with large cheek pouches, like all the genuine monkeys. The face, ears, and posteriors are naked, and, when tamed, it evinces a fondness for tobacco, snuff, and mustard, which it eats without apparent inconvenience. In this respect, it comes nearer to man than any of its species! These animals, unlike the apes, usually travel in companies, and their thieving and libidinous propensities constitute the prevailing features of their character. They are also natives of Africa, and live principally on grains, fruits, and succulent herbs. The monkey tribe is very numerous, and differs from the apes and baboons in being furnished with very long tails, which are useful in allowing their suspension from the limbs of trees. Cunning, trickery, thieving, and capering are their leading characteristics; and, excepting that their faces are moulded somewhat after the human type, they partake of the nature and features of squirrels. The Mico is much the handsomest specimen of the tribe. It has a rather handsome face, with a red flesh-color, is covered with a rich coat of white hair, and a tail nearly twice the length of its body. Its habits are also more agreeable than most of the species. It is found in large numbers in South America, particularly in Brazil. The green monkeys are found in Northern Africa and India. They are light and fragile in form, and, like the mico, have tails of extraordinary length; some of which are naked and others clothed with hair. The human aspect of the face gradually disappears in this group, and takes the form, in some of them, of the squirrel, the fox, and the rat. The ring-tailed monkey has the head of a fox, the body of a cat, and the gait of the kangaroo. The yellow Macauco approaches the opossum. They both belong to Africa, although found in numerous other localities, and are remarkable for their playfulness. The remains of monkeys have lately been found in the upper strata of the Tertiary in all the countries where they now live, and in some where they have long since become extinct. Those found in the Tertiary of India have been referred to a species of *Semnopithecus*, and indicate animals of large size. Those discovered in the South of France are smaller; while again, those of South America point to species more than twice the stature of those now living there. The fossil remains of this order establish the interesting fact that, before the close of the Tertiary era, four leading divisions of the monkey family existed; that these divisions were as distinctly separable by structural features and general habits as are the living apes from the monkeys; and that, moreover, they had a wider geographical distribution, and existed in regions of country where the climatical changes which have occurred in the mean time have utterly extinguished the ancient species, and forbid the residence of those now living under the torrid zone— hence their limitation to Africa and South America, and portions of Asia; and their invariable exclusion from regions of glacial cold.

The last order of the mammalia consists of man. It was toward the evening of this great geological day— between the closing of the post-Pliocene and the fragrant and peaceful dawn of the Present, that the *ne plus ultra* of the creative work had been attained. After the most elaborate preparations had been made for his reception; after all the animals had contributed their share, directly and indirectly, toward the attainment of that harmonious equilibrium in nature by which governmental dominion was to be forever maintained over them; after the earth had been clothed with herbs, and fruits, and shady groves, and the valleys and the hills strewn with blooming flowers; after the terrestrial stage had been furnished with "new and gorgeous scenery," and the vast over-arching dome hung with brilliant chandeliers, the light of which dazzled with the varying lustre of millions of celestial Koh-i-noors; —it was then that the great versatile actor, *Man*, made his first appearance on the stage, and entered at once upon the round of characters which usually distinguish his brief and arduous engagement. We have no disposition to speak of him in the spirit of levity; but we may say with a sigh, but with entire truth, that the characters he assumes upon the stage of life are often but indifferently performed. His most lamentable failure is in the rendition of the sterling character of an *honest, upright, God-fearing man!* It is one often assumed, but alas! how complete and overwhelming are the failures! The canting hypocrite—the selfish miser—the envious, shallow-pated pretender—the unfeeling destroyer—the Aminidab Sleeks—the arch-nosed Shylocks—the plausible, slippery, crawling, insidious Iagos—these constitute the staple characters of life, and they are almost everywhere played off before indulgent audiences. But to go back:

The order of *Bimana*, which embraces all the varieties of the human species, was the last and the greatest work of the Creator. This fact is not only apparent in the statement of Moses, but it is rendered perfectly indisputable by the revelations of geology. As to the exact time of man's appearance, there is room for a little doubt; I shall, however, endeavor to show that it must have been toward the close of the Tertiary, in which his fossil remains have been sparingly found. Like all the inferior animals, the species of mankind comprise a large number of distinctive varieties, which physiologists arrange under five leading groups or families, viz., the *Caucasian*, the *Mongolian*, the *Malayan*, the *Negro*, and the *American Indian*. Each of these principal divisions comprises a large number of sub-groups, nearly all of them speaking different languages, and characterized by local habits and mental idiosyncrasies as diverse as could well be conceived. Emigration, intermarriage, and commercial and social intercourse between individuals, tribes, and nations, have constantly tended to increase the lines of separation and the complexity of type, or to gradually fuse old into new ones. From the earliest history of mankind, however they all appear to resolve themselves into five principal

23

families, and still more remotely into *three*. Thus, the Caucasian is everywhere distinguished from the others by his white skin, his tall stature, and the general proportion, harmony, grace, and beauty of his person, while his moral and intellectual faculties are also better developed, and appear to be of an infinitely finer temperament than those of all the others. His nose is straight or aquiline, his face and ears small, the forehead broad, and the complexion roseate, and expressive of thought. His face often sustains a beard on a line with the cheek bone —a feature, which, though not invariable, is yet peculiar to none of the others. While it is far from being the most numerous, the Caucasian race may be regarded as the *governor* of mankind, and as the head of all animated nature, because it has attained a higher degree of civilization, refinement, power, and mental culture. It includes most of the inhabitants of Western Asia, the Tartars, the Caucasians, Georgians, Armenians, and Circassians; the Turks, Persians, Arabians, Affghans, Egyptians, and Abyssinians, and with the exception of the Finlanders, it embraces all the Europeans, and their American descendants. The nobility of England may be regarded as its highest type, both in physical and mental development.

The Ethiopian, or Negro, though a native of all parts of Africa, is yet distinct from certain tribes inhabiting the northern and central portions, some of which, according to recent discoveries, appear to have attained a considerable degree of civilization, and are no less remarkable for the purity of their complexions, and the European aspect of their features. The Negro is the very converse of the Caucassian—his skin being perfectly black, and his hair and eyes generally of the same color. But his hair, instead of being long, straight, or curling into ringlets, is woolly and knotted. His nose is broad and flat, his lips thick and protruding, and his forehead low. There are many minor features by which he is everywhere distinguishable; but there is one which, though I have never seen it mentioned by physiologists (except by Mr. Jefferson, in his *Notes on Virginia*), appears to be universal even with mulattos and half-breeds. The skin of the Negro secretes a perspiration which emits a peculiar and unpleasant musky odor, somewhat in the manner of those animals of the family *Mustela vulgaris* already described.

The *Malayans* are natives of the numerous islands of the South Sea, as New Zealand, New Guinea, New Holland, and the Malayan Archipelago. Their skin has a dark brown color, hair long and black, nose full and broad, mouth and face long, large, and prominent. They can very easily be traced back to the Ethiopian type.

The Mongolian race is well represented by the Chinese, the Japanese, the Mongols, and other Asiatic tribes, together with those of Tonquin, Siam, and Thibet, and the Laplander and the Esquimaux. Their color is that of yellow or olive. They have but little beard, generally none at all; low forehead, broad face, rather flat nose, large ears, thick lips, projecting round cheeks, black hair and eyes, and stature inferior to the Caucasian.

The American Indians present several groups, but they all resemble the Mongolian race so closely, especially in the structure of their heads, that they may properly be regarded as primitive off-shoots from it. Their color varies from a dark red to an olive—as a whole, they may be termed *copperish*, with an infusion of yellow sufficient to identify them with the Mongolian. Their hair is also black, long, and coarse, but they are destitute of beard. Their leading characteristics are essentially Mongolian, while their individual features sometimes approach the Caucasian. As a race, they are tall, straight, and graceful, and in natural sprightliness of intellect, dignity of carriage, and eloquence of speech, are hardly inferior to any.

After the creation of man, toward the close of the Sixth Day, the world seems, for the first time, to have been lulled into a serene and calm repose. Creative action had attained the *ultimatum* in man. All the previous work ended and blended harmoniously in him, as the days themselves blended into the great Sabbath of Nature which ensued. All the angels of heaven participated with God in the joyful work of creation. Even the little children-angels, that play around the opal-throne of the Creator, seem to have contributed their little works in the annelids, insects, and flowers that peep out from their secret retreats in the earth; while God himself was mainly engrossed in infusing the *coup de grace* to the whole—in concentrating and embodying all the animal functions, under the sceptre of Reason, in the brain and body of man! How wonderfully and fearfully he is made! What an infinite number of delicate vessels traverse his frame! How nicely all his bones fit into their varied and complicated sockets! How gracefully and elastically his limbs bend and move under the guiding tendons, muscles, and fibres with which they are padded! How God-like is his *Mind!* Composed of a variety of organs, hitched to the car of Thought, Reason holds the reins, and drives the body through the earth—sometimes over rough and dangerous roads, sometimes in happy thoroughfares and along broad avenues, strewn with flowers and sweetened perfumes! Man is the Governor—the Monarch of Creation! In him all things earthy end—in him all things return back to God! After his creation, therefore, God ceased from his work. Day after day, each filled with its appropriate work, passed in succession; but the *Seventh* day, instead of being ended, is only begun. Moses does not speak of the Sabbath as having any *end,* so far as man is concerned, it is an eternity, during which God will rest from his creative labor! Upon the close of the Sixth Day, after a terrestrial equilibrium—a systematic organization of all the types, creatures, and circumstances of created life had been secured, the whole was placed under the dominion of *man*. It was expressly consigned to his custody. He became the god of earth—the recognized agent of the Almighty, to take care of the creatures and the varied works he and his angels had made!

Man's creation was the preliminary step to the Sabbatical repose which ensued. When the sky began to gleam in vast lakes of liquid azure, and bold promontories of fleecy drapery—when the mountains rejoiced in spreading foliage, and hid their hoary peaks in the clouds, to conduct angels to the paradise below; after the sloping plains had rolled out their carpets of velvet, and crystal streams glistened in the sun, or leaped in cascades, or waltzed around in eddies and gurgling pools to the soft cadences of the musical air;—in a broad vale, fringed with thymy terraces, and boquetted with flowers and blossoming orchards, whose fragrance infused ethereal intoxication;—in a garden, by nature formed to please, laden with fruits, and vines, and gums, and juices nectarial—amid the joyous songs of birds, and the approving throbs of animated life; surrounded with the winged ministers of the planetary universe, and the angelic hosts of the court of Jehovah—there, as the last and crowning act of creation, God "formed man in his own image, male and female created he them;" and breathing into his nostrils the breath of life, he became a living and an immortal *soul*. Adam and Eve, thus created and united in wedlock, became flesh of one flesh, and bone of one bone. In the presence of the angelic throngs of heaven—with the approving smiles of all the representatives of the ethereal worlds—the powers of space, of Eternity and Immortality

—the nuptials of man's progenitors were solemnized, and as their endowment, God consigned to Adam the sole custody of the great work thus completed; crowning him with unrestricted dominion over the boundless earth—the fishes of the sea, the varied creatures of the land and air! He then blessed and sanctified the Sabbath that ensued; and now, the joyous train showering their gratulations on the wedded pair,

> The heavens, and all the constellations rang!
> The planets, in their places, list'ning stood,
> While the bright pomp ascended, jubilant!
> Open! ye everlasting gates! they sung;
> Open! ye heavens! your living doors—let in
> The great Creator from his work returned,
> Magnificent! His six days' work a World!
> Open! and henceforth oft, for God will deign
> To visit oft the dwellings of just men
> Delighted! and with frequent intercourse
> Thither will send his winged messengers
> On errands of supernal grace!

THE SEVENTH DAY—THE SABBATH.

1 Thus the heavens and the earth were finished, and all the host of them. 2 And on the seventh day God ended his work which he had made; and he rested on the seventh day from all his work which he had made. 3 And God blessed the seventh day, and sanctified it, because that in it he had rested from all his work which God created and made. 4. These are the generations of the heavens and of the earth, when they were created, in the day that the Lord God made the earth and the heavens. 5 And every plant of the field before it was in the earth, and every herb of the field before it grew; for the Lord God had not caused it to rain upon the earth, and there was not a man to till the ground. 6 But there went up a mist from the earth, and watered the whole face of the ground. 7 And the Lord God formed man of the dust of the ground, and breathed into his nostrils the breath of life, and man became a living soul.

I REMARKED at the outset of the present work, that I had no desire to enter the field of theological speculation. Indeed, it was a part of my original plan expressly to avoid it, except only in those cases where there seemed to be a direct conflict between the Bible and Geology. In these conflicts, I have ventured to explain the geology, rather than the theology of the dispute. I have, in fact, not written *as a* theologian at all—I have not even confined myself within the sectarian discipline of the church, I have simply taken up the *Cosmogony* of the Bible, as an humble but original and independent investigator of nature, and endeavored to sustain its integrity against the insidious assaults of the most distinguished geologists. This work, I am well aware, could have been better performed, so far as the Church itself is concerned, by some one more closely identified with its tenets and policy than myself. But while the field of theology is comparatively well filled with expounders of every degree of merit and intelligence, it will be conceded that but few of them have devoted attention to the department of practical geology. Instead of dedicating a portion of their time and energies to a proper investigation of the field, so as to protect the Church from its assaults, many preachers have really diverted their pulpits from their more legitimate duties, to the promulgation of what may be termed "sensational" sermons, or have partially abandoned them in favor of the lecture-room and the political arena, for the poor and questionable compensation of popular applause and notoriety. The pulpit, thus neglected and weakened, has been left unprotected from the attacks of pernicious geological theories and speculations; and the result is, that we now see its authorized guardians and defenders driven to miserable special pleading, or compelled, in many cases, to abandon the explicit statements of the Bible in favor of mere subterfuges and unmanly evasions and apologies.

Notwithstanding the enthusiasm—not to say fanatical zeal, which a large number of ministers evince in the abstract political questions of the day, the great majority, no doubt, deserve respect for *good intentions*, at least. There is no character on this broad earth more entitled to our kindly sympathies, our warm regard and admiration, than the conscientious, open-hearted, and devoted minister of the gospel. Their social influence is, and ought to be, unlimited, because they are necessary to the preservation of society, government, and civilization. When, therefore, we see men compromising their high position as ministers of the gospel—neglecting its immediate and legitimate objects, or prostituting them in the strife and angry tumult of partisan politics, the cause of religion is not only weakened, but the Church itself suffers degradation. Diversity of opinion upon doctrinal points may safely be tolerated, and is perhaps unavoidable; but a diversion of the obvious functions of their office from its high social, moral and spiritual, to a sectional and political purpose, can only lead to the embitterment of popular prejudices, to the disruption of society and of governments, and to the *ultimate extinction of religion itself.*

But the very fact, paradoxical as it may seem, that scarcely any two ministers can agree upon *all* the doctrinal points of the Bible, is a powerful argument in favor of its divine authenticity. Like all the works of God, it is sometimes difficult to understand. But all that which we see around, above, and beneath us, although quite familiar to our senses, is yet only one great, stupendous, unfathomable mystery. We ourselves are no less so. And yet we see in the daily operations of nature, things that we *know* to be fixed and certain; and from things thus *known*, we can easily pursue those that are unknown, or involved in obscurity. We can thus trace God through *all* his works. And it is precisely so with the Bible. The great bulk of it, although mysterious, we know to be true and fixed, because self-evident to our senses—and by the *faith* we thus have in the known, we are justified in our inferences of the unknown. Geology proves, for example, that there *was* a beginning;—that there was a time when man did not exist;—that there was a time when animals and vegetation did not exist;—that there was a time when the earth itself could not have existed. Now, knowing this, we must infer that a *cause* produced the effects which we see and feel—that the operating cause must be one of superior intelligence, and that originally it must necessarily have been in close communion with the creatures of its volition. Hence, we cannot but believe that man, in the beginning, was in friendly intercourse with his Maker—that he received from him the means whereby to sustain and develop his nature, and the admonitions and advice by which happiness, love, and peace would be forever secured.

God is said to unite three persons in one—as the Father, the Son, and the Holy Ghost. This combination of personality is incomprehensible to us, but not incredible. As Father, he is the undoubted Creator of all things—the personification of worlds, of matter, elements, and force. We may recognize in him the imponderable and indestructible elements—carbon, oxygen, hydrogen, electricity, light, gravitation. These are omnipresent in space and matter. Besides these, there are sixty-two simple minerals that cannot be destroyed or changed from one to another, as gold, iron, mercury, etc. The elements may be presumed to constitute the instruments of creative action, order, and law, because all matter and life are subject to their control. Some of these elements and minerals may have affinities in other worlds, or in space, and the whole may have originally formed a unit, or *one*. This primary unit, existing in unimaginable chaos, evolved a governing principle—a principle inheriting all the properties of elements, as subsequently diffused in worlds and life. Now, this vital principle was and is none other than *God*. It may be said that God himself was thus *created*. No: he primarily existed; he now began to *create*. His will was manifested by diffusion—irradiation. The elements incorporate in him were invisible creatures, like the ideal visions of man's dreams, but capable of assuming *fixed forms in matter*. God is thus the source from which *all* things flow.

But this personality of Creator is distinct from that of *Son*, in which he also became, by his irradiated elements in worlds, *the governor of all things*. That is to say, having *created*, he now also governed by *Law*, which Law may be regarded as the embodied *Son or offspring*. He is thus Father *and* Son; or the source of elemental matter, and the diffused law that *controls* created elemental matter. As Holy Ghost, he further exists in man in the form of *mind*, or as the subtle spiritual principle. As *Father*, he creates; as *Son*, he is law; as *Holy Ghost*, he is the impulse to guide the creatures of law. God said, "Let *us* now make man in our *own image*—after our likeness;" that is, let us endow man with some of our qualities. He was accordingly created to govern the world, and was therefore supplied with *mind*, a function of the Creator. Man, therefore, comprises *two* persons in one—a mental and a physical being. His spiritual part is immortal, and exhibits the governing principle under the guidance of the Holy Ghost. Being the creature of Law, man can neither modify nor evade it. God, therefore, combines the attributes of three distinct personalities—Father, Son, and Holy Ghost.*

Even before the creation of worlds, God was surrounded by angels or spirits representing the different irradiated offsprings of his Volition. When he began to create worlds, and to confer authority on his Son (embodying *Law*), certain of these angels rebelled, and because they sought to subvert Law, or harmony, or order, they were expelled into the regions of unreclaimed darkness. The chief of these was Satan. Under every form of disguise, he has since sought to subvert the Son in his administration of law. Although forever expatriated from heaven, he still preserves his immortality, with which he was clothed in the beginning, and which is consequently irrevocable, since God, having finished his work of creation, now *only governs by law*, and has transferred the

---

* These are but ideal speculations, and will of course be so regarded by the reader.—E. B.

custody of the earth to his embodied Son. Satan, therefore, is a rebel against the Son of God; and having known the original decrees of heaven, and that the earth formed a portion of the plan of the Universe, he undertook its subjugation. Such was his subtilty, that he not only deceived man (a creature of law), but also the angels guarding the gates of Paradise. Effecting an entrance into Eden, the beauty of the garden recalled visions of that ethereal paradise which he had forever lost. He traversed its broad and sinuous avenues, paved with sands of gold "unnumbered as the dust of Barca, or Cyrene's torrid soil." With stealthy eye, he traced its grotesque walls of alabaster, sinking into dim perspective. He stole into its secret grottos, walled with crystal amethyst, and ribbed arches of gnarled topaz. He contemplated its diked porphyry, rising upon the level plateaux in monumental piles to the glory of the Creator, glittering with intermingled clusters of massive diamonds, opals, and sapphires, like the stellar diadem of Night. Trees bowed low with the offering of their luscious fruit; and the vine clung to the spreading oaks, tempting with her pulpy nectar. Leaping cascades, gushing fountains, expanding lakelets diffused their refreshing draughts to the thirsty sun; while liberal Flora, from exhaustless stores, fed the fragrance-seeking breezes to balmy repletion. In the deep recesses of the wood, surrounded by their brooding mates and their unfledged young, myriads of plumaged warblers prolonged their happy strains, and bore higher and higher the universal anthems which animated Nature sent up, in harmonious accord, to the throne of the divine Creator.

Thus stealthily stalked the angel of Evil from one object and scene to another, until all their varied features had been duly scanned; while he secretly gloated over the impending gloom (decreed in the bituminous caverns of hell), which his machinations were now suspending over the young earth, like the web of a spider. Anon, he espied the secret bower, sacred to love and wedded bliss. On an elevated step, overlooking the glories of the paradise which stretched out in profuse luxuriance beneath, a Parian nook opened its chaste portals to the nuptial feast. From the top, clusters of amaranth diffused immortality upon the rosy couch; while dahlias, geraniums, heliotropes, myrtles, and olives, sacred to eternal love, devotion, holiness, and peace, twined their tendrils and their new-blown buds in long-drawn aisles and foliaged arches, to screen them from the peering sun. There, on sheets of silken down, newly hackled from gossamer cocoons, the progenitors of mankind resigned themselves to rest, locked in the arms of holy love;

> "The loveliest pair
> That e'er in love's embraces met;
> Adam, the goodliest man of men, since born
> His sons; the fairest of her daughters Eve."

In the very citadel of earthly bliss the tempter had already entered, in a dark disguise. Drawing near to the pillow of Eve, he deduced from their whispered speech that the *tree of life* was to them forbidden fruit; and he infused the spirit of unrest in her dreams. He obtained a clew—the artful spider here found a beam from which to stretch his subtle web; and now the angels of Sin sent up a shout, in anticipated triumph, that echoed through the parched and blistered gloom of their infernal caverns. He left the sleepers ere Uriel's sunbeams had arrived, and again transforming himself into a new shape, wandered among the cattle of the fields;

> "For spirits, when they please,
> Can either sex assume, or both; so soft
> And uncompounded is their essence pure;

> Not tied or manacled with joint or limb,
> Nor founded on the brittle strength of bones,
> Like cumbrous flesh; but in what shape they choose,
> Dilated or condensed, bright or obscure,
> Can execute their airy purposes,
> And works of love or enmity fulfill."

In his present purpose, however, he assumed the body of the serpent, which he knew to possess "more subtilty than any beast of the field." In this guise, he approached Eve, while she was alone, and engaged in training the flowers of Paradise. Struck with the beauty, and the extraordinary faculty of speech and intelligence which the serpent possessed, Eve unfortunately became interested in its discourse. Inquiring whence it had obtained such varied accomplishments and wisdom, it replied that they flowed from a certain tree in the garden, the fruit of which it had freely eaten! The innocent curiosity of the mother of her race was at once aroused, and following the serpent to the spot, discovered that the tree in question was the *forbidden tree of life*. The serpent, with the freedom of levity, and with an air of affected sympathy and incredulity, inquired, "Yea, hath God said, Ye shall not eat of *every* tree of the garden?" To which poor Eve replied, half ashamed that she should be *inferior* to the serpent in the gratification of her tastes; "We may eat of the fruit of the trees of the garden; but of the fruit of the tree which is in the *midst* of the garden, God hath said, Ye shall not eat of it, neither shall ye touch it, lest ye die." To this the subtle villain again, with mock pity and commiseration, and with a half-suppressed smile at her innocent *naïveté*, replied, "Ye shall *not* surely die; for God doth know that in the day ye eat thereof, then your eyes shall be opened; and ye *shall be as gods*, knowing good from evil." This certainly looked plausible, and the *experience* of the serpent seemed to confirm it;

for if *it* could eat with impunity, and inherit wisdom, and *live*, why should not *Eve?* Alas!

> "Neither man nor angel can discern
> Hypocrisy, the only evil that walks
> Invisible, except to God alone,
> By his permissive will, through heaven and earth;
> And oft, though wisdom wake, suspicion sleeps
> At wisdom's gate, and to simplicity
> Resigns her charge, while goodness thinks no ill
> Where no ill *seems!*"

The woman fell, and brought down her husband with her; for when Adam perceived that she had violated the direct and express command of God, he was confounded with conflicting emotions—love for the lost woman, whose crime might perhaps be palliated by his own participation, or his happiness jeopardized in her death; fear of the just retribution of God, and yet a longing desire for the wisdom which the fruit could impart; credulity in the experience of the serpent, because manifest to all the senses, and yet a lurking suspicion of its honor and integrity—all these struggled in the excited mind of Adam, and he grasped the forbidden fruit as with uncontrollable desperation. They ate! For a time the effect was exhilarating, like the juice of the grape. Their carnal natures were inflamed—they burned with voluptuous desire—their veins swelled, and their cheeks flushed, and staggering, they fell to the ground! They slept, but it was *now* the sleep of remorse and exhausted nature, instead of innocence and peace. They awoke—not to the smiles of hope, and virtue, and harmony; but to mutual shame, recriminations, bickerings, distrust, and reproach. In short, all the dark and damning passions of sin—disobedience, lust, falsehood, jealousy, revenge, and all their horrid train, marshaled by Sickness, Disease, and Death;—all the polluted spirits of the infernal regions arose from the earth like spectres, and became installed in the world—raising their arms against law and harmony—against peace and good-will—against virtue and love—and asserted dominion over Adam and his captive race!

Why, it may be asked, did not God place them beyond the *power* of disobedience? Why involve them in a position in which they would be exposed to the assaults of sin? To this it may be suggested that it was not the *plan* of the Creator to people worlds with absolute Gods. He had a purpose to subserve, decreed in the beginning. Man was his *servant*, owing to his *law* implicit obedience, but endowed with reason to obey or not. He was provided with every earthly comfort—he was warned of his danger—and thus left *free* to act! Had he been endowed with higher qualities, *heaven* would have been his proper sphere at once, *not earth*; but made as a servant of God, his subsequent *promotion* depended on his implicit obedience to law.

Now, after Adam's transgression, God placed enmity between the seed of the serpent and that of Eve, and doomed Satan to crawl upon the earth in the character or form he had thus assumed. The serpent, therefore, is the personification of evil—*first:* because when coiled, it is emblematic of eternity, or the original immortality of angels; *second:* when wounded or dismembered in its exterior parts, it has the power of renewing the lost flesh; *third:* it changes its skin during the vernal season, thus renewing its youth with increasing age; *fourth:* its poisonous secretions, and needle-like fangs, are often fatal to man, and significant of his original fall; and *fifth:* its movement is stealthy, noiseless, and slimy, eluding grasp, and readily accommodating its body to the most contracted retreats—all of which are significant parallels of those subtle spirits which haunt, annoy, and destroy the peace and happiness of mankind. Besides all this, there is an inherent dread of serpents in the human species, notwithstanding that it is one of the most beautiful, graceful, and agile animals that God has created. Its very appearance occasions an irresistible nervous tremor—a secret shudder, which it is impossible to control. We thus perceive that the enmity between the two races, is universal and palpable; but this is not all. Man has the power of bruising its *head*—its vital and only dangerous part. He, therefore has the power of destroying *evil*—while in return, the serpent only bruises man's *heel*, or his *fleshly* part, thus indicating that no part of man is liable to its attacks if he treads *not in the paths of evil*, where only the serpent is concealed. It is a singular and extraordinary fact, that what *seem* to be the most simple and commonplace expressions in the Bible, are in truth invariably pregnant with volumes of meaning, all of which are developed upon the most casual investigation of the fundamental laws of nature. It is these, among thousands of other coincident features, that establish its divine inspiration beyond all doubt or cavil, while, at the same time, much that still remains obscure and dubious to our imperfect comprehension, will ultimately be revealed by investigation of nature's laws.

In decreeing Adam's punishment, God cursed the *ground* for his sake—he did not impair, or blight, or afflict any of his physical functions. The punishment may appear severe, but no one can question its wisdom. Had he inflicted personal injury to the race, it would certainly have savored of cruelty; but he merely cursed the ground. And how, and why? He caused noxious weeds, and plants, and trees, to grow spontaneously, and with extraordinary prolificacy, at the same time that blight, barrenness, disease, and death, attacked the fruits. In Adam's fallen condition, bodily and mental exercise became absolutely essential. Idleness is the prolific mother of sin, and the great Creator saw that *to save the race of man*, he must be furnished with employment. Hence he *cursed the ground*, and consigned Adam and his children to the task of subduing and tilling it. Under the circumstances, it was no punishment at all—it was a benevolent, kind, and fatherly act, for it saved mankind from the *ennui*, the idleness, remorse, and languor that would inevitably have resulted in the ultimate extinguishment of the race, or in its effeminate degeneracy to that abject weakness, in which it would have fallen a prey to the predaceous animals. To poor deceived Eve, he simply multiplied the sorrows of her conception—a punishment which, after all, is scarcely greater than that visited on the inferior animals. And yet, who can fail to detect Almighty wisdom and benevolence in this? Had her conception brought with it no pains and pangs, no sorrows and troubles, and apprehensions, what a world of beastly profligacy and prostitution we should have had! It is absolutely terrible and horrible to contemplate. The holy institution of marriage, the divine sentiment of love, of parental affection, and domestic virtue would not, and *could not*, have existed on the earth; but unbridled licentiousness, beastly sexual intercourse, harlotry, and debauchery would have reigned unchecked and supreme. *Punishment*, indeed! It is just such punishment as a loving parent, knowing the weakness of his child, would inflict under the *pretext* of a terrible corrective, but in reality a wise and benevolent protection from impending calamity, already brought down upon its head by unrestrained liberty. The whole history of mankind and of individuals, reveals nothing but the utmost benevolence of the Creator—love, forbearance, anticipation of his needs and rational desires, forgiveness, partiality, and parental fondness. There is hardly a creature, a vegetable, a jewel, or a substance of any sort whatever, above, beneath, or in the bowels of the earth or seas, that is not made tributary to his wants or to his childish fancies and desires. He is an infant that tumbles into every sort of mishap, and not the watchful care of all the angels of heaven can restrain him from mischief, or from the injurious contact of the prowling cormorants of sin and folly!

But, we are told, God put enmity between the seed of the serpent and that of the woman. What further meaning has this sentence, in addition to that already presumed? Wherefore seed of the *woman*? Satan seduced woman; he leveled his poisoned shafts at the *weaker vessel*, not at the strongest. To render *his* punishment the more poignant, therefore, God redresses the wrongs of woman through the *strength or seed of woman*. He caused the Virgin Mary to conceive in holy immaculation—the spirit of God descended in her, and she brought forth Christ in the image of man. Christ assumed the cause of his fallen race, and armed in the holy panoply of heaven, lived, suffered, and died like a god. Thus the human race was cleansed, purified, *reborn, as it were, through his spiritual incarnation;* and man again stands redeemed and disinthralled from the manacles of Satan. But as a servant of God, he owes obedience. He must not transgress his laws—he must not again eat forbidden fruit. He must believe in Christ, that is, he must believe in and exemplify the doctrines taught and practically illustrated in the whole life, actions, and doctrines of Christ. He is at once the head and front, the very incarnation of law; and by pursuing the course he lays down, everlasting peace and good-will on earth, and felicity in heaven will be secured. He did not come upon the earth to destroy the law previously established, but rather to exemplify it in his own incarnation; he came not to destroy, but to fulfill, and thus it is that *Christianity will ultimately govern the whole earth*, and by its equitable operations, rescue all mankind from the doom that would otherwise befall them. Much has been said and written regarding the divinity of Christ. One thing is certain,—he was a pure man, and this character is little inferior to that of a god, since it inherits God's kingdom, and secures immortality. But every thing tends to confirm his divinity. If he were not divine, he himself was a deceived man—nay, worse; he lived and died a *wicked man*. But where was the *motive* for deception? He did not assume worldly pomp—he did not covet regal splendor—nor princely luxury—nor wealth, nor temporal applause, honor, or station! No! His whole life was directly the reverse. He was poor and despised. He was jeered and scoffed. He was a wanderer—without a home—without courtly friends—without a pillow to lay down at night, or a roof to shelter him from the storm. He was poor, indeed! And yet, what philosopher of any age or clime has promulgated doctrines more in harmony with the higher decrees of heaven? What man, since the fall of Adam, has lived a purer, a more holy, and angelic life? He was without reproach. Like the lamb, he could "lick the hand just raised to shed his blood!" He was the personification of wisdom. His words were more cutting and more powerful than swords! And yet his mission was not to *destroy*, but to *save*. His object was *love*—not *gain*. He wanted *souls*—not *wealth*.

Either, then, he was divine or he was *not*. It will not suffice that he was a *monomaniac* in Religion, gifted with extraordinary accomplishments and wisdom, and powers of meek suffering and endurance. A diseased mind is always arbitrary, incoherent, feverish. His was not; on the contrary, it was strong, flexible, persistent, and logical. It was controlled by god-like reason and wisdom, and no devices or arguments of man or devil could withstand its pointed barbs. If he was not divine, and especially appointed by heaven, he deceived himself and his devoted apostles, and added to the enormity of a life of imposture, eternal fraud, falsehood, and perjury. But in the face of his spotless life, his holy work, his unrecompensed suffering, persecution, and death, is it not *absolutely sacrilegious to doubt his divinity?* Can any one study his character, motives, and actions, and discover any thing like *human* frailty, pride, vanity, selfishness, ambition, or folly? Cæsar, Pompey, Xerxes, Alexander, Hannibal, and all the great captains of the earth, before and after him, sought the glory of States and of National power;—but he, without arms and munitions of war; without Senates or pompous oracles; without games, feasts, statues, paintings, or idolatrous images, yet sought no less an object than the *redemption and conversion of the world*, past, present, and future,—and that, with no other instrument than his *inspired words and deeds!* Living in the most critical, learned, and voluptuous age that had yet dawned upon the earth,—in the very zenith of Roman, Grecian, and Persian renown—an age of Poetry, Philosophy, Science, Art, Literature, Architecture, Oratory, Music, Diplomacy, and Civil Government—an age whose heroes stand before time in long lines of historical statues;—living, preaching, suffering, and dying in such an age, and at such a time, yet not a single speck or blemish could be detected in his person, manners, doctrines, or motives—exposed, as he constantly was, to all the assaults and blandishments of the wicked! Not the shadow of a defect could be pointed out in his proffered scheme of salvation. His doctrines were the sublimated essence of all wisdom and philosophy, based as they were, and are, upon faith, obedience, love, and virtue.

The fall of Adam and his species, it is claimed, was thus compensated by the *implied promise of this very Saviour*

—hence the enmity between the serpent and the descendants of *Eve.* Adam saw in this arrangement the future regeneration of his race, and by its *retroaction upon the past,* that of himself. This consoling promise saved him from again relapsing into the arms of sin, and under it he died a good and holy man. Had no such hope existed, he might again have embarrassed the plans of the Creator by self-destruction, or that of his doomed issue. But God, it would seem, was merciful, and did all for the best, but in a manner peculiarly his own. Alll his acts and works seem dim to our feeble and limited vision; but when we view them through the glass of familiar Nature, they are all sufficiently plain and simple. He not only cheered the unfortunate couple with this promise, but furnished himself the skins to clothe them. With sad and subdued hearts, he then led them out of the Paradise they had desecrated, lest they should be tempted to eat of the fruit of the tree of life, and thus *live forever* in their fallen state.

Although but three of Adam's sons, and none of his daughters are named in the Bible, it is very likely that his family was numerous. These, from the necessity of the case, must have intermarried, and extended their settlements in various directions, though not in the more remote continents of the earth. The Bible deals only with such names and circumstances as are essential to the prominent facts of its text—leaving minor details to be inferred. Therefore, considering the early habits of mankind,—their pastoral occupation and general non-commercial and non-manufacturing character, we may conclude that but a small portion of the earth's surface was populated during the sixteen hundred years that intervened between the marriage of Adam and the birth of the three sons of Noah. Such, too, was the longevity of the antediluvian race, that but a very few generations could have flourished during this period; and as there could have been no necessity for their *distant migration,* so we have no right to infer that, up to the time of the flood, any of Adam's descendants wandered far off from the original centre of creation. But such was the inherent wickedness of the race, that God determined to wash it from the earth — preserving, however, its concentrated virtue and wisdom in Noah and his family. From the fact that Noah is represented as "*perfect in his generations,*" it is concluded that the promised seed of the Saviour had descended to *him,* and thence through *Abraham* to *Mary,* the immaculate *mother* of Christ. It is impossible for us to dwell upon the religious aspect of the history of the antediluvian race; we merely purpose to consider the phenomenon of the Noachic flood, and the remarks thus far offered are only intended as a *connecting link* between the close of the Sixth Day and the dawn of the ensuing Sabbath. The enmity which God placed between the seed of Evil and that of Eve, it will be thus observed, was brought to another crisis in the deluge, and to its culminating point in the birth of Christ.

In reference to the flood, nothing appears more certain than its universality; and it is therefore a matter of astonishment that men standing high as Theologians should undertake to dispute it. Among the most recent, and perhaps the most plausible doubters, was the late lamented Hugh Miller, who brings to the support of his views the testimony of some of the most distinguished Doctors of Divinity. The whole premises of these distinguished Christian-doubters can be demolished with a single word: if the flood was only *partial,* and confined (as they allège) to a small area, where was the *necessity of the ark?* Why could not God have removed Noah and his family, and the animals of the earth, to the adjacent districts or continents that *remained unsubmerged?* Yet these plausible

Christian philosophers (embracing bishops, doctors, vicars, and laymen without number,) fill volumes to explain and apologize for the supposed errors of Revelation, as if the word of God were a series of bungling errors. I confess I have no respect either for the learning or the religious integrity of such men. They would have the world believe that the Bible, as it stands, is *true;* and yet they write unnumbered scientific homilies to prove that it is a fable, a farce, and a cheat.

God communicated to Noah his reasons for decreeing the destruction of the animated earth. He became disgusted with the folly and wickedness of man—"seeing that every imagination of the thoughts of his heart was only evil continually." He repented that he had made man, or beast, or bird; for on looking upon the earth, he could see nothing but corruption, in which all flesh was alike involved. To his faithful old servant he said, "The end of all flesh is come before me; for the earth is filled with violence through them; and behold, I will destroy them with the earth" (that is, he will *use* the earth which they were abusing to effect their destruction). He ordered Noah, therefore, to build an ark, and gave him the directions as follows: "The length of the ark shall be three hundred cubits, the breadth of it fifty cubits, and the height of it thirty cubits. A window shalt thou make to the ark, and in a cubit shalt thou finish it above: and the door of the ark shalt thou set in the side thereof; with lower, second, and third stories shalt thou make it."

The Scripture cubit, according to Sir Isaac Newton, is a fraction over twenty inches; but, according to Bishop Wilkins, it is over twenty-one inches. This may, or may not, be true; but assuming either of them to be correct, or approximating the true length, the dimensions of the ark would be as follows: Length between perpendiculars, 515 or 547 feet; breadth, 85 or 91 feet; depth, 51 or 54 feet; keel, or tonnage capacity, 464 or 492 feet, which would give a tonnage of from 18,500 to 22,000 tons. A parallel to Noah's ark is presented in the steamship *Great Eastern*, which recently visited our shores. The length of this *Leviathan* of the deep is, I believe, 680 feet, breadth 83 feet, depth 58 feet, keel 639 feet, and capacity 23,000 tons. But this is exclusive of her engines and propelling machinery, and of her elaborate furniture, fixtures, and properties—none of which encumbered the plain vessel of Noah. Besides, all these estimates are made for *live freight;*—the actual tonnage of the Great Eastern, if laden with iron, would probably be over 26,000 tons, and that of the ark would have been in proportion. Her machinery and properties occupy at least one-half of her space, and one-half of her carrying capacity, which would thus raise her tonnage to nearly fifty thousand tons. And in this view, according to the cubic standard of Newton or Wilkins, the carrying capacity of the ark would be considerably augmented, and might be safely computed, on the basis already presented, as nearly, if not fully equal to the Great Eastern, *without her machinery.* This would give at least *forty thousand tons burden for live freight.* An idea of the enormous dimensions of the Great Eastern may be formed from the fact, that she presents as much available room, and could carry the combined freight, of *ten* of our largest war vessels. Thus, the tonnage of the steamer Pennsylvania, which used to be regarded as the giant of the ocean, is 3,211 tons; that of the Columbus, 2,489 tons; the Ohio, 2,757; the North Carolina, 2,633; the Delaware, 2,633; Vermont, 2,633; New Orleans, 2,805; Alabama, 2,683; Virginia, 2,633; and the New York, 2,633—making a total of over 28,000 tons. These steamers are all of the largest class, and *yet all combined* are barely equal to the Great Eastern! And after removing their machinery, the whole would present an aggregate of storage surface very little superior to the great ark of Noah. Now, if this great steamer subserves no other purpose, it will at least demonstrate that it is sufficiently capacious *to constitute a menagerie of all the terrestrial animal species now living!*

Hugh Miller, however, with characteristic special pleading (in his *Testimony of the Rocks*), brings down the dimensions of the ark to 450 feet in length, 75 feet in breadth, and 45 in height. He accomplishes this *feat* by an ingenious analysis of the modes of measuring adopted by some of his provincial countrymen. "There," says he, "is the *span*, the *palm*, the *hand-breadth*, the *thumb-breadth* (or inch), the *hair-breadth*, and the *foot*. The simple fisherman on our coasts still measures off his fathoms by stretching out both his arms to the full; the village seamstress still tells off her cloth-breadths by finger-lengths and *nails;* the untaught tiller of the soil still estimates the area of his little field by *pacing* along its sides." According to this system of measurement it is obvious that every thing depends on the physical proportions of the measurer. The fathoms of a big Scotch fisherman would certainly exceed those of Tom Thumb; while the pedal extremities of the "untaught tiller of the soil" would enlarge or decrease the area of his "little field" exactly as they happened to be large or small. The Irish clod-hopper, by this rule, might easily obtain broader acres than the Chinese rat-catcher. But even if this standard were adopted, Noah's ark would rather be enlarged than decreased. The same chapter that specifies the dimensions of the ark in cubits, informs us that "there were *giants* in the earth in those days; and also after that, when the sons of God came in unto the daughters of men, and they bare children unto them; the same became mighty men, which were of old men of renown." Now, if the ark was built during an age of giants, of men *renowned* for their mighty proportions, and the cubit was a measure of some portion of the human frame, (as Miller says it was, and still is,) instead of its having been 450 feet in length, it was most *likely six or more hundred feet*, and had a tonnage capacity in exact proportion! And this, after all, is just as likely to be correct as any of the estimates since proposed. Miller, however, adopts the dimensions according to the cubit already given; and then proceeds to consider how the animals could be accommodated in the ark. He calls to his aid Sir Walter Raleigh, an experienced seaman, who proceeds to pack them in after the following manner. "If in a ship of such greatness"—(it was a square, flat-bottomed *ark*, without masts, spars, or rigging!)—"if in a *ship* of such greatness," says Sir Walter, "we seek room for eighty-nine distinct species of beasts, or, lest any should be omitted, for a hundred several kinds, we shall easily find place both for them and for the birds, which in bigness are no way answerable to them, and for meat to sustain them all. For there are three sorts of beasts whose bodies are of a quantity well known; the beef, the sheep, and the wolf; to which the rest may be reduced by saying, according to Aristotle, that one elephant is equal to four beeves, one lion to two wolves, and so of the rest. Of beasts, some feed on vegetables, others on flesh. There are one-and-thirty kinds of the greater sort feeding on vegetables, of which number only three are clean, according to the law of Moses, whereof seven of a kind entered into the ark, namely, three couples for breed, and one for sacrifice; the other eight-and-twenty kinds were taken by two of each kind; so that in all there were in the ark one-and-twenty great beasts clean, and six-and-fifty unclean; estimable for largeness as *ninety-one beeves:* yet, for a supplement (lest, perhaps, any species be omitted), let them be valued as a *hundred and twenty beeves*. Of the lesser sort, feeding on vegetables, were in the ark six-and-twenty kinds, estimable, with good allowance for supply, as four-score sheep. Of those which devour flesh, were two-and-thirty kinds, answerable to three-score and four wolves. All these two hundred and eighty beasts might be kept in one story or room of the ark, in their several cabins; their meat in a second; the birds and their provision in a third, with space to spare for Noah and his family, and all their necessaries."

This estimate of Raleigh's was made nearly two centuries ago, and instead of there being but eighty or one hundred and twenty species of animals on the earth, as then estimated, discoveries made in the mean time have raised the number, large and small, of all the different species, to several thousand. Up to this time, every known region of the earth has been visited and fully explored, and the animal kingdom is found to embrace the number of species as follows: Quadrumana 170; Marsupialia 123; Edentata 28; Pachydermata 39; Terrestrial Carnivora 514; Rodentia 604; Ruminantia 180; Birds 6,266; Reptiles 657; Turtles, etc. 15—making a grand total of living species of 8,596. This embraces all the animals of the globe, large and small, from the minute mouse up to the elephant. It must be borne in mind, however, (as we have all along warned the reader,) that much diversity exists in the systems of classification adopted by Naturalists. What one Zoologist would divide into three classes, another arranges under two: what one divides into five orders or tribes, another will arrange in three; what one separates into eight families or sub-families, another can dispose of in four or five; and what one will spread out into ten, twenty, or thirty *species*, another will embrace in five, ten, or fifteen. Thus, the feline animals comprise many species, the number of which might, with propriety, be greatly reduced. And so with nearly every division of the animal kingdom. Species, as ordinarily used, implies a *general*, not an *exact* resemblance to the parents. It could easily be demonstrated that climate, external circumstances, and the controlling exigencies of necessity have created more species in the classifications of Naturalists than *primarily existed* in the original progenitors of the Animal Kingdom. But to enter upon a discussion of this point here, would appear hypercritical, and might be regarded as an *evasion of the main question*. We despise any such pretexts or subterfuges, and should still rely on the abundant capacities of the ark to accommodate them all, if the number of species were twice as great as is now claimed—for the question involved is *not as to the number of animals*, but simply as to the dimensions and capacities of the *ark*. And this, singularly enough, and in the face of the exact and explicit language of Scripture, turns upon the value of the cubit. In ordinary, the cubit is the *ulna*, or bone which extends from the elbow to the wrist; but in modern mensuration, it represents the length of a man's arm from the elbow to the extremity of the middle finger. Webster observes that the standard length of the ancient cubit varied with different nations, as we know the length of the hand and the ulna vary with different individuals. With most men, the *ulna* seldom exceeds eleven inches, and the hand, from the wrist joint to the point of the middle finger, rarely exceeds seven and a half inches. The two combined, in our largest individuals, would perhaps reach nineteen inches. The Roman cubit, according to Dr. Arbuthnot, was seventeen inches and four-tenths, and that of England eighteen inches. The Scripture cubit is defined as being a fraction less than twenty-two inches; but as to the *antediluvian cubit*, we have no positive information. If, however, the average length of the ulna and hand of men now living be estimated at nineteen or twenty inches, we are entitled to infer that, during the race of giants which flourished before and after Noah's era, they averaged from twenty-one to twenty-two inches. The cubit, thus graduated, would make the ark quite as large as the Great Eastern, and, leaving out her machinery, masts, and ordinary sailing and propelling properties, the tonnage capacity of such a vessel might be estimated, in round numbers, at *fifty thousand tons*. But, not to appear unreasonable in our estimate, we will say—according to the twenty-two inch cubit of Newton—that the tonnage of the ark, for live stock, was at least equal to *twenty thousand tons*, of twenty-two hundred and forty lbs each, or to 44,800,000 lbs. This is considerably *less* than the capacity of the Great Eastern, even including her five or six steam engines, her steam pumps and lifeboats, her sails, masts, and rigging.

As to the structure of this enormous boat, but little need be said. The timber was close at hand, and its architecture, though simple and rude, was yet substantial. It was, simply, a gigantic flat-bottomed boat, with three stories, instead of one, exclusive of the roof or deck. Like the small plank and lumber arks that annually descend our rivers, its joints were carefully caulked and pitched; and this was all that the circumstances required. It was not *intended* to ride the waves and storms of ocean, but rather to be borne gradually up along the sinking land, like a vessel during the flowing in of high tide.

A difficulty has been suggested as to how the animals came to Noah; but if the other end of the proposition were presented, there would be no difficulty at all. The animals did *not* come to Noah; Noah went to *them*. His orders were specific, and he obeyed them. But how? Did he or his agents wander over the earth, armed with spears, and lasso, and traps to hunt down and capture the animals? or, like a man of sense, taking a practical view of the enterprise committed to him, did he merely seek the *young* of each species, and arrange them in a general cosmopolitan menagerie, to be trained to the voyage they were to undergo? We have no right to suppose that Noah was an ignorant, simple-minded, old man; on the contrary, he was eminent for his wisdom and virtue. He was, consequently, fully equal to the great enterprise with which God intrusted him. Being a man of practical sense, therefore, his obvious policy was to obtain *young animals; first,* because they would be more tractable; *second,* because they would occupy less space in the ark; *third,* because they would not encumber the ark with brood; and *fourth,* because their powers of recuperation would afterward be superior to those of adult animals; and *fifth,* because they would require less forage for their keeping. Would not the cubs of the bear, the lion, the tiger, the elephant, rhinoceros—the calves of the herd, of camels and horses, and the young of all quadrupeds, answer his purpose *better,* in every point of view, than the full-grown animals? Would snakes a week old, not suffice as well, (or rather better,) than boas and vipers *twenty feet in length?* Was it necessary to fill his ark with antiquated oxen, and elephants, and camels, that had done service in the plow or the caravan? Must he select poor old spavined horses, toothless lions, and tigers, and bears, when the little cubs would best correspond with the object of his mission? The idea is too absurd to be entertained.

That the ark, indeed, was filled with the young of each species, is manifest in the fact that no *births* occurred on board, although the vessel was afloat for one hundred and fifty days. This remark, of course, applies only to the larger quadrupeds; for among that somewhat numerous and diminutive class, whose existence is limited to a few months, the propagation of the species must have continued uninterrupted. A similar exception may be made for birds, among whom the process of incubation must have been constantly maintained. As it required many years to *build* the ark, there was abundant time to collect the animals. And as these occupied but a comparatively small geographical area, and were still under the dominion, and many of them in the service of man, there was really little difficulty in bringing them together. The lion and the tiger had little of the ferocity which, in their native jungles, distinguishes them now. The whole animal creation, like man himself, was originally of a more subdued and tractable nature than it has since become. And, mainly confined to the area then occupied by the human species, it required no extraordinary efforts to bring them together, or to collect a menagerie of specimens. The species that may have wandered off into other continents, if they left no descendants at home, necessarily *became extinct after the flood*. The Bible says so, and Geology proves it. The earth teems with their fossil remains, and the extinct are buried in the same strata with those of the representatives of living species. Their number, however, was comparatively small; for the great bulk of *all* the animals of the antediluvian earth continued within the range of the original circle of the Adamite Creation. But while some of them may, and did wander off, and even reach distant continents, most of the original types, or progenitors of species, remained behind, and came into the ark of Noah. There is no fact in the entire range of Geological discovery, more palpable and overwhelming than this. It forms a line of separation between the fauna of the past and the present, as broad as the continents themselves. From facts that are thus clear, it is easy to follow the foot-prints of others more obscure, and bring order out of a seeming chaos.

But let as now inquire how the animals can be accommodated. The ark, we will suppose, on the basis of a twenty-two inch cubit, to have been 600 feet in length, and 100 feet in width. This would give us an area of 60,000 square feet to each floor, making an aggregate of 180,000 square feet for the three stories, exclusive of the cabin under the roof for Noah and his family, and for the storage of such articles as could occupy the deck of the vessel. We, of course, cannot assume to understand the real interior plan or structure of the ark, but presume it was extremely simple. We can suppose it to have been divided into stalls, running lengthwise with the ark, and separated by narrow alleys for drainage, and to allow the passage to and fro of the grooms of the animals. The first two ranges of stalls, extending along both sides of the vessel, we shall devote to the cubs of the larger animals—as the elephant, the rhinoceros, the camel, the horse, the domestic cattle, including the buffalo, the bison, and the elk, moose, and deer. These, although extremely numerous in all parts of the earth, comprise but few species; scarcely more, in fact, than originally estimated by Sir Walter Raleigh. He computed them at eighty-nine, but his magnanimity extended the number to one hundred, which he estimated in bulk as equal to 120 beeves. We, however, will *double* the liberal estimate of the gallant chevalier. We will suppose that there are now living 200 of the larger animal species, counting from the elephant to the horse. Of these animals that went into the ark, according to the Mosaic law, some were clean and others unclean. For the sake of facilitating our estimate, we will suppose 200 of the former, and 400 of the latter—in all 600 animals provided for in the ark. Adding 20 per cent. to equalize their bulk, we have a number equal to seven hundred and twenty head of cattle. This, of course, is based upon the theory of *full-grown* animals—a proposition which cannot be allowed for a moment, because it is at war with *all* reason and common sense. We take it for an absolute certainty that *all* the larger animal species were represented by cubs or young, for the reasons already stated. The young of all classes of the large animals, but especially those of the elephant, rhinoceros, bison, etc., are very slow in their growth and development, and for *several years* do not exceed the proportions of an adult bear. Those of most of the others are even smaller, as cattle, the moose, deer, etc. It would be safe, therefore, to compute the average size of *all* as equal to the bear, agreeably to the basis of Raleigh and Aristotle. Instead, therefore, of 720 head of *cattle*, we now have 720 *bears* to provide for, the aggregate of whose bulk occupies the same relation to the former as their weight. That is, if the 720 cattle averaged 1,200 lbs. each, the bears would not average more than 300 lbs., or one fourth. This, we think, is a liberal allowance, and contemplates cubs more than one-half developed (for weight generally increases only with the *adult* animal) Instead of 864,000 lbs., we thus have but 216,000 lbs., or instead of the bulk of 720 cattle, we now have that of 720 *bears*. Now, if we suppose the first floor of the ark to have been apportioned into ranges of stalls, corresponding in size to the animals to be accommodated, an estimate of the quantity of superficial feet occupied, can readily be formed. Thus, the stalls ranging along the sides of the vessel, might be partitioned into apartments *averaging* 8 feet in lateral depth, and 6 feet in width. The ark being 600 feet long, would thus afford two hundred stalls of this uniform width. The animals being reduced to the average size of bears, three and a half of them could be accommodated in each stall, and leave abundant room for their movement backward and forward. The Royal Bengal tiger and his mate, or the African lions, are often confined in the cages of menageries, to areas relatively very much smaller. Allowing three and a half animals, therefore, to each stall, we can provide for seven hundred in the *two ranges along the sides*. We thus dispose of all the *large animals*, except twenty, which can be hereafter provided for in other stalls.

We have thus taken sixteen feet from the one hundred feet width of the ark. Allowing two passages or alleys, each 3 feet wide, we now erect two other ranges of *double stalls*, running parallel with the first-mentioned, the entire length of the ark. These, instead of eight feet in depth, will be but *five;* and instead of six feet in width, will be but *four*. Two double ranges would thus give 150 to each range, or 600 stalls in all. In these could be accommodated the cubs of what may be classed as *second-class* animals—such as the bear, the lion, the tiger, the leopard, the panther, the jaguar, sheep, swine, goats, etc. Four of the cubs of these animals could be placed in each stall, and we should be able to provide comfortably for 2,400 individuals, or twelve hundred progenitors of species. Allowing 250 species for the 720 large animals, we should thus far have a total of 3,120 animals, and 1,450 species; while the room thus far occupied is but 48 per cent. of the width of the ark, and perhaps not *one-sixtieth* part of its actual tonnage capacity.

Leaving two other alleys, each $2\frac{1}{2}$ feet wide, we shall again extend two double rows of stalls the whole length of the ark. Instead of five feet in depth, we shall have these but *four;* and instead of four feet in width, we shall make them *three*. We should thus have two hundred stalls in each range, making for the four ranges, eight hundred in all. These stalls would suffice for the cubs of *third class* animals—such as dogs, cats, armadillos, sloths, chinchillas, foxes, rabbits, pigs, dogs, hares, beavers, wolves, etc. These could be stalled in fours, but we will say an average of three—making a total of 2,400 individuals, and 1,200 species, and summing up an aggregate, thus far, of 5,520 animals, and about 2,650 species. These stalls and alleys have absorbed 21 feet of the width of the ark, which, added to the previous 48 feet, make 69 feet in all. Of the tonnage, not over one-fiftieth part has been absorbed.

We now again leave two alleys, each $2\frac{1}{2}$ feet wide, and then arrange two more sets of double stalls. These, instead of four, will be $3\frac{1}{2}$ feet deep, and instead of three feet in width, will be but two. Consequently, each range will have 300 stalls, or in all, 1,200. In these we shall place the young of the animals of the *fourth class*—such as civets, weasels, ferrets, martins, pole-cats, sables, otters, moles, Guinea-pigs, squirrels, jerboas, marmots, rats, mice, —though many of them, being aquatic, need hardly be provided for. Four of these could again be accommodated in each stall, making an aggregate of 4,800 animals, and, say, 2,400 species. Added to the others, we have thus far arranged for 10,320 individuals, and 5,050 species. We have occupied 88 feet of the 100 feet width of the ark, leaving 12 feet to our credit. We will appropriate this space to two alleys, each two feet wide, with a double row of stalls down the centre, for a distance of 500 feet, leaving the remainder to be occupied by the cubs of the twenty large animals hitherto unprovided for. The stalls thus erected would be $3\frac{1}{2}$ feet in depth, by two in width, making altogether 500 stalls. These could be occupied, three, or four, or half a dozen to the stall, by such animals as were hitherto unprovided for—provided, however, *that any remain on hand!* The stalls would accommodate at least 2,000, which, added to the previous number, would make in all 12,320 animals, and 6,050 species, or kinds.

According to Newton, the ark must have been from fifty-five to sixty feet in perpendicular height or depth. The lower story must therefore have been at least twenty feet in height; and the second and third, nearly, if not exactly the same. Now, if this were so, there is no reason why cages for the feathered animals should not have been arranged *over all the stalls here mentioned.* Allowing a height of eleven feet for all the stalls of the animals, there would remain nine feet for the bird cages; and arranging these on the basis of the different ranges of stalls, at least twenty thousand birds, and ten thousand species, could be comfortably provided for. This, added to the other, makes 32,320 animals, and 16,050 species! And yet, not more than one-fortieth part of the tonnage capacity of the ark is thus far occupied.

The second story, being a mere copy of the first, needs no amplification here. The only animals that yet remain unprovided for, are those of turtles, snakes, lizards, toads, crocodiles, and the numerous species of flies, insects, spiders, scorpions, etc. All these, however, it will be readily admitted, could be accommodated in a single row of stalls. The marine and semi-aqueous species, of course, could shift for themselves—including many fowls, insects, and mammals. As not more than half of the actual capacities of the ark could have been occupied by the animals, it may be inferred that a large space was devoted to forage and provisions. But when we bear in mind that all, or most of the animals were young—mere cubs, calves, or lambs—I am unable to perceive the necessity for such an inference. The fact is, comparatively little space was required for such storage. Few of the animals required even straw to lie down upon;—for if the stalls had a gradual slope into the alleys, covered gutters would convey away all the natural excrements, and they could be cleaned as circumstances required. A little hay would be relished by some, grains by others, and flesh by all the carnivora. A large number, however, would be fed on milk, to supply which there must have been a special stock of adult cows and goats, as well as flesh for the strictly carnivorous. But one half of the area of the second story would suffice to store all the forage necessary for a year's voyage. The whole of the third story, and the roofed deck, would thus remain comparatively unoccupied; but we can well suppose that, in addition to animals, Noah had also supplied himself with a stock of plants, vegetables, seeds, and fruits, so that, on his return to the impoverished land, he would not long remain without the customary necessaries of life. We have thus shown, that, *according to the data furnished by the Doubters themselves*, there was room for more than thirty thousand animals, representing nearly half that number of species; and that, if there were more, more could easily have been accommodated! Zoologists may therefore go on, and ransack with impunity every corner of the earth, to multiply species; but they will find the ark prepared to receive all they can bring forward, and still *have abundant room to spare!* I will merely add, in leaving this branch of my subject, that whatever may be thought of the space allotted to the animals by my arrangement of the ark, it far exceeds that usually devoted to the animals in traveling menageries, where, as before remarked, the Royal Bengal Tigers and African Lions are often confined in cages barely large enough for them to turn round. Moreover, all my arrangements are predicated on the basis of Aristotle and Raleigh, and if any mistake has been made, it is theirs, not mine. But, if any exist, it will be found to militate against the doubters of the ark's capacities, and not against those who regard it as sufficiently large to have subserved all the purposes contemplated by the Creator.

SUBTERRANEAN STREAMS. 339

We now come to consider the flood itself. The animals, whatever their number and variety may have been, were safely disposed in the ark, and it only remains for us to institute some inquiries touching the flood. Here, again, the Bible is explicit: " In the six hundredth year of Noah's life, in the second month, the seventeenth day of the month, the same day were *all the fountains of the great deep broken up*, and the windows of heaven were opened. And the rain was upon the earth forty days and forty nights. And the waters increased, and bare up the ark, and it was lifted up above the earth. And the waters prevailed exceedingly upon the earth; and all the high hills, that were under the whole heaven, were covered. Fifteen cubits upward did the waters prevail, and the mountains were covered. And the waters prevailed upon the earth one hundred and fifty days."

"The fountains of the great deep broken up"—what is meant by this? What *are* the fountains of the great deep? They are, in part, the springs of water which gush out from the surface of the ground, and underneath rivers, lakes, and oceans. The crust of the earth is permeated by water. Almost everywhere, by digging a hole ten, fifty, or a hundred or more feet deep, according to the position of the strata, water will gush forth—the deeper the excavation, the greater the supply. Now, in general, the water thus obtained is accumulated in the fissures of the earth from surface drainage. This is evident from the fact that, during seasons of drought, many springs are temporarily exhausted. But there are also lakes and great rivers of water in the *bowels* of the earth, *not* directly due to surface drainage. In France, in boring for water, the auger suddenly fell a foot or more, when a stream of water rushed forth—thus proving that a reservoir had been struck. In another instance, after boring 375 feet, fine sand, vegetable matter, and shells of species living in the vicinity were brought up, indicating that they must have descended by some passage in the bottom of a river. In Germany, fish were thus brought up, although no river existed within many miles. Salt springs, although they are thought to emanate from deposits of rock salt lying at a great depth in the earth, may, in some cases, be due to subterranean drainage from the ocean. The borings for salt, in nearly all the Western States, are scarcely ever less than eight hundred, and many of them are more than fifteen hundred feet. The Artesian wells of Paris are over eighteen hundred, and that of Charleston, S. C., (if we are not mistaken,) is nearly two thousand feet deep. That recently sunk by Mr. Lauer, of Reading, is nearly two thousand feet deep. So, also, another, in Columbus, Ohio. These deep borings very often pass through the upper formation or system of rocks, into the adjacent ones below, where the water is usually found in anti-synclinal basins; and as it cannot, therefore, be derived from the surface immediately overlying, it follows that it must emanate from a distance, which may be great or small; and, in the case of salt springs, may proceed from the ocean, or distant salt lakes, or salt deposits. In the State of Michigan, subterranean lakes have been found but a short distance from the surface; and, in one instance, in excavating the track of a railway in that State, a large and deep basin was struck, into which the surface earth disappeared so rapidly, that many of the laborers barely escaped with their lives. The waters were beauti fully clear, and contained myriads of fish of the finest flavor. Such lakes, however, generally occur at a great depth, and their waters are only forced up through the fissures of the rocks by powerful hydrostatic pressure. Artesian wells are borings made through rock, or a stratum of clay, into a more porous or cavernous formation beneath, which, in consequence of such porosity, becomes a storehouse of water. Thus, a valley between two hills or mountains, may be covered by clay, through which no water will drain. But the slopes of the mountain may be sandy, in which case the water will *soak through the sand*, and accumulate in the valley, under the layer of clay. Now, by boring through the clay, a column of water will rise to the surface, and sometimes shoot up in a fountain, to the height of ten or fifty feet. These are called Artesian wells, because the experiment of sinking them was first made in the district of *Artois*, in France. Limestone formations are almost invariably of a stratified and cavernous structure, and consequently often contain immense stores of water. These reservoirs are generally supplied by surface drainage, and then tapped by springs in the valleys and plains, which form the sources of rivulets, creeks, rivers, lakes, and oceans. Where a broad surface of water is presented to the sun, evaporation takes place. The heat of the air absorbs it from the ground, as well as from lakes and oceans. The heat of summer, in a few hours, will lick up all the water of a shallow pond, and bestow it upon vegetation ; or it will gather it into the atmosphere in clouds of vapor, and again spread it over the earth in genial showers.

On the summits of all the higher mountains, as those of the Alps, the Himalaya, and the Rocky mountains, the rain congeals, and with the snow, forms stratum upon stratum of sleet and ice. The mass constantly increases until its ten thousand peaks are buried in the clouds— forming, as it were, gigantic stalagmites or steeples, piercing the regions of eternal cold. On the summit of the Alps, the ice and snow thus accumulated, forms plateaux two or three hundred square miles in extent, and varying in thickness from one to three hundred feet. The table-lands are traversed by rivers of melted snow, which are filled with fragments of ice, cut loose from the adjacent cones, pyramids, and canoned shores. During the summer, especially after long-continued rains, the rivers and lakes are swollen, and immense bodies of ice are undermined by the water, and borne slowly down the mountain slope. Before they proceed far, however, winter again sets in, and the isolated masses are united by new congelations. While constant additions are thus made to the aggregate bulk, the descending movement, although very slow, and rarely exceeding from one to five hundred feet per year, still continues. The summer always relaxes the icy grasp of winter, for a short season, and the canon streams cut immense incisions into the mountains of snow, or undermine the huge cliffs. As they thus become detached, the downward movement proceeds, when finally, reaching the precipices of the mountain, the enormous masses slide down with irresistible force. These are termed avalanches. They invariably carry with them immense rocks, trees, gravel, and earth; and entire villages, and hundreds of people have been destroyed by their unlooked-for visitations. The narrow valleys of the Alps teem with little villages, all of which are more or less exposed to the contingencies of avalanches, which, although very slow in their downward movement until they reach the fatal precipices, are yet certain, if sufficient time be allowed. The glaciers of the Alps vary in their altitude —some being as low down as 3,000 feet, while others ascend to a height of from 7,000 to 8,000 feet. Some of them are fifteen miles long, and are so permanent in their features and characteristics, that they are known by specific names. When these glaciers or avalanches descend the mountain, they sometimes stretch across the narrow valleys, and form dams which arrest the water. The water is thus backed, until finally a sluice is cut, and the whole gradually is swept away. If, however, the avalanche descends into the ocean, it becomes an iceberg.

and it is then floated about, until the ice gradually melting, the rocks, trees, gravel, and mud, are deposited in irregular heaps or lines over the bottom of the sea. Such deposits, when they occur in heaps, are styled *moraines*.

Great as are the effects wrought by these high mountain avalanches and glaciers, they are mere trifles compared with the stupendous icebergs annually sent into the temperate zones, from the Artic regions. The whole Artic circles are covered with snow and ice, and the atmosphere presents a perpetual winter.

Every one who has crossed the Atlantic ocean, must have more or less experience and knowledge of icebergs. Very often the reminiscences connected with them are of an unpleasant nature—for many a noble vessel has gone down in the fathomless deep by too close a contact with them. A passenger on the steamer Persia, writing to a journal in Mobile, details some of the incidents of such a voyage, and the allusion to icebergs being *apropos*, I will here introduce an extract:

"Then came fog, fog for three days and nights, until one thousand miles were passed—Cape Race and the banks of Newfoundland. Here the air began to be very cold, requiring thick winter clothing, and indicating that we were approaching the region of icebergs. Sure enough, on the afternoon of the 9th, while we were at dinner, the cry of "*icebergs*" was heard through the cabin (a convenient excuse for many to leave the table, as the sea was a little rough)! We rushed on deck, and there, far away, was a dim mass of white substance, which we could not distinguish from land; then another came, very large and grand, about ten miles distant—a grand mountain of ice, like a huge, bold promontory, jutting out into the wild waste of waters, while the waves dashed in foam and spray upon its cold and barren sides. Then the sunlight flashed over its glassy heights with a dazzling brilliance, reflecting all the colors of the rainbow, from peak to peak, until the mass passed into a shadow, and then appeared like a great mountain of snow, of the purest whiteness, untouched by that which defiles and darkens. To one we passed within half a mile, and could with great distinctness see its huge sides cut into ridges and gullies, by the streams that were trickling down to the ocean; on the summit there seemed the form of a house; indeed, there was any thing there that the imagination could picture out of such fantastic shapes and strange appearances. The cold gushes of wind that swept the ice-fields came over our vessel like wintry blasts, producing the most intense cold. These are the strange visitors from the unknown regions of the inhospitable North that break away from the icy fetters of their frozen continents, drifting down by current and breeze, seeking the warm gushes of the treacherous sunlight of a milder clime, which deceives their confidence, and before the huge ice-monsters become conscious of this snare into which they have incontinently drifted, the warm embraces of an enemy are around them, melting away their proud significance into the common level of the waters."

When icebergs contain imbedded rocks, and are thrust against the edges of rocky promontories and cliffs, they polish the surface, and leave long parallel grooves and scratches. Many of the rocks of previous formations, occupying the sides and summits of mountains, having been submerged during the flood, have their peaks rounded and their surface smoothed for many miles by the masses of ice and debris which floated over them.

"The recent polishing and striation of limestone by coast-ice, carrying boulders even as far south as the coast of Denmark, has been observed by Dr. Forchhaumer, and helps us to conceive how large icebergs, running aground on the bed of the sea, may produce similar furrows on a grander scale. An account was given, so long ago as 1822, by Scoresby, of icebergs seen by him drifting along in latitudes 69° and 70° north, which rose above the surface from one to two hundred feet, and measured from a few yards to a mile in circumference. Many of them were loaded with beds of earth and rock, of such thickness that the weight was conjectured to be from fifty to one hundred thousand tons.* A similar transportation of rocks is known to be in progress in the Southern Hemisphere, where boulders included in the ice are far more frequent than in the north. One of these icebergs was encountered in 1839, in mid-ocean, in the antarctic regions, many hundred miles from any known land, sailing northwards, with a large erratic block firmly frozen into it. In order to understand in what manner long and straight grooves may be cut by such agency, we must remember that these floating islands of ice have a singular steadiness of motion, in consequence of the larger portion of their bulk being

---

* Lyell's Elements of Geology, p. 122.

sunk deep under water, so that they are not perceptibly moved by the winds and waves, even in the strongest gales. Maury had supposed that the magnitude commonly attributed to icebergs by unscientific navigators was exaggerated; but now it appears that the popular estimate of their dimensions has rather fallen within than beyond the truth. Many of them, carefully measured by the officers of the French Exploring Expedition of the Astrolabe, were between one hundred and two hundred and twenty-five feet high above water, and from *two to five miles in length.* Captain D'Urville ascertained one of them, which he saw floating in the Southern Ocean, to be thirteen miles long and one hundred feet high, with perfectly vertical walls. The submerged portions of such islands must, according to the weight of ice relatively to sea-water, be from six to eight times more considerable than the part which is visible, so that the mechanical power they exert, when fairly set in motion, must be prodigious."

Persons living in the temperate zones can entertain but a feeble idea of the extent and power of the glacial influences of the polar regions. Much information has recently been obtained from the explorations of the late Dr. Kane, who, for three years, sailed through towering icebergs, or was hemmed in during the winter by frozen seas, and glacial precipices, and lands whose dust was drifted snow, and whose rocks were massive ribs of ice. His little crew were sometimes reduced to the last stages of human endurance, suffering, starvation, and disease, in a climate where the breath of man's nostrils would congeal, and his blood stagnate into torpidity. Even in the month of August, they were frozen in for the ensuing winter while attempting to reach the settlements of Greenland, after having delineated nearly one thousand miles of coastline. The amount of travel to effect this exploration exceeded two thousand miles, and was performed solely on foot and by dog-teams.

"On one occasion, during this coast-exploration," says Doctor Kane, "we were made aware of a remarkable feature of our travel.* We were

* Arctic Explorations, vol. i. p. 92.

on a table or shelf of ice, which clung to the base of the rocks overlooking the sea, but itself overhung by steep and lofty cliffs. Pure and beautiful as this icy highway was, huge angular blocks, some many tons in weight, were scattered over its surface; and long tongues of worn-down rock occasionally issued from the sides of the cliffs, and extended across our course. The cliffs measured one thousand and ten feet to the crest of the plateau above them. They were," adds Dr. K., in a note, "they were of tabular magnesian limestone, with interlaid and inferior sandstone. Their height, measured to the crest of the plateau, was nine hundred and fifty feet—a fair mean of the profile of the coast. The height of the talus of debris, where it united with the face of the cliff, was five hundred and ninety feet, and its angle of inclination between 38° and 45°." "We pushed forward on this ice-table shelf as rapidly as the obstacles would permit, though embarrassed a good deal by the frequent water-courses, which created large gorges in our path, winding occasionally, and generally steep-sided. We had to pass our sledge carefully down such interruptions, and bear it upon our shoulders, wading, of course, through water of an exceedingly low temperature." . . . "On the 1st of September, still following the ice-belt, we found that we were entering the recesses of another bay but little smaller than that in which we had left our brig. The limestone walls ceased to overhang us; we reached a low fiord, and a glacier blocked our way across it. A succession of terraces, rising with symmetrical regularity, lost themselves in long parallel lines in the distance. They were of limestone shingle, and wet with the percolation of the melted ice of the glacier. Where the last of these terraced faces abutted upon the sea, it blended with the ice-foot, so as to make a *frozen compound of rock and ice.* Here, lying in a pasty silt, I found the skeleton of a musk ox. The head was united to the atlas; but the bones of the spine were separated about two inches apart, and conveyed the idea of a displacement produced rather by the sliding of the bed beneath, than by a force from without. The paste, frozen so as to resemble limestone rock, had filled the costal cavity, and the ribs were beautifully polished. It was to the eye an imbedded fossil, ready for the museum of the collector. . . . I am minute in detailing these appearances, for they connect themselves in my mind with the fossils of the Eischoltz cliffs and the Siberian alluvions. I was startled at the facility with which the *silicious limestone, under the alternate energies of frost and thaw, had been incorporated with the organic remains. It had already begun to alter the structure of the bones, and in several instances the vertebræ were entirely enveloped in travertin.* The table-lands and ravines round about this coast abound in such remains. Their numbers, and the manner in which they are scattered, imply that the animals made their migrations in droves, as is the case with the reindeer now. Within the area of a few acres, we found seven skeletons and numerous skulls; these all occupied the snow-streams or gullies that led to a gorge opening on the ice-belt, and might thus be gathered in time to one spot, by the simple action of the water-shed." In a note, Dr. K. adds, "A reindeer skull found in the same gorge was *completely fossilized*. That the snow waters around Rensselaer Harbor held large quantities of carbonate of lime in solution, was proved not only by the tufaceous deposit which incrusted the masses, *but by actual tests*. The broken down magnesian limestones of the upper plateau readily explain this." . . On the 4th of September, in detailing the incidents of the day or two previous, the Doctor again remarks in a note: "This halt was under the lee of a large *boulder of greenstone*, measuring fourteen feet in its long diameter. It had the rude blocking out of a cube, but was rounded at the edges. The country for fourteen miles around was of the low-bottom series; the nearest greenstone must have been many miles remote. Boulders of *syenite* were numerous; their line of *deposit* nearly *due north and south*."

"Our progress on the 5th was arrested by another bay much larger than any we had seen since entering Smith's Straits. It was a noble sheet of water, perfectly open, and thus in strange contrast to the ice outside. The cause of this, at the time, inexplicable phenomenon, was found in a roaring and tumultuous river, which, issuing from a fiord at the inner sweep of the bay, rolled with the violence of a snow-torrent over a broken bed of rocks. This river, the largest probably yet known in North Greenland, was about three quarters of a mile wide at its mouth, and admitted the tides for about three miles; when its bed rapidly ascended, and could be traced by the configuration of the hills as far as a large inner fiord. I called it Mary Minturn river, after the sister of Mr. Henry Grinnell. Its course was afterward pursued to an interior glacier, from the base of which it was found to issue in numerous streams, that united in a single trunk about forty miles from its mouth. . . I shall never forget the sight, when, after a hard day's walk, I looked out from an altitude of eleven hundred feet upon an expanse extending beyond the eightieth parallel of latitude. Far off on my left was the western shore of the Sound, losing itself in distance toward the north. To my right, a rolling primary country led on to a low, dusky, wall-like ridge, which I afterward recognized as the *great Glacier of Humboldt;* and still beyond this, reaching northward from the north-northeast, was the land which now bears the name of Washington. . . The great area between was a solid sea of ice. Close along its shore, almost looking down upon it from the crest of our lofty station, we could see the long lines of hommocks dividing the floes like the trenches of a beleaguered city. Further out, a stream of icebergs, increasing in number as they receded, showed an almost impenetrable barrier; since I could not doubt that among their recesses the ice was so crushed as to be impassable by the sledge." In a note, in explanation of the previous remark that Humboldt Glacier connected the two continents of America and Greenland, the Doctor observes that all "Arctic America, north of Dolphin and Union Straits, is broken up into large insular masses, and may be considered as a vast archipelago." "Grinnell land," he says, "cannot be regarded as part of the continent of America; while Washington land seems, in physical character and position, to be a sort of middle ground."

In the second volume of Dr. Kane's Arctic Explorations, page 146, some interesting phenomena touching glaciers, are described.

"The bend of this glacier, (that named after Humboldt,) is a few degrees to the west of north. We followed its face afterward, edging in for the Greenland coast, about the rocky archipelago which I have named after the Advance. From one of these rugged islets, the nearest to the glacier which could be approached with any thing like safety, I could see another island, larger and closer in shore, already half covered by the encroaching face of the glacier, and great masses of ice still detaching themselves, and splintering as they fell upon that portion which protruded. Repose was not the characteristic of this seemingly solid mass; every feature indicated activity, energy, movement. The surface seemed to follow that of the basis-country over which it flowed. It was undulating about the horizon, but as it descended toward the sea, it represented a broken plain, with a general inclination of some nine degrees, still diminishing toward the foreground. Crevasses, in the distance mere wrinkles, expanded as they came nearer, and were crossed almost at right angles by long continuous lines of fracture, parallel with the face of the glacier. These lines, too, scarcely traceable in the far distance, widened as they approached the sea, until they formed a gigantic stairway. It seemed as though the ice had lost its support below, and that the mass was let down from *above in a series of steps.* Such an action, owing to the heat derived from the soil, the excessive surface drainage, and the constant abrasion of the sea, must in reality take place. . . . The indication of a great propelling agency seemed to be just commencing at the time I was observing it. These split-off lines of ice were evidently in motion, pressed on by those behind, but still widening their fissures, as if the impelling action was more and more energetic nearer the water, till at last they floated away in the form of ice-bergs. Long files of these detached masses could be traced slowly sailing off into the distance, their separation marked by dark parallel shadows—broad and spacious avenues near the eye, but narrowed in the perspective to mere lines. A more impressive illustration of the forces of Nature can hardly be conceived.

Regarded upon a large scale, I am satisfied that the iceberg is not disengaged by *debacle*, as I once supposed. So far from falling into the sea, broken by its weight from the parent glacier, it *rises from the sea.* The process is at once gradual and comparatively quiet. The idea of icebergs being discharged, so universal among systematic writers, and so recently admitted by myself, seems to me now at variance with the regulated and progressive actions of Nature. Developed by such a process, the thousands of bergs which throng these seas, should keep the air and water in perpetual commotion, one fearful succession of explosive detonations and propagated waves. But it is only the lesser masses falling into deep waters which could justify the popular opinion. The *enormous masses of the great glacier are propelled, step by step, and year by year, until, reaching water capable of supporting them, they are floated off to be lost in the temperature of other regions.*

"The crevasses bore the marks of direct fracture, and the more gradual action of surface-drainage. The extensive water-shed between their converging planes, gave to the icy surface most of the hydrographic features of a river system. The ice-born rivers which divided them were margined occasionally with spires of discolored ice, and generally lost themselves in the central areas of the glacier before reaching its foreground. . . . The height of this ice-wall, at the nearest point, was about three hundred feet, measured from the water's edge; and the unbroken right line of its diminishing perspective showed that this might be regarded as its constant measurement. It seemed, in fact, a great icy table-land, abutting with a clean precipice against the sea."*

We have made these extracts from the work of the lamented Kane, our indomitable countryman, for the purpose of impressing upon the reader the extraordinary extent and nature of the Glacial mountains, seas, and deserts of the North and South poles. No conception of this can well be formed by persons who have paid no previous attention to the subject—they are not prepared to realize the fact that, for *thousands and thousands of miles, nothing but these stupendous glaciers, drifted snows,* and *ice-covered seas* meet the eye of the explorer, and the navigator. The glaciers of the Alps and of other mountains

* Arctic Explorations, vol. ii.

that rear their peaks into the upper atmosphere, although extensive, when considered as belonging to latitudes strictly temperate, are yet the merest nothings when compared with those of the Arctic zones.

Now, it is a singular fact, that after volcanic action had comparatively exhausted itself in the middle belts of the earth, it was *transferred to the poles*. Whether this was due to the fact that, after the deposition of the great continents, the polar circles became the weakest parts of the earth's crust, or whether they still retained their bituminous inflammability, after other districts had parted with it, is a question which it would be useless for me to discuss here. But it is nevertheless true, that toward the close of the Tertiary period and the dawn of the present geological epoch, both the polar regions of the globe simultaneously became the theatres of the most violent and terrific volcanic action, the effects of which we propose briefly to consider.

Dr. Kane, in the extracts we have given, has sufficiently indicated the *primitive geological character* of the arctic country. Whenever he found boulders or erratic blocks, they proved to be *greenstone, porphyry, syenite,* or *amygdaloid*. We have before remarked, in the earlier part of this work, the primary origin of the entire northern portions of the continents of America and Asia. Most of these regions remained undisturbed, covered only with accumulating snows and ice, during all the subsequent geological periods; but after the Tertiary, the whole northwest coast of Greenland, and the northeast border of Baffin's Bay, became the theatre of the most violent volcanic eruptions. An uninterrupted chain of extinct volcanoes stretch all around, and over the entire frozen country explored by our great geographer—a country which presented to his feet the very antipodes of previous heat and present cold. Nor was the volcanic action *con-*

THE ERA OF VOLCANIC ACTION. 401

fined to that particular region. Nearly every portion of the earth, except America, came more or less under its influence; but, as we observed before, the Arctic Zones were in this instance the chief centres of its operations. Along the Asiatic coast of the Pacific, including all the islands outside of the sea of Okhotsk, that of Japan, and of China, many thousands of small islands were elevated above the ocean, most of which were wholly unknown until recently, and every day reveals new ones. Thousands of islands, and groups of islands, some of them by no means inconsiderable in size, have been thus redeemed from the waste of waters since the Tertiary period, or *since the creation of man*. The American coast is scarcely less prolific, but they are all confined to that of the Pacific. The extensive chain of mountains known as the Andes, the Cordilleras, and the Rocky mountains, extending through South America to Oregon, are for the most part volcanic; and although they were in action during previous geological eras, they again poured forth their streams of lava after the Tertiary, and at irregular intervals of time, continue to do so even now. The greater portion of North America—in fact the whole of it, with the exception of the extreme northern parts, and a narrow belt along the Atlantic slope, already mentioned, was unaffected; but nearly the entire surface of Europe was simultaneously convulsed and disruptured from one end of it to the other. Scotland, Ireland, France, Germany, Italy, Austria, and all the northern coast of the Mediterranean, into Asia Minor, and thence east to Persia, *was an almost uninterrupted théatre of volcanic fury*. The Ural mountains, and those of the Altai ranges in Central Asia, were affected to some extent, while nearly the whole of India, and the southeastern slopes of the Indian Ocean, shared in a common and almost universal disturbance. This was emphatically the *era of volcanic action*, both on the land and beneath the seas; and Moses says truly, that "the fountains of the great deep were broken up." With the single exception of North America, the volcanoes were distributed all over the world; and in many portions entire regions were engulfed in their simultaneous action. While thousands of islands were thus rising from the ocean, what was the effect in the polar circles—among the mountains, deserts, and seas of ice? Volcanic eruptions in these regions would, of course, produce peculiar and extraordinary effects. They would not only crack and split the rocks, but inject upward the usual streams of red-hot liquid lava. This, coming in contact with the exterior ice and snow, would hurl the loftiest glacial mountains from their base, and send up columns of boiling water into the freezing air. The very opposites of elemental qualities would thus meet in angry combat. Glaciers, that reared their hoary beards into the clouds, would be precipitated into the adjacent seas—while the ordinary water courses would be swollen into mighty and resistless torrents—carrying with them the red-hot cinders and boulders thown in their way. Where the heat was greatest, the ice and snow would mingle in their original fluidity; but in those points less affected, *mountains of glaciers would slide from the heated rocks,* and be borne off into the seas. The mighty volume of waters, now bearing thousands of these massive icebergs, loaded with earth and debris, are now borne off by the currents of the ocean, which sweep around the continents of Asia, Africa, and America. Those of the North Pole move toward the south; while those of the South Pole move toward the north. The waters rush on, and as they proceed, encounter thousands of terrific volcanoes, bursting up from the oceans—"the fountains of the great deep." The waters still increase—the gigantic-rock-freighted icebergs still multiply The atmosphere of the temperate zones becomes suddenly refrigerated, while that of the poles is, for a time, moderated. The dense moisture of the icebergs envelops the world in fogs and vapor. The clouds of thick sulphurous mist, rise higher and higher, while the icebergs still pour in with every succeeding wave. The moisture is rarified in universal auroras; electricity is generated; heaven's artillery prepares to enter the elemental strife, and detonates its threatening thunder. Anon the "windows of heaven are opened," and then descends rain in ceaseless torrents. Still the waters rise higher and higher—still the long islands of icebergs, laden with rocks, and trees, and mud, pour in from the north and the south—still the waters are enveloped in dense vapor—while in terrific and overwhelming showers—

> "The impetuous rain descends,
> Nor from his patrimonial heaven alone
> Is Jove content to pour his vengeance down;
> But from his brother of the seas he craves
> To help him with auxiliary waves.
> Then with his mace the monarch struck the ground,—
> With inward trembling earth received the wound,
> And rising streams a ready passage found.
> Now seas and earth were in confusion lost,—
> A world of waters, and without a coast."

The crust of the earth now quakes and shudders, while bleeding red-hot scoriæ, from ten thousand gaping pores. The expansive force of the exploding gases—the continuous *upward pressure* of the volcanic currents, draws the water secreted in the bowels of the earth to the surface; all the reservoirs are broken up, and the strata become *disjointed*, *twisted*, and *melted*. The *fusion* of the cavernous and disjointed rocks, occasions their subsequent contraction and solidification, while the waters themselves having suddenly passed from icy coldness to that of boiling heat, *expand to nearly twice their usual volume.* The disrupturing of the earth, and of the floor of the sea, thus liberating the *boiling water and liquid lava of the volcanoes*, added immeasurably to the extent of the flood, while the subsequent subsidence and solidification of the interior crust of the earth, *caused the whole surface of the earth to be overflowed!* The "fountains of the great deep are broken up" indeed, and water, in its most powerful, excited, and irresistible form, again sweeps over the land, and asserts its dominion in the air! The mountains, none of which had yet attained their full-grown height (the Alps, the Himalaya, the Altai, the Urals, Andes, Alleghanies—all were growing in altitude with the varying movements that originally raised them up), were now covered by the triumphant waves. Neptune, enthroned in clouds of rain, with his brother Jove, reigned supreme over the sea, earth, and air!

The people of the Eastern and Middle States are all familiar with the action of ice-freshets upon their rivers—such as the Ohio, the Potomac, the Susquehanna, the Penobscot, Connecticut, etc. The sources of these rivers are in the mountains, which, during the winter, accumulate deposits of snow. In the spring, when the sun sends forth his rays to waken nature from her torpor, these snows are melted. The water, pouring into the main river from a thousand swollen rivulets, produces a sudden enlargement. The ice that covered the streams during the winter is also suddenly broken up by the influx of waters, and then hurried down the agitated stream. Where the river contracts between high banks or hills, the ice gorges or wedges itself across the channel. For a time, the ice and water are both arrested, and then flow back, overleaping the shores, and forming a succession of steps and terraces. When the river has accumulated sufficient water to break through or override the narrow obstruction, the whole mass again moves forward, and bears along with it trees, rocks, fences, and sometimes stables, houses, and bridges. These freshets occur in nearly every river on the globe. "The effusion of only a part of the ices of the Cordilleras in Peru," says St. Pierre, in his Studies of Nature, "is sufficient to produce an annual overflow of the Amazon, of the Orinoco, and of several other great rivers of the New World, and to inundate a great portion of Brazil, of Guiana, and of the *Terra Firma* of America; that the melting of part of the snows on the Mountains of the Moon, in Africa, occasions every year the inundations of Senegal, contributes to those of the Nile, and overflows vast tracts of country in Guinea, and the whole of Lower Egypt; and that similar effects are annually reproduced in a considerable part of Southern Asia, in the kingdoms of Bengal, of Siam, of Pegu, and of Cochin-China, and in the districts watered by the Tigris, the Euphrates, and many other rivers of Asia, which have their sources in chains of mountains perpetually covered with ice, namely, Taurus and Imaus."* It may be assumed that what the occasional extraordinary overflows of these rivers are to the particular regions which they drain, are the still rarer overflows of *oceans* in respect to the *continents* which they drain. The parallel between them may be wide, but it is nevertheless correct; for while the floods of rivers are more *frequent*, those of oceans, we are well assured by geology, have been no less regular and periodical.

But what evidence have we of the submergence of the mountains? We have, *first*, the evidence of the *Bible*, which is reliable beyond all other evidence—beyond the evidence even of our eyes and senses; because all these organs may and do deceive us, but the Bible *never*. Sec-

---

* Studies of Nature, by Bernardin St. Pierre.

ondly, we have the evidence of *Nature*, also a reliable witness; and *thirdly*, the *traditions of man himself, in all parts of the earth*. All these declare the *universality* of the flood, and that all the mountains were alike overflowed.

The Bible, however, does *not* aver that all the mountains overflowed by the deluge were as *high as they are now;* nor, aside from the flood, have we any *reason* to suppose that they were. The Himalaya range, we have reason to believe, did not *exist* as mountains before the Tertiary; neither did the Alps. The Alleghanies rose after the Carboniferous era; but their rise was also gradual: and so, with but a few exceptions, was that of nearly *all* the lofty mountains of the earth. The earth itself was not made in a *day;* and it would be the height of nonsense to claim that the mountains, which form its ribbed axes, were.

But granting, for the sake of argument, that all the mountains *before* the flood were as high as they are *now*—what is gained? All the mountains are or were covered with diluvion—most of them with erratic boulders, and the moraines of icebergs. How are we to explain this phenomenon? Geologists, with universal accord, assume that the mountains have been sunk *under* the level of the sea, and then again *elevated*. In some instances, indeed, this supposed subsidence and elevation has occurred *repeatedly*, at long geological intervals—for in no other way can the alternation of *marine, land*, and *fresh-water strata* be accounted for. In truth, most geologists assign greater stability to the seas than to *terra firma*. "While we have no evidence," says High Miller, "that the sea-level has changed during at least the ages of the Tertiary formations, and absolutely know that it could not have varied more than a few yards, or at most a few fathoms, we have direct evidence that during that time great mountain chains, many thousand feet in height, such as the Alps, have *arisen from the bottom of the ocean*, and that great continents have *sunk beneath it and disappeared*. The larger parts of Northern Europe and America have been covered by the sea since our present group of shells began to exist. In 1819, a wide expanse of country in the delta of the Indus, containing fully two thousand square miles of flat meadow, was converted, by a sudden depression accompanied by an earthquake, into an inland sea. About three years after this event, a tract of country between the Andes and the Pacific, more than equal to all Great Britain in area, was *elevated* from two to seven feet over its former level, and rocks laid bare in the sea, which the pilots and fishermen of the coast had never before seen."*

The sea appears to be receding from the shores of Sweden at the rate of four vertical feet per century; while on the coasts of Greenland (as also remarked by Dr. Kane) it appears to be advancing at a rate somewhat more considerable. But while Miller and all cotemporary geologists thus assume the *repeated* subsidence and submergence of elevated plateaux, mountain ranges, and vast continents, none of them attribute any operating or auxiliary effects to the *flood*. The sea, they maintain, is stationary, and, with the exception of its tidal variations, always maintains its level. Both geologists and Christian-Infidels (if I may coin a severe term) therefore unite in elaborating very ingenious but very flimsy pretexts to utterly destroy the *idea of a universal deluge*, and with it all the co-ordinate ideas of a primary *central creation of man and animals*, and several other leading doctrinal pillars of Revelation. They would thus pull down the whole fabric of Christian faith under the pretext of "*har-*

---

* Hugh Miller, Testimony of the Rocks.

monizing" Revelation with Geology! The foundation of their argument against the universality of the deluge amounts to this: *that there was an insufficiency of water.* They all tell us that the whole earth *originated from the deep*, and that continents were afterward repeatedly *submerged*; but, in the case of this *particular flood*, there could not have been sufficient water to cover the high mountains! This, to be plain, is nothing but the argument of fools! Besides which, it betrays the most consummate hypocrisy.

In his account of Canada and the United States, in 1845, Sir Charles Lyell announced the conclusion to which he had then arrived, "that to explain the position of the erratics, and the polished surfaces of rocks, and their striæ and flutings, we must assume first a *gradual submergence of the land in North America*, after it had acquired its present outline of hill and valley, cliff and ravine, and then its *re-emergence from the ocean.* When the land was slowly sinking, the sea which bordered it was covered with islands of floating ice, coming from the north, which, as they grounded on the coast and on shoals, pushed along such loose materials of sand and pebbles as lay strewed over the bottom. By this force all angular and projecting points were broken off, and fragments of hard stone, frozen into the lower surface of the ice, had power to scoop out grooves in the subjacent solid rock. The sloping beach, as well as the floor of the ocean, might be polished and scored by this machinery; but no flood of water, however violent, or however great the quantity of detritus or size of the rocky fragments swept along by it, could produce such long, perfectly straight and parallel furrows, as are everywhere visible in the Niagara district, and generally in the region north of the 40th parallel of latitude."

The principles of continental subsidence here presented,

the distinguished writer also applies to Europe, and argues the case at considerable length. But, it will be observed, he could not restrain a passing *kick* at the Noachian flood—"no *flood of water*, however great the quantity of its detritus, could produce such long and parallel furrows in the subjacent rock." "This," he says, "was the work of icebergs;"—and who, *for one moment, disputes it?* Does the Bible say or intimate, that there could have been no *icebergs?* Still less does the Bible intimate that there was no *sinking down of continents.* Water was *wanted;* a universal deluge was *decreed;* and it was therefore *necessary* to "break up the fountains of the great deep," not only to obtain *icebergs*, but water in *every form* in which it existed on the earth or in the air.

The subsidence of entire continents is thus admitted, not only by one, but by all geologists, properly so called. Many of them, as I said before, demand the submergence of continents and of particular districts, again and again, so as to account for the alternation of strata. But the submergence in this case was the most recent—the very *last* that occurred on the earth. We *know* that it occurred during or toward the close of the Tertiary, and we *know* that it has *not* occurred during the present historic era of man. The highest mountain ranges on the globe, as the Himalaya, the Alps, and the Andes, are strewn with *Tertiary remains*, and are scratched with *Diluvial icebergs and glaciers*, while their slopes and valleys are *filled with boulders and moraines*. We can thus trace the flood, or the "breaking up of the fountains of the deep," to the age of *Noah*. All co-ordinate circumstances lead directly to *that particular time;* and when thus hemmed in, on all sides, there is *no longer any other alternative but to believe that it occurred exactly as the Bible describes it.* There is no other alternative.

How, then, are we to account for this last submergence of continents? The geologists universally demand their subsidence and submergence, but they do not volunteer to enlighten us as to the *modus operandi*. Lyell presumes that it was "gradual;" all his theories, indeed, are on a graduated scale. I have already intimated the cause to which I refer it, namely, the renewal of *volcanic action, simultaneously, in nearly every portion of the globe.* Volcanic eruptions, such as have already been described, would break up the "fountains of the deep," and this was the preliminary *commencement* of the flood that ensued. Expelling the water from its subterraneous caverns, and consolidating the strata in a general fusion and contraction, would have the effect of *lowering the surface of the earth*, and especially of the higher mountains. We must constantly bear in mind that, while the rocky material omposing the crust of the earth is from ten to fifty or a hundred miles in thickness, the whole mass is extremely porous, and like a sponge, becomes the receptacle of immense reservoirs of water, as already described. We find caverns near the surface, like those of the Mammoth Cave of Kentucky, or Weyer's, or Madison's in Virginia, extending from ten to fifty miles in length. There are many thousands such on the earth, as yet unexplored. It is only when they occur *near the surface* that we obtain any knowledge of them; but we know, from the phenomena of mineral and thermal springs, that many others of a similar character exist at great depths in the earth—far below the system of rocks prevailing near the surface. The effect of volcanic action is to expel the water of such caverns and fissures, compress and solidify the strata, and absorb their gases. At the same time, volcanoes form the nucleus of all mountains, and throw up islands in the midst of the ocean. Mountains are, in fact, nothing but protuberances on the crust of the earth, and in extent, bear the same relation to it that a mere cutaneous pimple or boil bears to a man's body. When the blood becomes stagnant and corrupt, the system endeavors to expel the foul humors through the pores of the skin. This engenders fever or heat; the body is covered with eruptions, expelling putrid matter; but after such eruptions, the cuticle again resumes its natural position. It is thus with volcanic mountains. They are *protuberances*, emitting gaseous matter from a distempered earth, —upon the discharge of which they relapse, or the crust subsides, until new gaseous secretions have accumulated. The natural springs of the earth may be compared to the pores of man's skin—since their obvious office is to *discharge the humors of the interior;* and whenever these become inefficient or inadequate, eruptions are certain to follow, and the *fluid vessels of the earth are proportionally depleted.* Thus whole continents are lowered, and again elevated by expansion; and this, we have every reason to infer, was the condition of the earth during the Noachian deluge.

But, say the doubters: if continents were submerged, during the era of man, the diluvium should contain his remains and those of his associate animals; whereas, there is a very evident scarcity or absence of fossils, of every description, in that formation. This, however, instead of being an objection, is only a *confirmation* of the truth of a previous remark of mine, viz., that the land animals outside of the original circle of the Adamite creation, were extremely few, but that all which were outside, and had no specific representatives in the central creation, necessarily became utterly extinct *after* the deluge. The universality of the flood is not only indicated in this fact, but it also effectually disproves the theory of the *plurality* of creative centres, of which some writers claim from three to six for man, ten to fifteen for animals, and from twenty to thirty for vegetation! If the fossils are scarce or totally absent, (as they generally are,) in the diluvial strata, and confined mainly to species extinct, what becomes of the great bug bear of Geology, as to the *antediluvian fauna having extended all over the earth?* If it really extended over all the continents, *where are its remains to be found?* If man inhabited America *before* the flood, let the geologists show us his fossil remains, or traces of his work. If the cotemporary animals of man, such as were necessary to his sustenance, as the sheep, cattle, fowls, horses, camels, dogs, etc.,—if these inhabited America *before* that event, be kind enough to *let us see their fossil bones.* I am aware that, in South Carolina, a pretended discovery of such bones has recently been made; but it will require something more than isolated and scattered teeth to demonstrate that they really belonged to the domestic animals of man, and that they lived here *anterior* to the Adamite creation. Their remains, if genuine, are found in deposits, and under geological circumstances too closely related to the *present*, to raise them to the dignity of a position from which to contradict the order of physical cosmogony and the revelations of the Creator. There is a large class of "sensation" geologists, who are ever on the *qui vive* for something new and startling. Their little pamphlets, printed at their own expense, but circulated under the ostensible auspices of learned Scientific Academies, are distributed with profuse liberality, and are almost invariably directed against the common target—Moses and the order of divine creation. If the Bible had not been an inspired work, it would have been buried in contempt and oblivion many centuries ago; but fortunately, all the assaults of the devil, and the insidious pamphlets of his Scientific coadjutors, only excite the spirit of *investigation*, which invariably discloses its solemn and benignant truths like the effulgent beams of the sun, peering through the dark and flimsy mists of the horizon.

The extinct species of animals belonging to the diluvial, are well known. They are of a character distinct and separate from most of those now living, although their bones sometimes occur side by side. The gigantic Mastodon, the Castoroides Ohioensis, the Myladon, Capabara, Megalonyx, and others, are strewn over the diluvial plains of the South and West, and sufficiently attest the destructive effects of the deluge. Lyell, however, having found the bones of the Mastodon in fluviatile beds in New York, containing shells of the genera Melania, Lymnea, Planorbis, Cyclas, Unio, etc., all of recent and existing species, goes on to show that the Straits of Niagara cut through these shell-deposits, and that, *therefore*, the Mastodon must be of cotemporary age with the shells, and that all are more ancient than the Falls, the erosion or retrogression of which he estimates at 36,000 years. All this is very astute—very ingenious; but extremely absurd. Has it never occurred to the distinguished geologist, when contemplating the varied and magnificent phenomena of the Falls, upon which he dwells, in all his books, with so much care and seeming pleasure, that the narrow, deep, perpendicular chasm which affords a grudging outlet to the river, has been repeatedly choked up with the ice pouring down from the cold region of the great lakes which it drains? When thus choked up, the surrounding country would be overflowed, and the lake of Erie would expand greatly beyond its ordinary extent. During these repeated cataclysms, the sand and clay in which the shells are found were deposited, and the river, in point of fact, *never cut through them at all*—its narrow channel having been previously cut through the Silurian lime and shale. The surface is alluvial, derived from the lakes in the form of sediment, and scattered over an extensive area by means of small icebergs and muddy slush, in which the shells themselves were transported and deposited. The Mastodons, and other extinct animals of the diluvian era, wandering over the sandy plains, laid down to die, and their bones were afterward covered up by new accumulations of comminuted mud and sand. Who has not observed the sand-banks deposited along the sloping shores of rivers after a recent flood or ice-freshet? Sometimes they are of great extent, and in level districts invade the adjacent country for many miles. The Mississippi often pours its waters, to the depth of eight or ten feet, over the adjacent prairies, not unfrequently twenty or thirty miles from the main channel. The sediment thus deposited is invariably strewn with the characteristic testacea that inhabit the river or its numerous sources. Now, if an old horse should lie down to die on the prairies thus covered with shells and sediment, and his bones afterward be covered over by new accessions of fluviatile silt, no one except geologists of the Lyell school, would have the hardihood to assert that the horse was of the same *age* as the channel of the Mississippi!

The scarcity of animal species, therefore, in the diluvial sands of the flood, proves, if it proves any thing, that there was but *one* original centre of creation, from which but a *few types* had wandered, all of which became extinct immediately afterward, or were reproduced by the progenitors left behind in the ark. Man, and all his works, were confined to their originally limited geographical sphere—the exact location of which it is now impossible to determine. It seems, however, to have been located somewhere between the Red Sea, the Caspian, the Black Sea, and the Mediterranean. Some suppose it to have been further southeast, between the Red Sea and the Persian Gulf, in Syria or Arabia; while others, again, remove it still further east, into India. The greater portion of these countries is a Tertiary and modern formation—having been but recently redeemed from the seas and lakes that drain it. Much of it is sandy desert, or alluvial silt, poured from its great rivers, and whatever remains of the antediluvial era were deposited over the surface, have been obliterated by subsequent races, whose bones and monuments form of themselves the most attractive objects of antiquarian research. In such a region, forming one uninterrupted and exhaustless museum of historical treasure, it would be impossible to identify the relics of ages so nearly joined together; and hence the obscurity which must forever surround the remains of the Adamite race. The present age, with all the appliances which highly developed and inquiring mental capacities could bring to bear, has yet found sufficient material upon which to exercise its antiquarian skill, in the buried ruins of Egypt, in its catacombs, pyramids, palaces, cities, and works of art and internal improvement. When this field shall have been exhausted, another still more ancient, and perhaps equally interesting, may be exhumed from the arid deserts that once formed the estuaries of the Mediterranean, and the Black, and the Caspian Seas.

It may, however, be worth while here to observe, that the hope of ever finding the remains of the antediluvian race, aside from the difficulties already suggested, are rendered still more remote from another cause. Before the flood, the race of man was necessarily limited in number, as well as in geographical range. But sixteen centuries had elapsed from the creation of Adam—all of which, judging from those mentioned in the Bible attained an extraordinary age. Consequently there could not have been many distinct generations, nor could any of them have reared works of art of an enduring and imposing character. Their lives were pastoral, and massive buildings and monuments were unnecessary. But few and limited as the race was in individuals and in works, it is probable that the whole was swept into the sea. If, indeed, the deluge had been stationary and permanent, man's remains might be looked for in the alluvial silt covering his original habitat; but it came on suddenly, and then receded, like the ebbing of a tide. The bodies, obeying the same law which exists now, under similar circumstances, would float off with the tide, and would finally lodge in the bottom of the adjacent seas, or perhaps be borne to great distances in the ocean. Decomposition would ensue with rapidity—for it is a great mistake to suppose, as St. Pierre has shown, that because the waters of the ocean are *salt*, they are therefore preservative of animal matter. Sea-water is saline, but it is not a pickle. If a bottle be filled in a tropical climate, decomposition will soon occur, when it becomes nauseous and putrid. The very saltness of the sea is supposed to proceed from the decomposition of animal and vegetable matter, pouring into it from all quarters of the earth. The ocean is a common sewer or cess-pool, into which all the filth of the earth is emptied, and from which, by the movement of its waves, salt is distilled. It is properly a lixivial, and can, therefore, readily decompose any organic substance introduced into it. If the Adamite race had been wafted into *such* waters, it is plain that all traces of it would soon disappear, and be forever lost to the curious speculations of modern paleontologists and comparative anatomists.

Before leaving this branch of the subject, I shall notice one or two additional points suggested by the anti-Noachians. "Of the creatures that live on vegetables," says Hugh Miller, "many are restricted in their food to single plants, which are themselves restricted to limited localities, and remote regions of the globe. Though these were estimated in 1842, to consist of no fewer than five hundred and fifty thousand species, they might yet be accommodated in a comparatively limited space. But how extraordinary an amount of miracle would it not require to bring them all together into any one centre, or to preserve them there! Many of them, like the myriopoda and the thysanura, have no wings, and but feeble locomotive powers; many of them, such as the ephemera, and the male ants, live after they have got their wings only a few hours, or at most a few days; and there are myriads of them that can live upon but single plants that grow in very limited botanic centres. Even supposing them all brought into the ark by miracle as eggs, what multitudes of them would not, without the exertion of further miracle, require to be sent back to their proper habitats, as wingless grubs, or as insects restricted by nature to a few days of life! Or supposing the eggs all left in their several localities to lie under water for a twelvemonth, amid mud and debris—though certain of the hardier kinds might survive such treatment, by miracle alone could the preponderating majority of the class be preserved. And be it remembered, that the expedient of having recourse to supposititious miracle in order to get over a difficulty insurmountable on every natural principle, is not of the nature of argument, but simply an evidence of the want of it. Argument is at an end when supposititious miracle is introduced."*

The closing sentence of this paragraph is somewhat anomalous. First, we are told, "that by miracle *alone* could the preponderating majority of certain classes of the five hundred and fifty thousand species of insects be preserved in the ark;" and *then* we are reminded that, if we establish their preservation by calling in the aid of miracle, or miraculous agency, (the same thing,) "there *is an end*

* Hugh Miller's Testimony of the Rocks.

to argument. It seems to me that there is an end to argument in the very terms of the quoted sentence. This whole discussion involves miracles. Mr. Miller himself, in all his books, deals in nothing but miracles; for all God's works are such. He never consummated a greater miracle than when he created these very insects. "Man the wonderful" scarcely displays nicer or more miraculous functions than the butterfly, the bee, the ant, or the spider. The earth and the heavens are merely stupendous miracles; and how could the Bible, if it be a faithful revelation of God's acts and commands, be otherwise? It is this very feature, so perfectly consistent with all that we see or know of the Creator, that gives the stamp of *authenticity to the sacred volume.* Mr. Miller, and all the geologists of his school, profess to *believe* in that book; and yet we find them constantly assailing its integrity, *because, forsooth! it daals in miracles!* If its statements were based upon truths such as man would dictate—such as writers like Miller himself would expound—a new and revised edition would have to appear annually—future generations would have to erase the errors and absurdities of the present, and the present those of the past. But the Bible, based upon the solid truths of God's eternal laws, is the authentic record of miracles—miracles connected together as a chain, any single link of which proving false, would precipitate the whole Christian fabric to oblivion. But God's

> "Creation is no less
> Than a capacious reservoir of means,
> Formed for his use, and ready at his will."

But all this is perhaps foreign to the main question. We are to determine, without resorting to "supposititious miracle," how five hundred and fifty thousand species of minute animals, such as flies, spiders, beetles, locusts, etc., were provided for in the ark. *First,* we have no reason to believe that there are, or ever were, five hundred and fifty thousand *species*—the largest and most recent estimate that we have seen not exceeding four hundred thousand. This number, however, is sufficient, and would be equally as formidable as the first, were it not susceptible of very considerable reductions. More than one-half of them are aquatic, *spawning* in mud and water, and therefore require no more attention than fishes, or the cetacean mammalia. This brings us down to 200,000 species, of which number more than one half are burrowers in the earth, and have the faculty of lying dormant during seasons of cold, no matter how long continued, as worms of every description, grubs, ants, and thousands of other minute and nameless creatures, which it would be out of the question here to describe. The whole earth, however, teems with them, and a moment's reflection on the part of my reader will recall myriads of different species to his recollection. Of the remaining 100,000 species, nearly all of them are natives or frequenters of the forest, the orchard, or the garden. They secrete themselves or their invisible larvæ (visible, for the most part, only by the microscope) in the bark of trees, or bore into the solid trunks or roots, and in the partings of decayed stumps and rotten logs. No one who has not had some practical experience of forest life, can form the remotest conception of the almost unlimited extent and variety of the creatures which live and generate upon the trees. They are inseparable from vegetation, and are perhaps essential to terminate the life of the giants of the forest; for there can be no doubt but that their rapid increase ultimately attacks the vitality of trees, and hastens their death. This is familiarly illustrated in the case of fruit and ornamental trees—the ravages of insects, in a single season, destroying not merely single individuals, but entire gardens and parks.

But we have made no provision for that numerous class of insects which infest animals, as fleas, lice, flies, gnats, and various other parasitic bugs and vermin. This class, also embracing many varieties, is equally inseparable from animal life, no matter under what circumstances it may exist. But besides these, there is still another class, by no means less varied or numerous, which are to be found in the crevices of furniture, houses, ships, barns, granaries, and among stores of every description—embracing spiders of all kinds, stylops, moths, ants, wasps, bed-bugs, roaches, and the whole family of Myriopoda cited by Mr. Miller. All these, I repeat, are inseparable from the substances in and around which they are found. Maggots exist in putrid animal matter—in cheese, in vegetables, in earths and soils. They cannot be eradicated, for their larvæ, like that of common flies, is generally invisible. Besides these, there are many species of locusts, caterpillars, beetles, cochineals, dragon-flies, etc., etc., the larvæ of which, like that of all the others, is diffused in the secret pores of vegetable, animal, and earthy substances, and generally lie dormant until wakened into life by heat or moisture.

The simple truth is, that the great majority of these minute creatures, in their larval condition, exist primarily in the *juices of animals and vegetables*, precisely as the microscope shows thousands of distinct creatures inhabiting a *single drop of water*. I have already remarked, in the earlier pages of this book, the microscopic revelations of Ehrenberg and others, (among whom I may include my neighbor, Dr. Wythes, of Port Carbon, who has written a valuable book on the subject, and is perhaps one of the most scientific entomologists of the present age,) by which it appears that a greater number of inhabitants occupy a cubic inch of water than there are human beings on the face of the globe! Water is water, whether it be found in the body of animals, or in trees, plants, and fruits; and it has been ascertained that no matter how carefully it may be distilled, animalcules still exist in it in one form or another. The author of the Vestiges of Creation refers, at considerable length, to experiments made some years ago, in England, by Messrs. Weeks and Crosse, by which insects of the species *acari* were elaborated from solutions of ferrocyanate of potassium, obtained by boiling salt in distilled water, and then subjecting the solution, under closed vessels, in an atmosphere of pure oxygen, to long-continued electric currents. These experiments are cited to prove the soundness of the development hypothesis, which is based upon the assumption that animal life can and has originated otherwise than *ex ovo*. Instead, however, of proving any such nonsense, they prove incontestably what has already been mentioned, and what Ehrenberg, Wythes, and others have demonstrated, that a drop of water is as much a globe to microscopic animalcules as *terra firma* is to man; that while water may be changed, mixed, distilled, or heated, the animals or their larvæ still remain instinct with life; and that they perpetuate their respective species by spawn so infinitesimally minute, that the highest powers of the microscope cannot detect them, but which are *proven to exist in the ovo of the very insects thus generated by the chemical experiments of Messrs. Weeks and Crosse!* But it is an extraordinary fact, that many of these animalcules, as well as certain species of lice which infest vegetation, are not only gifted with fecundating properties to perpetuate themselves, but *transmit such properties to their female descendants for many succeeding generations.* The minute spawn thus generated has, within its individuals, the spawn for *other* generations; and so they may continue to multiply, under the most diverse circumstances that the mind can conceive, without incurring the remotest liability of the ultimate extinction of their species. The celebrated Reaumer has proved, that one single female may become the parent, in five generations, of more than *five thousand nine hundred and four millions of descendants!* The higher species of insects are less prolific, but their nature is better understood. A single fly will generate, in a period of three or four months, a brood of from seven to eight hundred thousand flies; the wasp will deposit thirty thousand eggs, the queen-bee from forty to fifty thousand, and the moth of the silkworm from five to six hundred. It is the wonderful *fecundity* of these animals, and their adaptation to the most obscure and imperceptible retreats and crevices, and the ability of their *infinitely minute spawn to resist all the adverse circumstances of earth, air, and water*, that render the extinction of their species a physical impossibility, so long as earth, air, water, vegetation, and animals remain!

> "Why has man not a microscopic eye?
> For this plain reason: *man* is not a *fly*

And for this very plain reason, it is impossible for *man* to fathom the deep and wonderful mysteries of the great Creator. His works are miracles—large or small, animate or inanimate, gaseous, liquid, or solid;—*all* are wrought with a mechanism at once skillful, mysterious, and incomprehensible.

I am astonished that a writer of the reputed intelligence and scientific acumen of the late Mr. Miller, should have advanced a proposition so perfectly destitute of significance, and reflecting so seriously upon his judgment as a Naturalist. He might as well have undertaken to separate animals or vegetables from the *diseases* to which they are liable, as to separate them from the parasitic insects and vermin which they absorb in water, nourish in their vital juices, and yield up in their putrefying bodies! Yet we find him exclaiming, with a dogmatic and triumphant air, as if his philosophic vagaries had dealt a stunning blow to the blunt and unaffected statements of the Bible: "But how extraordinary an *amount of miracle* would it not have required to bring them *together* into any one centre, or to *preserve* them there!" On the contrary, the *amount of miracle* (if it is estimable by bulk) necessary to bring them together, was extremely small—for nearly all the animals of the Adamite creation, as we have already shown, *occupied their original centre*. The insects dwelt and moved with animals, and fed upon the vegetable and other stores in the ark; while their spawn remained behind in seas and rivers; under shelving rocks, and their burrows in the earth, and in the pores of roots and trees; in perforated stumps and decomposing rubbish, and in nuts, fruits, and plants. Of the four or five hundred thousand estimated species of insects and vermin, it was hardly probable that *special* attention was necessary to the preservation of a *single species*. Even the bee could shift for herself, locked up in the cavities of trees, amid her stores of manufactured sweets. But this, it may be objected, was in conflict with the divine decree, which ordered *all* animal life to be extinguished from the earth —"all flesh wherein there is life." So far as the *living creatures* themselves are concerned, the order was obeyed; but it could have had no application to life in embryo. The myriads of spawn *in* the earth, and in the recesses of vegetation, were not living animals—they required the nourishing care of nature for their future development, and would only spring into existence with the renewal of vegetable and animal life. But the great bulk of the higher class of insects swarmed to the ark with the animals and their forage, and had their larvæ secreted long in advance of its embarkation. But inasmuch as nothing was said of fishes and cetacea, we have a right to assume that insects stood in a similar relation, especially in view of the fact that the great bulk of them are aquatic and amphibious. Knowing their inseparable identity with vegetation and animals, the Creator was incapable of giving an order to his servant Noah, which it would have been difficult for him to carry out in any other way than that already suggested. The object was to annihilate *man*, and the order of mammalian animals more directly associated with the economy of the earth. But " myriads of them," continues Miller, " can live upon but *single plants*, that grow in very limited botanic centres." If they can *only live* on certain plants, where would be the propriety of their *removal?* And does it not follow, as a logical sequence, that if they can *only* live on such plants, they derive their vitality *from* them—that the spawn exists in the sap of such trees or plants, while the adult lives as a parasite upon their exudations? Most trees were not destroyed by submergence; or, at all events, their roots remained, even though their trunks were detached, and scattered over the ground. No injury, therefore, could result to the larvæ secreted in their pores or beneath their bark. But besides all this, *water alone* would in most cases not injuriously affect the ova or spawn of the great bulk of insects and vermin. Many of them are incubated by moisture and heat; and the large family of fleas, musquitoes, and various bugs, seem to prefer such situations, and are often found swarming in countless myriads over stagnant pools or along the sea coasts, and in marshy swamps and meadows. " But," he continues, "even supposing them all brought into the ark by miracle as eggs, what multitudes of them would not, without the exertion of *further miracle*, require to be sent back to their proper habitats as wingless grubs, or as insects restricted by nature to a few days of life!" It is, indeed, painful to notice such innuendos, because they reveal the melancholy truth that, under the pretext of *harmonizing God's law with "science,"* the writer has no proper conception of one or the other. When men betray such consummate ignorance of the ordinary economy of nature, and of the wonderful powers of recuperation possessed by its microscopic creatures, it is hardly a matter of astonishment that they should, with high-sounding words, shows of superficial learning, and with affected zeal and boldness, undertake to question the plainest decrees of the Creator. Although "wingless grubs" exist in cheese, walnuts, fruits, and animal putrefactions, and millions of other similar worms and crawling grubs in filth of every description; yet this Naturalist is puzzled to know how Noah could have brought them into the ark, and then how he afterward distributed them over the earth! If Noah himself was puzzled at any thing, it was to know how he could keep them *out* of his ark; and if he had a moiety of the common sense usually allotted to human nature, he never gave himself the least uneasiness as to their subsequent geographical distribution. I may here remark, what has already been noticed in my observations on coal, that every successive stratum of the earth appears to be impregnated with the seeds of its own peculiar species of vegetation; and on consulting the Bible, I find nothing to contradict such a theory, but much to confirm it. The coal measures, wherever they outcrop, are prolific in ferns; and I find precisely the same varieties imbedded in the rocks below! It may be assumed that the Carboniferous strata are the true native soils of ferns and all coniferous trees; and that wherever these soils are brought to the surface, and unaffected by artificial culture, such trees will spontaneously spring into existence, and constitute the prevailing vegetation. It may be objected that the seeds could not be *preserved* in the earth for such an indefinite and incalculable space of time, and that they would become fossilized with the limbs and trunks of the trees. But even in the hardest argillaceous rocks, and more especially in soft slates and shales, what we regard as *stone* is merely baked comminuted *mud;* and we know from experience how the crevices and the interior surfaces of rocks clothe themselves in the most remote and loftiest situations, surrounded by water or by glacial mountains, with vegetation corresponding in every particular to the ancient species. And since God nowhere alludes to new creations of vegetation until he "planted the garden eastward in Eden," and thus introduced the varied fruits and flowers which the world now enjoys, we have abundant reason to believe that every geological formation has its peculiar species of plants. I therefore lay it down as a theory, based upon my personal observation, that every system of rocks furnishes now, as it did in the past, its peculiar plants, affected only by changes of climate; and it is in consequence of this fact that we have the *unlimited diversity of species and genera* which now distinguishes the earth's crust. And should further investigation corroborate this hypothesis, the varied centres of vegetable creations claimed by nearly all botanists, like many other similar propositions that conflict with the Bible, will be forever *nailed to the counter*. After the creation of man, when the strata of the earth were thrown together upon the surface of the earth, in consequence of previous disturbance, God foresaw the intermixture of the seeds thus primarily imbedded, and hence decreed man to the *task of cultivating the ground and subduing it*. This intermixture of seeds must forever continue, so long as water runs, winds blow, or birds and animals migrate. It is a *law*— a miracle of nature, decreed from the beginning.

Although I have already dwelt on this branch of my

subject at greater length than I intended, another proposition in Mr. Miller's theory of the deluge, and the one upon which he evidently most relies, common courtesy—*ad hominem argumentum*—requires me to notice. "The great continents," says Cuvier, "contain species peculiar to each; insomuch, that, whenever large countries of this description have been discovered, which their situation had kept isolated from the rest of the world, the class of quadrupeds which they contained has been found extremely different from any that had existed elsewhere. Thus, when the Spaniards first penetrated into South America, they did not find a single species of quadruped the same as any of Europe, Asia, or Africa. The puma, the jaguar, the tapir, the cabai, the lama, the vicuna, the sloths, the armadillos, the opossums, and the whole tribe of sapagos, were to them entirely new animals, of which they had no idea. Similar circumstances have recurred in our own time, when the coasts of New Holland and the adjacent islands were first explored. The various species of kangaroo, phascolornys, dasyurus, and perameles, the flying phalangers, the ornithorhynchi, and echidnæ, have astonished naturalists by the strangeness of their conformations, which presented proportions contrary to all former rules, and were incapable of being arranged under any of the systems then in use."

Mr. Miller quotes this paragraph of the Baron Cuvier, and then adds: "And it is a most significant fact, that both in the two great continents and the New Zealand Islands, there existed, in the later geologic ages, extinct faunas that bore the peculiar generic characters by which their recent ones are still distinguished. The sloths and armadillos of South America had their gigantic predecessors in the enormous megatherium and mylodon, and the strangely-armed glyptodon; the kangaroos and wombats of Australia had their extinct predecessors in a kangaroo nearly twice the size of the largest living species, and in so huge a wombat, that its bones have been mistaken for those of the hippopotamus; and the ornithic inhabitants of New Zealand had their predecessors in the monstrous birds, such as the dinornis, the optornis, and the palapteryx—wingless creatures like the ostrich, that stood from six to twelve feet in height. In these several regions, *two generations of species*, of the genera peculiar to them have existed—the recent generation by whose descendants they are still inhabited, and the extinct generation, whose remains we find locked up in their soils and caves. But how are such facts reconcilable with the hypothesis of a universal deluge?"

Upon the premises thus surveyed, Mr. Miller proceeds to argue against the probability of these animals having been collected into the ark by Noah; or, if they had been, by some miraculous means unknown, he considers it inadmissible that they should *afterward* have been *returned* to the several islands and continents previously inhabited, without the exercise of still greater miracle. In short, he regards the whole thing as physically impossible, but *die novello tutto par bello*. The immortal bard has said that "Truth is stranger than fiction;" to many persons it is not only a *stranger*, but it is not half so fascinating *as* fiction. If, indeed, the animals now inhabiting these regions were in the ark at all, it needs no argument to prove that they, or their progenitors, afterward *returned*. But this is not the point; the main question at issue divides itself into three parts: *first*, were the extinct *species* (of the genera) really identical, and the "predecessors" of those now inhabiting those continents; *second*, did the extinct *species* really live *before* the flood, in the localities where their remains are found; and *third*, if they lived *before* the flood, does it follow, and must we therefore believe, that they were *unrepresented* in the original Adamite creation?

" The sloths and armadillos of South America had their gigantic predecessors in the enormous megatherium (and mylodon), and the strongly-armed glyptodon." This sentence is evidently intended to pass current for more than its exact value. They had their "predecessors"—that is to say, they were preceded by animals of *a similar nature*. It implies a *near relation;* because, if there *were* no such relation, the fact of the one class preceding the other would have no significance. We say it *implies* a near relation, but the writer has omitted to point it out. He says nothing of the habits or anatomical structure of any of them. Where, then, is the propriety of creating inferences not sustained by *facts?* The animals may or may not be allied by generic features; if they are, they should have been pointed out. A dog resembles a horse *because* it walks on four legs. Here we see the *exact amount* of the resemblance—its beginning and its end. The lion and the tiger, the bear and the cat, the panther and the leopard, are equally *members* of the feline tribe, the *generic* features of which embrace a very large number of different species, and *all* of them distinct from each other. Now, (according to the implication of Miller's proposition,) if a menagerie traveling through South America, and including in its collection the Bengal tiger, the African lion, the chetah, the cat, and the leopard, the zoologist would be justified in asserting that the native puma and the jaguar, living or dead, were the "*predecessors*" of these animals—and that, in *consequence of their visit to South America*, they attained a *nearer relation* to the puma and the jaguar than that *previously* held when separated by the Pacific ocean? In other words, according to Miller's premises, the fact of their meeting on *common ground* is sufficient to assimilate species! Accordingly, it is only necessary to remove the South American tapir to Africa, in order to establish its *identity* with the elephant! But let the tapir be extinguished, and its bones be scattered in caves and alluvial silt, and then bring the elephant to South America, it will be easy to show, (by this kind of reasoning,) that the tapir was the "predecessor" of the elephant; and, agreeably to the *development hypothesis*, (or *vice versa*, the *degradation theory*,) it would follow that the one was preceded by, and descended *from* the other!

Having already referred to and briefly described all these animals, it would be a useless repetition to perform the same task again. The armadillos bear considerable resemblance to the manis, a lizard-like ant-eater, though they themselves also eat roots and vegetables. They are distinguished *as burrowers in the ground*, a characteristic utterly incompatible with the extinct Glyptodon, in consequence of its enormous dimensions. The Sloth, on the contrary, *lives on trees*, and travels in its native forests, from one to the other, by means of the interlacing branches; but unlike squirrels and other climbers, it suspends *itself from the limbs of trees, and travels like a fly suspended from a ceiling!* It is a comparatively small animal, covered with long hair, and presenting some remote resemblance to a baboon! Yet this is the worthy descendant of the great *Megatherium* and *Mylodon*—animals that were fully as large as existing elephants and rhinoceri! *They* would make a very beautiful figure in traveling through the forests upside down, suspended from the bending limbs of trees! The resemblance in habit and structure of the armadillo to the extinct *Glyptodon*, is equally striking and pleasing to contemplate! The bones of the *Glyptodon* are so enormous that they were long supposed to be those of the *Megatherium*. We can accordingly imagine how it would burrow in the ground, and bury itself, like its supposed pigmy descendants, in the bowels of the earth!

The resemblance between the living animals and the extinct species is about as striking as that between the hog, the tapir, and the elephant. The hog has a thick skin;—so has the elephant. The tapir has a proboscis; —so has the elephant;—*ergo*, they all belong to the same *general family*. The *Glyptodon*, like the existing armadillo and the manis, was covered by a scaly coat of mail, but in this respect they all resemble alligators or marine turtles as much as they do each other! Their anatomical structure exhibits few features in common; and it would require the highest powers of the imagination to detect the smallest identity between *any* of them, either as respects instincts, habits, or structure! From these simple facts, the reader will perceive how extremely obscure are the inferences of comparative anatomists and zoologists when dealing with the scattered remains of extinct animals; and how perfectly unauthorized and reckless some writers are in predicating the most stupendous speculations upon such unsubstantial data. The classifications of anatomists, botanists, and geologists, we have already shown, are very arbitrary and diverse—based as they are, upon dental formula, or that of the hoof, the skin, the skull, or other general or specific features. But notwithstanding this, a certain class of writers, whenever the Bible is to be assailed, rear theories high as heaven—employing no other basis than an old tooth, a jawbone, a fossil fern, or the tail of a fish or a serpent! Under the technical flummery of science, the most monstrous absurdities—(absurdities, in the present case, only equaled by the giant proportions of the extinct animals,) pass current in the world, and are as eagerly swallowed and believed in as the panaceas, sarsaparillas, and life-pills of the **medical** empirics! They are all gulped down with **avidity,** and the only difference between them is, that while one disorders the *body*, the other corrupts the *mind;*—one spreads contagion in the *blood*, the other diffuses it into the very *soul!*

But did the extinct species really live *before* the flood, in the several localities where their fossil remains are found? Miller intimates very broadly that they did, by saying that "their gigantic remains are locked up in their soils and caves." After the examples we have had of looseness and recklessness of expression, leading to erroneous conclusions, it is absolutely necessary to be on our guard. "Locked up in *their* soils and caves." What, "in the names of all the gods at once," can this mean? Can any human being determine *what* particular soils are here meant; and as for caves, he may as likely contemplate those of the Paleozoic as the Secondary or Tertiary, and either of them would be equally *theirs*. But were they locked up in *caves* of any kind? If so, they must have been enormous ones, with openings infinitely larger than a Pennsylvania barn-door! But as Mr. Miller has not pointed out the *kind* of soil and caverns in which their fossils are locked up, we shall have to consult some other writers, who may have been a little more explicit. Prof. Richardson, in his "Introduction to Geology," Buckland, in his "Bridgewater Treatise," and Lyell, in his "Elements," may throw a glimmer of light on the subject, or lend us a key by which to unlock these mysterious "caves" and "soils" of Geology. From these respectable writers we learn that not only the Megatherium and Glyptodon, but the *Chlamydotherium, Hoplophorus, Pachytherium, Euryodon,* and *Xenurus,* as well as the *Scelidotherium,* and *Platyonyx,* and the *Cœladon, Sphenodon,* are all "locked up," some of the smaller animals in the caverns, and the larger ones in the soils of South America, answering to the upper strata of the *sub-àpen-*

nine group of the Mediterranean. The sub-apennine group is a term bestowed by Brocchi, an Italian geologist, who investigated the argillaceous and sandy deposits, replete with shells, which form a low range of hills, flanking the Apennines on both sides, from the plains of the Po to Calabria. The deposits embrace strata of different ages, the *oldest* of which are *newer* than the Tertiary basins of London and Paris. The upper strata, as well as some intermediate, contain shells of *recent species*, both of marine and fresh-water origin. The shores of the Mediterranean often exhibit the *washings and silt of rivers*, intermixed with the debris of the sea. In South America, the same formation is represented along the great rivers discharging into the Atlantic and the Pacific, like the adjacent plains of the Mississippi, in North America. "The seven hills of Rome are composed partly of marine tertiary strata of the older Pliocene period, and partly of superimposed volcanic tufa, on the top of which are usually cappings of a fluviatile and lacustrine deposit. Thus, on Mount Aventine, the Vatican, and the Capitol, we find beds of calcareous tufa with incrusted reeds, and recent terrestrial shells, at the height of two hundred feet above the alluvial plains of the Tiber. The tusk of the mammoth has been procured from this formation, *but the shells appear to be all of living species.*"* It is a singular coincidence that, before the building of Rome, the Almighty should have strewn the humble but highly sculptured shells of the sea, and the remains of terrestrial life upon the spot since made famous and classical with the works of man! Beneath the broad dome of St. Peter's and the massive walls of the Vatican, with its five hundred grand stairways, its spacious saloons, and wide avenues—its walls animated with the *chef d'œuvres* of ancient art, and its niches glit-

---

* Lyell's Elements of Geology.

tering with statuary and anaglyphs of marble, and brass, and gold;—beneath all this, the Creator had written *his* sermons in stones, as if to teach his vicegerent a practical lesson in Natural Theology. The very rock upon which the Church of Peter was built, is a rock full of cosmical *witnesses*—all of which can bear testimony more ancient than the bulls of the Popes, to the goodness, grandeur, power, and unbounded wisdom of the great Creator.

Prof. Lyell, after stating that an analogy exists between the skeletons of the Megatherium, Megalonyx, Glyptodon, Toxodon,* and other extinct forms, and the living sloth, armadillo, cavy, capybara, and lama of South America (from which statement Miller undoubtedly borrowed his ideas, and took the liberty of enlarging very considerably upon them), goes on to remark:

"That the extinct fauna of Buenos Ayres and Brazil *was very modern*, has been shown by its relation to deposits of marine shells, agreeing with those now inhabiting the Atlantic; and when in Georgia, in 1845, I ascertained that the Megatherium, Mylodon, Harlanus Americanus (Owen), Equus Curvideus, and other quadrupeds allied to the Pampean type, were posterior in date *to marine shells belonging to forty-five recent species of the neighboring sea. However modern, in a geological point of view, we may consider the Pleistocene epoch, it is evident that causes more general and powerful than the intervention of man have occasioned the disappearance of the ancient fauna from so many extensive regions. Not a few of the species had a wide range: the same Megatherium, for instance, extended from Patagonia and the river Plata, in South America, between latitudes 31° and 39° south, to corresponding latitudes in North America, the same animal being also an inhabitant of the intermediate country of Brazil, where its fossil bones have been met with in caves (of the modern era).* The extinct elephant, likewise of Georgia (*Elephas primigenius*), has been traced in a fossil state northward from the river Alatamaha, in latitude 33° 50' north to the polar regions, and then again in the Eastern Hemisphere from Siberia to the south of Europe. If it be objected that, notwithstanding the adaptation of such quadrupeds to a variety of climates and geographical conditions, their great size exposed them to extermination by the *first*

---

* Lyell's Elements of Geology, and Principles of Geology.

hunter tribes, we may observe that the investigations of Lund and Clausen in the ossiferous limestone caves of Brazil have demonstrated that these large mammalia were associated with a great many smaller quadrupeds, some of them as diminutive as field mice, *which have all died out together, while the land shells, formerly their cotemporaries, still continue to exist in the same countries.* As we may feel assured that these minute quadrupeds could never have been extirpated by man, so we may conclude that all species, small and great, have been annihilated one after the other, in the course of indefinite ages, by those changes of circumstances in the organic and inorganic world, which are always in progress, and are capable, in the course of time, of greatly modifying the physical geography, climate, and all other conditions on which the continuance upon the earth of any living being must depend."

I have thus taken the pains to contradict the unauthorized inferences contained in the propositions of Mr. Miller, *by the very authority from which he seems to have drawn his ostensible facts* to destroy the idea of the universality of the flood! It will be seen that not only are the remains of these extinct animals found in the *modern alluvial strata*, but that Prof. Lyell deems it *necessary to dispute the theory of their extinction by man*—a supposition which is by no means groundless, as I shall soon demonstrate. The evidences of the flood are found in nearly every instance *intermediate between these strata*— that is to say, between the post-Pliocene and the more modern of the geologists. The fact is, that nearly *all* the extinct animals that ever breathed upon the *land* are embraced within *these very strata*; and as the boulder formation, or the deluge, occurs *during* this period, the fact is sufficient and irresistible in establishing the *universality of its prevalence.* Sir Charles Lyell wrote without any reference to theological bearings; but he has unconsciously accumulated an amount of evidence which tends powerfully and directly to the phenomena of the Noachian deluge, notwithstanding the perversions to which cotemporary writers have diverted his facts.

Whether man had any agency in the extinction of the race of giant animals which inhabited the earth in the more remote ages, is a question upon which we can do little more than speculate. One thing, however, is certain, that the aborigines of America have traditions, handed down from father to son, of the existence of a race of quadrupedal monsters. I have in my library Jefferson's "Notes on Virginia, for the use of a Foreigner of Distinction"—being a copy of the original edition as printed for private circulation. I need not say that I value this book highly, in consequence of its direct emanation from one of the greatest patriots and sages that the world has ever produced. Jefferson, while President of the American Philosophical Society, made a collection of the bones of the Mastodon, as found on the banks of the Ohio, and, conjointly with Dr. Franklin and a few others, may be said to have been the pioneer in planting the Natural Sciences in the New World. On page 69 he says, "Our quadrupeds have been mostly described by Linnæus and Mons. de Buffon. Of these, the Mammoth, or big buffalo, as called by the Indians, must certainly have been the largest. Their tradition is, that he was carnivorous, and still exists in the northern parts of America. A delegation of warriors from the Delaware tribe, having visited the Governor of Virginia during the present revolution, on matters of business, after these had been discussed and settled in council, the Governor asked them some questions relative to their country, and among others, what they knew or had heard of the animals whose bones were found at the Salt Licks, on the Ohio. Their chief speaker immediately put himself into an attitude of oratory, and with a pomp suited to what he conceived the elevation of his subject, informed him that it was a tradition handed down from their fathers, 'That in ancient times, a herd of these tremendous animals came to the Big-bone Licks, and began a universal destruction of the bear, deer, elks, buffaloes, and other animals which had been created for the use of the Indians: that the Great Man above, looking down and seeing this, was so enraged that he seized his lightnings, descended on the earth, seated himself on a neighboring mountain, on a rock of which his seat and the print of his feet are still to be seen, and hurled his bolts among them, till the whole were slaughtered, except the big bull, who, presenting his forehead to the shafts, shook them off as they fell; but missing one at length, it wounded him in the side; whereupon, springing round, he bounded over the Ohio, over the Wabash, the Illinois, and finally over the great lakes, where he is living at this day.'"

The bones of the Mastodon, thus referred to, occur in the same relative formation as those of the Megatherium, Glyptodon, and the other extinct species of South America. They are found scattered along the alluvial beds of rivers, in meadows and salt marshes,—sometimes covered over with sand, mud, and decayed leaves and mould to the depth of from three to ten and twenty feet. It was the opinion of Mr. Jefferson, after careful inquiry, that the Mastodon was still living, and inhabiting the northern regions of America. Several persons of intelligence, who had been taken prisoners by the Indians, and in their travels discovered bones of the Mastodon in remote regions of the country, learned from them that the living animals had been seen by their ancestors, and from their descriptions supposed it to present a close resemblance to the elephant or the hippopotamus. Mr. Jefferson also shows, with Mr. Lyell already quoted, that the remains of the animal have been found strewn over the earth very far to the north and west.

That all these animals lived within the recollection of the aborigines of America, is sufficiently clear in the fact that human bones and specimens of human art have been found in cotemporary geological formations—namely, the post-Pliocene and alluvial. In the West Indies, in the Island of Guadaloupe, human skeletons have been found imbedded in solid limestone. The stone is extremely hard, and chiefly composed of comminuted shells and corals, of species now living in the adjacent ocean. The coral reefs around the island are worn down by the action of the waves, and the detritus thus produced is swept upon the shore in the form of plastic mud, which afterward hardens on exposure to the atmosphere, or by a union with the waters of streams and rivers holding carbonate of lime in solution. The skeletons, strewn over or washed out of the land adjacent to the rivers and the ocean, have been thus infiltrated with the coralline mud and limestone, and in process of time the whole mass became hard and solid. The skeletons, from certain peculiarities of the skull, have been pronounced to be those of ancient Peruvians; and some of them have been found in a sitting posture, which was the usual mode of interment adopted by that race. With the skeletons, and around them, are found arrow-heads, fragments of pottery, and other articles of human workmanship. A limestone, with similar human contents, has been formed, and is still forming, in St. Domingo; and Prof. Lyell says that rocks still more ancient, referable to the Post-pliocene, are also to be found in the West Indian Archipelago, as in Cuba, near Havana, and in other islands, in which the shells are identical with those now living in corresponding latitudes.

The mounds so profusely scattered over the plains of the West, especially in the valleys of the Ohio and the Mississippi, bespeak a race of great antiquity, and of some mechanical skill, but of which the present race of Indians profess to be altogether ignorant. Some of these mounds, and the ancient fortifications near them, I have examined. One of the largest on the Ohio stands on a broad terrace of that river, twelve miles below the city of Wheeling, in Virginia. The borough of Moundsville derives its name from it. It is sixty-nine feet in height, and nine hundred in circumference at the base. It stands perfectly isolated, like a vast dome. It was overgrown with trees, one of which, according to the annual rings of growth, betrayed an age of over five hundred years. It was a white oak, and occupied a place on the very summit of the mound, which has a flat area of fifty feet in diameter. In driving drifts through the mound to ascertain its contents and the nature of the work, several chambers were found, in which were human bones. The necks of the skeletons were surrounded by many hundreds of ivory beads, and the wrists by copper bracelets. Sea-shells of the involute species were also strewn around, and were probably worn as ornaments. Isinglass or mica, in large plates, also occurred around and over the skeletons. Besides the bones, there were large deposits of what seemed to be the burned remains of bodies—indicating that the great bulk of those deposited therein had been consumed by fire before interment. In addition to the relics in this mound, a number of others have been found in the neighborhood and all over the western country, many of them associated with bones and skeletons more or less decayed. The proprietor of the mound land found, about two miles from it, a number of porcelain beads, in substance much resembling the artificial teeth of dentists. He had also an image carved in stone, which he found with other relics some eight miles distant. It is in human shape, sitting in a cramped position, the *face and eyes projecting upward*. The nose is what is called *Roman*. On the crown of the head is a knot, in which the hair is concentrated and tied. It is eleven inches in height, but, if it were straight, would be at least double that length. The head and features evince no inconsiderable skill and ingenuity, and, as a work of art, it far surpasses any similar efforts of the modern Indian.

But the most interesting object of antiquarian research, is a small flat stone, inscribed with alphabetic characters, which was disclosed on the opening of the mound. These characters, Mr. Schoolcraft believes, are in the ancient rock alphabet, of sixteen right and acute-angled single strokes, used by the Pelasgi, and other early Mediterranean nations, and which is the parent of the modern Runic, as well as the Bardic. The existence of this ancient art here, could hardly be admitted otherwise than as an insulated fact, without some corroborative evidence in habits and customs, which it would be reasonable to look for in the existing ruins of ancient occupancy. It is thought some such testimony has been found. Rude towers of stone, commanding a view of the adjacent plain and river, are to be found within the distance of a few miles back of the mound. They were, no doubt, used for look-outs to descry the approaches of an enemy. The towers were surrounded with circular walls, all of which have long since fallen to the earth, leaving only the merest traces of the ancient fortifications. Several polished tubes of stone have been found in one of the lesser mounds, the use of which is not very apparent. One of these is twelve inches long, one and a fourth inches wide at one end, and one and a half at the other. It is made of a fine, compact, lead-blue, steatite mottled, and has been constructed by boring, in the manner of a gun barrel. This boring is continued to within about three-eighths of an inch of the larger end, through which but a small aperture is left. If this aperture be looked through, objects at a distance are more clearly seen. Whether it had telescopic, or other powers, the degree of art evinced in its construction is far from rude. By inserting a wooden rod and valve, this simple siphon would be converted into a powerful syringe. Besides the mounds, numerous remains of ancient fortifications occur in various places in the West, especially near the city of Portsmouth, in Ohio, and on the upper Monongahela, in Virginia, and throughout the State of Illinois. They indicate a race of great antiquity, and of much higher cultivation than the Indians, who really appear to have no knowledge or traditions concerning them.

But one of the most interesting antiquarian discoveries of the present age, is that of the ancient burial grounds in Chiriqui, in the State of New Granada, on the narrow isthmus between the Atlantic and Pacific Oceans. With the bones of an ancient, and perhaps unknown race, this discovery reveals at once an apparently very rich auriferous and archæological treasure. A year since, the newspapers teemed with accounts of the curious golden images and metallic anaglyphs found interred with the bodies. The discovery was made in the month of June, 1859, and is thus described:

"A native of Bugalita, a small town in the district of Boqueron, in the province of Chiriqui, in New Granada, while wandering through the forest in the vicinity of his cabin, encountered a tree which had been prostrated by a recent tempest, and underneath its upturned roots he perceived a small earthen jar. Upon examination, this proved to contain, wrapped in swathing of half-decayed cloth, divers images, of curious and fantastic shape, and of so yellow and shining a metal that he at once suspected them to be gold. Knowing that he was in the midst of an ancient Indian *huaca*, or burial ground, he immediately commenced an exploration of the little burial mounds which were on every side, suspecting that they also might contain treasures of a like character. The result was that in three or four days, he succeeded in exhuming no less than seventy-five pounds weight of these golden images. Not entirely confident as to the quality and value of the metal, he disclosed his discovery to his neighbors, and in less than a fortnight, over a thousand persons were at work, and obtained over nine arrobas (or 225 lbs.) of images, most of which proved to be of the finest gold."

As yet, the antiquity of these sepulchral remains is involved in obscurity. That they date back to a very early period, there can be no question; for the present tribes of Central American Indians have no knowledge whatever of the huacas which abound throughout the whole country, and are alike ignorant of the art of making the images, or of the source from which the gold was obtained.

In a history of New Granada, by Colonel Joachim Acosta, mention is made of a similar discovery by the Spaniards in the fifteenth century. This was at Zenu, in the province of Antiochia, in New Granada—the golden images being in all respects similar. After speaking of the riches of the Indian burial grounds at Zenu, the historian observes: "The cemetery of Zenu was composed of an indefinite number of mounds of earth, some of a conical form, and others more or less square. When an Indian died, it was the custom to dig a hole capable of containing his arms and jewels, which were placed in the left-hand side of his grave, looking toward the east; and around these were placed earthen vases containing *chichi* and other fermented drinks; also Indian corn, and stones to pound the same; also his wives and slaves (if he was a principal man), which last thoroughly intoxicated themselves previously to the interment; and then the whole was covered over with a species of red earth brought from a distance. Then the mourning commenced, which lasted as long as there remained any thing to drink; while, in the mean time, the mourners continued to throw earth upon the grave; thus it was elevated according to the ability of the individual or family to provide a greater or less quantity of liquor." Jewels of gold, in large or small quantities, were found in all the tombs. In some were golden figures representing every species of animals, from man to the ant, and sometimes to the value of ten or twenty thousand dollars.

Mr. E. G. Squiers, who formerly represented the United States as *Chargé d'Affaires*, in some of the Central American states, and who is well known for his learning and antiquarian researches, has written a book on the antiquities of that remarkable region. In a recent communication, he says that it "is a mistake to suppose that the occurrence of these images, in considerable numbers, in the Indian graves of that district, is a late discovery. Large quantities have been taken out, from time to time, for many years past; and he was informed by the late Governor of the Bank of England, that several thousand pounds' worth were annually remitted from the Isthmus, as bullion, to that establishment. Most of the figures which he saw at the bank were exceedingly rude, but there were a few among them of relatively fine design and good workmanship. None, however, were quite equal in either respect to some relics dug up during the construction of the Panama railway, and which were found about seven miles from Panama, on the left bank of the Rio Grande, six feet below the surface of the ground. Trees between two and three feet in diameter were growing over them.

As to the origin and date of the golden relics of Chiriqui, Mr. Squiers thinks there can be no doubt. Columbus, when he discovered Chiriqui Lagoon, in his fourth voyage, found all the chiefs and important people decorated with these and similar ornaments, which, he says in his relation, gave him "great promise of the richness of the country in gold and silver." Hence he named the district *Castilla del Oro;* and hence the coast came to be known as *Costa Rica*, or rich coast—a name still preserved as that of the State of Central America adjoining the Isthmus. He mentions particularly, among the ornaments worn by the chiefs, great plates or *mirrors of gold*, suspended on their breasts, " which they would neither sell nor exchange." Columbus adds, that the Indians cast gold with some degree of skill, " but in no way equal to the Spaniards." He says also, that " in all the regions around Veragua, the Indians inter with their chiefs, when they die, all the gold which they possess. " Thus it is," he continues, in a moralizing strain, "that all men seek gold; they barter all they can spare of their produce for gold; gold is excellent; with it they lay up wealth here; and they even take it to their grave, as a comfort to their souls hereafter. Alas! for the follies of men who know not that gold is only valuable in its use, not in its accumulation."

The images and anaglyphs thus far brought to light, amount to several thousand. In their general character they bear a very striking resemblance to the designs and figures of the ancient Egyptians, Assyrians, and Babylonians—such as are found rudely carved in relief upon their obelisks, or on the alabaster slabs with which they paneled the apartments and broad avenues of their palaces. The fact of nearly all their artistic designs being based upon the *animal creation*, instead of the varied phases of the human countenance, and symbolical of human thoughts, acts, and events, seems to imply that, like the Egyptians and some other nations of the remote antiquity, most of the animals were held in *religious veneration*, and embodied some of the functions attributed to gods—hence they may have served as household idols, and were buried with the heads of the families who possessed them. We thus find the body of the South American tapir mounted on a bell; the snout of a shark furnished with the claws of a crab; the dog represented with a gigantic head, and a tail terminating in another head, partly canine and partly serpentine; the body of a crab having the arms of a man; beetles with human head and arms; vultures with outstretched wings, mounted on a pyramidal pedestal, etc. Besides these, there are unmutilated or uncombined figures of men, of toads, crocodiles, serpents, dogs, birds, etc. Judging from the specimens already found, there is scarcely an animal now living that has not been introduced into the curious religious mythology of this race. The resemblance between their works and those of the aborigines of North America is also peculiarly striking—though the latter would appear to have been inferior. But the accounts given by Col. Joachim Acosta, in his history of New Granada, of the mode of burial pursued by the Indians of Zenu, previous to the conquest of the Spaniards, would appear to be equally applicable to the predecessors of the present North American Indians. The graves, in both cases, were usually conical or pyramidal, and were large or small in proportion to the wealth, or the social, official, or religious standing of the deceased. The mound on the Ohio, must therefore have contained the remains of a person of the highest distinction, from its enormous dimensions; and this inference derives strength from the circumstance that but a few bodies were discovered in it—the first, or lowest skeleton, having been separated from the others by a considerable stratum of earth. It is probable that all the bodies in this mound were members of the same family, or held a similar official position in the community. Col. Acosta also mentions that the bodies in Zenu were surrounded by a peculiar *red earth*, brought from a distance; and it is singular that large quantities of such *red earth are found in the graves of the Ohio mounds*. The beads, pottery, images, inscriptions, and other works of art accompanying the remains of the bodies in the mound, are of a similar character to those of Chiriqui; but, instead of gold, they are fabricated of copper and other baser material. The ancient mining operations of Lake Superior were possibly prosecuted by these people, and furnished the supplies of that mineral, by which they wrought the trinkets and bracelets found in their graves. There can be no doubt, I think, of the identity of the two races; for, in addition to a similarity in their habits and works, the Indians that succeeded them are in both cases perfectly ignorant of their predecessors, and have no knowledge of the mechanical arts which distinguished them. The ancient race was, in every point of view, the superior of the existing Indians; and have left behind works that bespeak a considerable degree of civilization.

The palaces and works of art discovered by Stephens in Central America, (especially at Pelenque, and Cazmel, off the coast of Yucatan,) appear to belong to a similar race of people, but of a still higher degree of civilization. Among these ruins are gigantic pyramids and figures in bas-relief, together with hieroglyphic inscriptions. The pyramids are capped with buildings, leading to the inference that they may have been used for terrestrial or astronomical observations: but in this respect differ from those of Egypt. Mr. Stephens is of opinion that there is no identity between the Central American aborigines and the Egyptians, notwithstanding the similarity of their style of architecture, and especially in their sculptured figures. In this opinion, however, he stands alone; as nearly every other traveler who has described them, assigns an Egyptian origin to the race that produced them. As to their antiquity, no doubt is entertained on that point, some persons having even attributed to them *an antediluvian origin*. However this may be, there is no question as to the development of the race in the fine and useful arts; and all the works left behind attest the *Egyptian aspect of their civilization, manners, and idiosyn-*

crasies, to a greater extent than those of any other people, ancient or modern.

The present race of Indians was in existence at the first colonization of North America. Although it was successively explored by the French, Dutch, Spanish, and English, not a trace of the *previous inhabitants could be found*. The existing Indians were everywhere masters of the soil, and their habits and customs of life were such, that they presented no identity whatever with the extinct race interred in the mounds. And that it was at one time very populous, and spread over a vast extent of country, is sufficiently manifest in the number and geographical distribution of the mounds themselves, and the evidences of their mining operations. Their antiquity in North America is thus carried back beyond the fifteenth century, during which time the country was settled by missionaries, and explorers from several European governments; while in South America it was, at least, far anterior to the period of Columbus' visit to Costa Rica, shortly after which it appears to have degenerated, and been superseded by the present Indian natives. It is probable that those of South America, attracted by its gold, silver, and precious metals, absorbed the race from North America, and under the stimulus of such discoveries, it may have attained a somewhat higher civilization. This, however, is questionable; as it may be doubted whether the golden anaglyphs of Chiriqui exhibit more skill than the beads, pottery, and other ornaments of stone and copper found in the western mounds.

But many centuries before Columbus visited South America, portions of North America had been seen, and afterwards colonized to some extent, by the Northmen, *via* Greenland and Iceland.

"America was discovered in the year 1000, by Leif, the son of Eric the Red, by the northern route, and as far as 41° 30′ north latitude. (**Parts** of America were seen, although no landing was made, fourteen years before, in the voyage which Bjarne Hevpelfsson undertook from Greenland to the southward, in 986. Leif first saw the land at the island of Nantucket, 1° south of Boston; then in Nova Scotia, and lastly in Newfoundland, which was subsequently called 'Litla Helouland,' but never 'Vinland.' The gulf, which divides Newfoundland from the mouth of the great river St. Lawrence, was called by the Northmen, who had settled in Iceland and Greenland, Markland's Gulf.) The first, although accidental incitement toward this event, emanated from Norway. Toward the close of the ninth century, Naddod was driven by storms to Iceland, while attempting to reach the Färoë Islands, which had already been visited by the Irish. The first settlement of the Northmen was made in 875, by Ingolf. Greenland, the eastern peninsula of a land which appears to be everywhere separated by the sea from America proper, was only seen, although it was first peopled from Iceland a hundred years later, (983.) The colonization of Iceland, which Naddod first called *Snowland*, was carried through Greenlan I in a southwestern direction to the New Continent. Notwithstanding the proximity of the opposite shores of Labrador, 125 years elapsed from the first settlement of the Northmen in Iceland, to Leif's great discovery of America. So small were the means possessed by a noble, enterprising, but not wealthy race, for furthering navigation in these remote and dreary regions of the earth. The littoral tracts of Vinland, so called by the German Tyrker, from the wild grapes which are found there delighted its discoverers by the fruitfulness of the soil and the mildness of its climate, when compared with Iceland and Greenland. This tract comprised the coast line between Boston and New York, and consequently, parts of the present States of Massachusetts, Rhode Island, and Connecticut. This was the principal settlement of the Northmen. The colonists had often to contend with a very warlike race of Esquimaux, who then extended further to the South, under the name of the *Skralingers*. The first Bishop of Greenland, Eric Upsi, an Icelander, undertook, in 1121, a Christian mission to Vinland, (the name of this settlement at that time,) and the name of the colonized country has even been discovered in old national songs of the inhabitants of the Färoë Islands."*

The discovery and settlement of America by the Northmen, thus carries our knowledge of it back to the year 986. At that time there appears to have been *two distinct races* inhabiting it, besides the Norwegians themselves. The

* Baron Von Humboldt, quoting *Caroli* Christiani Rafu, *Antiquitates Americanæ*,—Cosmos, vol. ii. p. 230.

Skralingers, (whom they supposed to be Esquimaux,) told them of another race, living *still further southward*, beyond the Chesapeake bay. They were described as "*white men*, who clothed themselves in long white garments, and carried before them poles to which cloths were attached, and called in a loud voice." This account was interpreted by the Christian Northmen to indicate *processions*, in which *banners* were borne, accompanied by *singing*.

"An opinion," says Humboldt, "has been advanced by some northern antiquaries that, as in the oldest Icelandic documents the first inhabitants of the island are called *West Men*, who had come across the sea, Iceland was not at first peopled directly from Europe, but from Virginia and Carolina, (Great Ireland, the American White Men's Land,) by Irishmen who had earlier emigrated to America. The important work, *De Mensura Orbis Terræ*, composed by an Irish monk, Dicuil, about the year 825, and therefore thirty-eight years before the Northmen acquired their knowledge of Iceland from Naddod, does not, however, confirm this opinion." Humboldt makes various other suggestions to show the fallacy of this opinion,—which apply equally to the theory of Catlin of the descent of the Tuscaroras from the Welsh, or of those of various other antiquaries who, by linguistic characters, seek to ally the Indians with English, French, or other European nations.

"That the first discovery of America," adds Humboldt, "should not have produced the important and permanent results yielded by the re-discovery of the same continent by Columbus, was the necessary consequence of the uncivilized condition of the people, and the nature of the countries to which the early discoveries were limited. The Scandinavians were wholly unprepared, by previous scientific knowledge, for exploring the countries in which they settled, beyond what was absolutely necessary for the satisfaction of their immediate wants. Greenland and Iceland, which must be regarded as the Northern countries of the new colonies (in America), were regions in which man had to contend with all the hardships of an inhospitable climate. The wonderfully organized free state of Iceland, nevertheless, maintained its independence for three centuries and a half, until civil freedom was annihilated, and the country became a subject of Norway. The flower of Icelandic literature, its historical records, and the collection of the Sagas and Eddas, appertain to the twelfth and thirteenth centuries."*

The accounts which the Northmen give of the people called *Skralingers*, although represented as warlike, forbid the idea that *they* erected the western mounds, or that they dug copper, and wrought utensils and sculpture. The allusion to the *white people* south of them, is, however, significant; for all the relics hitherto found, betraying the least degree of civilization, are *south and west of the Chesapeake Bay.*

It has been suggested by some antiquaries that the original aborigines of America were derived from Asiatic stock, and that they effected an easy passage from one continent to the other at Behring's Straits, by means of the ice which generally blockades the channel during winter. These straits, at the narrowest points, are not over forty miles in width. If a passage was effected in vessels, it is equally probable that they may have crossed further south, from the coast of Japan, or from the Kubile Islands to those of the Aleutian, which stretch several hundred miles into the ocean from the Pacific coast of America. In this latter case, the distance from one island to the other would hardly exceed from fifty to one

---

* Humboldt's Cosmos, vol. ii. p. 237.

hundred miles, except in the main channel of the ocean where it would perhaps exceed two or three hundred. But as all the islands from Kamtschatka to, Borneo, on the one side, and from Behring's Straits to the Cordilleras of South America, on the other, are *volcanic*, it is at least *possible* that the *Aleutian chain at one time extended all the way across the ocean*—leaving but a few miles of distance intermediate between them. It is equally probable, however, that Behring's Straits, at a period not very remote, may have had no existence in fact—that it afforded a narrow isthmus which connected the two continents. The formation on both sides is the same—that of the hypozoic rocks, some of which, in the form of huge boulders or fractured masses, still lie scattered in the narrow channel in the form of ragged peaks and ice-clad islands. If the straits were thus traversed by an isthmus (or whether they were or not), there is little difficulty in referring the origin of the American aborigines to an Asiatic source; for they naturally resolve themselves into the Mongolian group. And we might, with equal propriety, fuse the Malayan into the African type, after which we should have but *three* leading divisions of the human species, corresponding with the three great continents of the old world, viz., the *Caucasian*, the *Mongolian*, and the *Ethiopian*, and with the descendants of the three sons of Noah,—Ham, Shem, and Japheth.

But, with the limited knowledge thus far obtained of these ancient people, it would be idle to speculate upon their origin, the probable date of their arrival, and the means by which they reached the American continent. The whole subject, as yet, is too obscure to justify lengthy antiquarian disquisitions. But it is nevertheless a very pertinent fact, that the characters inscribed on a stone slab, interred with the bodies in the mound at Moundsville, present a striking resemblance to the cuneiform or wedge-shaped letters of the ancient Pelasgians and Assyrians. Mr. Layard remarks, that it is not improbable that the cuneiform letters were originally formed by lines, for which the wedge was afterward substituted as an *embellishment;* and that the character itself may once have resembled the picture writing of Egypt, though all traces of its ideographic properties have been lost. The Assyrians, like the Egyptians, possessed at a later period a cursive writing, resembling the rounded character of the Phœnicians, Palmyrenes, Babylonians, and Jews, which was probably used for written documents, while the *cuneiform was reserved for monumental purposes.* There is this great difference between the two forms of writing, which appears to point to a distinct origin ;—the cuneiform runs always from left to right, the cursive from right to left. The cuneiform, under various modifications, the letters being differently formed in different countries, prevailed over the greater part of Western Asia to the time of the overthrow of the Persian empire by Alexander the Great.

The inscription on the Moundsville stone, I believe, has never been deciphered. Mr. Schoolcraft, however, identifies the letters as belonging to the alphabet of the ancient Pelasgi, and other early tribes inhabiting the shores of the Mediterranean. Their alphabet consisted of sixteen letters, formed by right and acute-angled strokes, and was no doubt derived, as they themselves were, from the Assyrians or Babylonians. If Mr. Schoolcraft is correct (and we know of nothing to impair his testimony since the time it was rendered), the people interred in the mounds were among the *earliest generations of the earth,* and must have reached this continent about the time that the Egyptians, under Cecrops, founded the Kingdom of Athens, which occurred 1556 *years before the birth of Christ.* Previous to this period, the southern corner of Europe, comprehended between the 36° and 41° of latitude, bordering on Epirus and Macedonia toward the north, and on other sides surrounded by the sea, was inhabited, above eighteen centuries before the Christian era, by many *small tribes of hunters and shepherds*, among whom the Pelasgi and Hellenes were the most powerful and numerous.

"The barbarous Pelasgi venerated Gracchus as their founder; and for a similar reason the more humane Hellenes respected Deucalion. From his son Hellen, they derived their general appellation, which originally denoted a small tribe in Thessaly; and from Dorus, Eolus, and Ion, his more remote descendants, they were discriminated by the names of Dorians, Eolians, and Ionians. The Dorians took possession of that mountainous district of Greece, afterward called Doris; the Ionians, whose name was in some measure lost in the illustrious appellation of Athenians, settled in the less barren parts of Attica; and the Eolians peopled Elis and Arcadia, the western and inland regions of the Peloponnesus. Notwithstanding many partial migrations, these three original divisions of the Hellenes generally entertained an affection for the establishments which had been preferred by the wisdom or caprice of their respective ancestors; a circumstance which remarkably distinguished the Hellenic from the *Pelasgic* race. While the former discovered a degree of attachment for their native land, seldom found in barbarians, who live by hunting or pasturage, the latter, *disdaining fixed habitations, wandered in large bodies over Greece, or transported themselves into the neighboring islands;* and the most considerable portion of them gradually removing to the coasts of Italy and Thrace, the remainder melted away into the Doric and Ionic tribes. At the distance of twelve centuries, obscure traces of the Pelasgi occurred in several Grecian cities; a district of Thessaly always retained their name; their colonies continued, in the fifth century before Christ, to inhabit the southern coast of Italy, and the shores of the Hellespont; and in these widely-separated countries, their ancient affinity was recognized in the uniformity of their *rude dialect and barbarous manners*, extremely dissimilar to the customs and language of their Grecian neighbors.'"*

The Noachian deluge occurred 1656 years after the creation of Adam. In less than 150 years after the flood, Nimrod, the son of Chus, grandson of Ham, and great-

---

* Dr. Gillas, History of Ancient Greece, p. 12.

grandson of Noah, founded the empire of Assyria, and built the city of Babylon, and afterward that of Nineveh. Moses speaks of him as a "mighty one in the earth," and as having distinguished himself as a *hunter*. Only six years after the founding of Babylon, (A. M. 1816,) Menes or Misraim, a son of Ham, settled in Egypt; and in six and a quarter centuries thereafter, (A. M. 2448,) colonies from Egypt, under Cecrops, founded Athens. Now, it is a very remarkable circumstance that, from the reigns of Semiramis and Ninyas, the third and fourth successors of Nimrod, to that of Phul and Sardanapalus, a period of over fourteen centuries, *absolutely nothing is known of the ancient empire of Assyria*. Although the ruins of Nineveh have been exhumed, and its buried sphinxes, human-headed bulls, winged animals, and sculptured panels, have been redeemed from the oblivious dust of centuries, but little has been added to our previous knowledge of its more ancient history. Some two or three centuries after the founding of Babylon, and the confusion of languages in the tower of Babel, we find the country overrun with *roving tribes of hunters and shepherds;* and under the circumstances, we are bound to infer that all or most of them were offshoots from the kingdom of Nimrod, and that they carried with them some of the principles, habits, and arts, which distinguished his race. Among these we include the Pelasgi—a people of varied habits, but like Nimrod, their ancestor, especially distinguished as hunters, and for their more barbarous, roving, and adventurous disposition as compared with the Hellenes. When, however, after the lapse of three or four more centuries, the territory of these roving tribes was invaded by colonies of Egyptians, and the kingdom of Cecrops founded, it is highly probable that their religion and habits were gradually absorbed into those of the more enlightened, enterprising, and powerful races. If this supposition be correct,—(and all the facts of history tend to confirm it in a most remarkable manner,) we at once obtain a *key* to the origin and meaning of the images and other works of art strewn over the American continent by its aborigines. They were the work of the *Pelasgic descendants of the* •*Egyptians.* Moved with• the spirit of emigration and colonization which prevailed at that early period, they carried with them to the new world the *religion of the Egyptians*, with the original instincts for hunting and adventure of the *Pelasgi.* Being thus a race of hunters, and moving about from place to place in quest of novelty, adventure, and game, they built but few cities or palaces, and hence have left little else behind except their conical mounds and pyramids; their rude towers and walls; their ornaments and idols of gold and brass; their utensils of stone and earth.

The Egyptians had a great number of gods, but only two of them were universal, viz., Osiris and Isis, supposed to represent the sun and moon. Both of these are typified in the Chiriqui anaglyphs, but in a manner somewhat obscure. But besides these, they worshiped nearly every known *animal*, as well as certain vegetables. Among those most popular were the ox, the cat, the wolf, the dog, the hawk, the owl, the crocodile, the ibis, the ichneumon, etc. While some of these were held in the highest estimation by certain tribes or cities, they were the superstitious abomination of others, and not unfrequently were the unconscious means of fomenting fraternal dissensions and bloody hostilities between tribes. Juvenal satirized these varied gods in the following lines:

> "Who has not heard where Egypt's realms are named,
> What monster-gods her frantic sons have framed?
> Here Ibis, gorged with well-known serpents, there
> The crocodile commands religious fear.
> Where Memnon's statue magic strings inspire
> With vocal sounds that emulate the lyre;

> And Thebes, (such, Fate, are thy disastrous turns!)
> Now prostrate o'er her pompous ruins mourns;
> A monkey-god,—prodigious to be told!—
> Strikes the beholder's eye with burnished gold.
> To godship here blue Triton's scaly herd,
> The river progeny is there preferred;
> Through towns Diana's power neglected lies,
> Where to her dogs aspiring temples rise;
> And should you leeks or onions eat, no time
> Would expiate the sacrilegious crime.
> Religious nations sure, and blest abodes,
> Where every orchard is o'errun with gods!"

Nearly every nation of antiquity had a superstitious veneration for certain animals or idols, which they worshiped under the pretext of symbolical gods. "Philosophers," says Plutarch, "honor the image of God wherever they find it, even in inanimate beings, and consequently more in those which have life. We are therefore to approve, not the worshipers of these animals, but those who, by their means, ascend to the Deity; they are to be considered as so many mirrors, which Nature holds forth, and in which the Supreme Being displays himself in a wonderful manner; or, as so many instruments which he makes use of to manifest outwardly his incomprehensible wisdom."

"Among us," says Cicero, "it is very common to see temples robbed, and statues carried away; but it was never known that any person in Egypt ever abused an ibis, a crocodile, or a cat; for its inhabitants would have suffered the most extreme torments rather than be guilty of such sacrilege." It is supposed that the celebrated Pythagoras derived his doctrine of the transmigration of the soul from the Egyptians; for their attachment and veneration for animals would appear to have been founded on the belief that, at the death of men, their souls transmigrated into other human bodies; but that, if they had been vicious during life, they were imprisoned in the bodies of unclean animals, to expiate in them their past transgressions. Who can doubt but that this idea formed the *basis of the religion*, whatever it may have been in other essentials, of the American aborigines, and especially of the Central American race, whose relics have been exhumed? Why surround their dead with the images of animals, unless, during life, they were regarded with *superstitious veneration*? Why repeat the follies of the Egyptians?—why rear *gigantic mounds and pyramids*, and carve in gold, and brass, and stone, the *forms of dogs, and vultures, and crocodiles, and crabs,* if there was not a connection, more or less direct, with the prevailing idiosyncrasies and superstitious religion of the original inhabitants of the Nile, the Euphrates, and the Mediterranean?

But we obtain, in the religion, and superstition, and the artistic works and evident *hunting proclivities of the Pelasgic-Egyptian race,* not only a clew to the origin of the American aborigines, but also to the introduction on the American continent and the various islands of the ocean, of *many of the animals belonging to the original Adamite creation;* but more especially those of the domestic kind, as the horse, sheep, cattle, goats, dogs, cats, etc. The remains of these animals, wherever found, are *cotemporary with man, and none of them*—no, not *one* of them—*existed here before the Noachian deluge!*

It would be foreign to the main purpose of this work, and would absorb altogether too much space, to trace the diffusion of the human race, *after* the flood, from Noah and his sons. After the landing of the ark on the mountains of Ararat or Armenia, the family of Noah remained for some time in that country, ranging between the Caspian Sea on the north, Asia Minor on the west, and the Red Sea on the south. The centre of population was doubtless in the vicinity of the Euphrates, which empties into the Persian Gulf. When God commanded them to disperse, and to replenish the earth, Nimrod, and the descendants of Ham, migrated northward, into what afterward became the Assyrian and the Babylonian empire. Shem and his descendants moved south, toward the Nile and along the Red Sea and the Mediterranean; while Japheth and his descendants settled along the northern slope of the Mediterranean, in Greece, Turkey, and the mountains of Caucasia. For many centuries, the human family, or rather, the great bulk of it, led a migratory and roving life—either subsisting by hunting, or by grazing sheep and cattle. As families increased, governments were instituted; and when, by marriages and social relations, families and tribes united for the common protection, governors were clothed with authority to regulate their domestic concerns. These little communities, by the natural law of increase, expanded into *nations*, and with such expansion came the ambition to build cities, and to surround themselves with the more permanent appliances and conveniences of life. With the increase of population and the growth of cities, the authority of rulers extended itself, whereupon new religions were adopted, new principles inculcated, and new and varied avocations pursued. Jealousies between tribes and nations were aroused; wars were carried on, and, in short, in a few centuries from the dispersion of Noah's family, millions of people were scattered over the vast region of country extending from the Atlantic coast of Britain eastward through Europe, Asia, and Africa, to the distant shores of India, China, and Japan. Nation after nation arose, some distinguished for their wonderful mechanical genius, as the Egyptians; some for their commercial enterprise, as the Phœnicians, and Carthaginians; some for their literature, poetry, and fine arts, as the Grecians and Romans; and all, more or less, for their wars, conquests, colonies, forms of government, science, philosophy, architecture, religion, etc., etc. They rose and fell, one after the other, until there is now little left of them, outside of Europe, but the sculptured remains of their ruined palaces, their superstitious idols, their works of art, and the names of their great kings, generals, poets, and philosophers.

In glancing over the history of the ancient nations thus distributed over Africa, Asia, and Europe, one fact will strike the reader with irresistible force, viz., that the *human species, four thousand years ago, possessed mental and physical properties fully equal*, if not superior in many respects, to the people of the present age. All their works were on a grand and magnificent scale, and have excited the wonder and admiration of all succeeding ages. Some of the palaces and temples of Thebes, presented a forest of lofty marble pillars, stretching out in long avenues and porticos, lined with innumerable sphinxes. A single hall, in one of these stupendous edifices, was supported by one hundred and twenty pillars, six fathoms round, of a proportionate height, and intermixed with obelisks which so many ages have not been able to demolish. Painting, too, had displayed all her art. "The colors themselves, which soonest feel the injury of time, still remained amidst the ruins of this wonderful structure, and preserved much of their original beauty and lustre—so happily could the Egyptians imprint a character of immortality on all their works."* Every portion of Egypt abounded in obelisks. They were for the most part cut in the quarries of Upper Egypt, where some were left half-finished. But the most wonderful circumstance is, that the ancient Egyptians should have had the art and contrivance to dig canals, from the very quarries to the river Nile, by means of which, during the high inundations of that river, they floated the obelisks to different parts

* Rollin, who also quotes Strabo.

of the country. This was accomplished by rafts or air-inflated skins; but when we consider the enormous proportions and weight of the obelisks, thus cut out of the quarries, the means would seem utterly disproportioned to the task. Thus, Sesostris erected in Heliopolis two obelisks of extremely hard stone, brought from the quarries of Syene, *each one hundred and eighty feet in length.* This, be it understood, was a single mass of rock, cut into a quadrangular shape, raised perpendicularly, and inscribed with hieroglyphics or mystical characters. The Emperor Augustus, after he made Egypt a province of the Roman Empire, caused these obelisks to be transported to Rome, one of which was broken into several pieces. Constantinus afterward removed a third one, even larger than the first. The pyramids also constituted a very remarkable feature among the monuments of the ancient Egyptians, and were scattered all over the empire. Those near Memphis, however, were the most considerable. The largest was built on a rock, having a square base, cut on the outside as so many steps, and decreasing gradually to the summit. It was built with stones of a prodigious size, the least of which were thirty feet, wrought with wonderful art, and covered with hieroglyphics. According to several ancient authors, each side was eight hundred feet broad, and as many high. The summit of the pyramid, which to those who stood below, seemed a point, was a fine platform, composed of ten or twelve massive stones, and each side of that platform sixteen or eighteen feet long. It is said that one hundred thousand men were constantly employed about this work, and were relieved every three months by the same number. Ten years were spent in hewing out the stones, either in Arabia or Ethiopia, and in conveying them to Egypt; and twenty years more in building this immense edifice, the inside of which contained numberless rooms and apartments.

These pyramids were tombs, and there is still to be seen, in the middle of the largest, an empty sepulchre, cut out of one entire stone, about three feet deep and broad, and a little above six feet long. Such were the pyramids, which, by their figure, as well as size, have completely triumphed over the injuries of time and the barbarians. Pliny gives us a just idea of their object, when he calls them a foolish and useless ostentation of the wealth of Egyptian kings, and adds, that by a just punishment, their memory is buried in oblivion—the historians themselves not agreeing about the names of those who first raised these vain monuments. Diodorus judiciously observed that the industry of the architects is no less valuable and praiseworthy, than the designs of the Egyptian kings are contemptible and ridiculous. But what we should most admire in these ancient monuments is, the true and standing evidence they give of the skill of the Egyptians in astronomy. M. de Chazeller, when he measured the great pyramid, found that the four sides of it were turned exactly to the four quarters of the world, and consequently showed the true meridian of the place.

What has been said concerning the judgment we ought to form of the pyramids, may also be applied to the labyrinth. This was not so much one single palace, as a magnificent pile composed of twelve palaces, regularly disposed, which had a communication with each other. Fifteen hundred rooms, interspersed with terraces, were ranged round twelve principal halls, and discovered no outlet to such as went to see them. There was the like number of buildings under ground. These subterraneous structures were designed for the burial place of kings, and also for keeping the sacred crocodiles, which a nation so wise and powerful in other respects, worshiped as gods!

But, in the estimation of many, the noblest and most

wonderful work of all the kings of Egypt, was the lake of Mœris. Its exact dimensions have been disputed; but according to modern travelers, it was twenty thousand paces, or over seven French leagues in circumference. Some of the ancient authors made it more than one hundred and eighty French leagues, but this is evidently an exaggeration. Two pyramids, on each of which was placed a colossal statue, seated on a throne, raised their heads to the height of three hundred feet in the midst of the lake, while their foundations took up the same space under the water. This lake had a communication with the Nile by a great canal, and was intended to store water for the supply of the country during seasons of drought, or when the river failed in its customary annual overflows.

We have cited these familiar examples of the works of ancient art, merely to remind the reader of the *almost unlimited resources, power, and mechanical skill of the earlier races of mankind*, and to support the proposition previously advanced, viz., that in many respects they far surpassed any subsequent age or nation. But we might cite many other examples—nations that equally surpassed the moderns in science, literature, and the fine arts; in mining, manufactures, and commerce; or in government, philosophy, and statesmanship. The present has the benefit of the discoveries of previous ages. It can detect the follies of the past, and improve upon the good that has descended from it. It enjoys the rich legacy bequeathed by its fathers, and nearly every thing that it accomplishes, is but the reflex of past ages, from whence our modes of thought and action were, in a great measure, derived. They were our predecessors in poetry, music, the drama—in war, the arts, sciences—in architecture, statuary, painting—and although we may, in some cases, have improved upon the original models, according to the idiosyncrasies of particular nations and ages, yet, upon the whole, we have never surpassed them, and probably never will.

Now, is it likely that nations thus distinguished—nations which, like the Phœnicians, Carthaginians, and Grecians, sent their fleets to every sea, and planted colonies in the most distant points, *should, for more than three thousand years, remain ignorant of the American continent?* Or is it to be supposed that, after their downfall, one following fast upon the other, it was finally reserved for *Spain*, (which had previously been but a colony of Carthaginia, from whose bowels she mined her gold, silver, copper, lead, and other minerals,) emerging from a long-protracted era of ignorance, superstition and darkness;—that it was reserved for her alone, in the fifteenth century, to make the discovery of America? I repeat: Can any one, after contemplating the extraordinary resources, enterprise, and learning of the previous ages, suppose that America remained unknown to the world until the discovery of Columbus? The idea is preposterous. The knowledge of this continent may indeed, *have died out* with the extinguishment of the ancient Asiatic and African nations themselves, as it did with the Icelanders, and in the darkness which enshrouded Europe during the middle ages, may have been completely lost to those people, if indeed, any knowledge of it had previously existed among them; —but it is utterly inconsistent with any estimate that we can form of the power, nautical enterprise, and learning of the previous ages, to suppose that they should have remained ignorant of it, when in an after age, notoriously effeminate and destitute of means, the discovery should still have been made, upon geographical grounds, supposed to be original and independent, but really far otherwise. Columbus derived his theory of the New World from the Icelanders, colonists from Norway, who had previously settled the Atlantic Coast from Boston to New York, as early as the tenth century, and it is well known that, when he actually landed in South America, four centuries after, he could not persuade himself that it was a *new* continent—every thing induced him to regard it as a portion of the far-off Indies, and hence the name which portions of it bear to this day!

There are so many things in the history and works of the more ancient races that we cannot comprehend, and so much to challenge our admiration and wonder, that it would be contrary to all reason and analogy to presume upon *their geographical ignorance,* in the face of monuments which our own continent is now revealing of an absolute identity between them and our aborigines. Such a presumption would not only be unreasonable, but it would be the sheerest folly when we remember that, while the works, physiological features, languages, and inscriptions of the aborigines bear no resemblance whatever to the nations of *Modern* Europe, *there is in everything a most remarkable similarity to the ancient inhabitants of the Mediterranean,* and of the adjacent countries around the original seat of the Noachian race. "The comparative study of languages," says the great and profound Alexander Von Humboldt, who maintains the unity of races, "shows us that races now separated by vast tracts of land *are allied together,* and have *migrated from one common primitive seat:* it indicates the course and direction of all migrations, and, in tracing the leading epochs of development, recognizes, by means of the more or less changed structure of the language, in the permanence of certain forms, or in the more or less advanced destruction of the formative system, *which* race has retained most nearly the language common to all who had migrated from the general seat of origin. The largest field for such investigations into the ancient condition of languages, and, consequently, into the period when the whole family of mankind was, in the strict sense of the word, to be regarded as one living whole, presents itself in the long chain of Indo-Germanic languages, extending from the Ganges to the Iberian extremity of Europe, and from Sicily to the North Cape. The same comparative study of languages leads us also to the native country of certain products which, from the earliest ages, have constituted important objects of trade and barter. The Sanscrit names of genuine Indian products, as those of rice, cotton, spikenard and sugar have, as we find, passed into the language of the Greeks, and to a certain extent, even into those of Shemitic origin."*

But the theory of the ancient colonization of America, is not only supported by the similarity existing between the trans-oceanic races and their works, but is strikingly corroborated by the animal creation. Cuvier, and other distinguished anatomists, from whom we have already quoted, pointed out the fact that each continent appears to have a fauna peculiarly its own; and that in America, the animals invariably belonged to a class very *ancient* in the scale of creation. When the " Spaniards first visited South America, they did not find a single species of quadruped the same as any in Europe, Asia, or Africa—the puma, jaguar, tapir, lama, sloths, armadillos, opossums, etc., were to them entirely new animals." He might have added that even the natives whom Columbus met, were an entirely new race—not different in species, for that was impossible; but like the animals themselves, *new and unknown.* And why was this? The races whom the Indians most resembled, had long before died out; and so even the animals had no longer any known representatives in Asia, Africa, or Europe. But on

* Humboldt, Cosmos, vol. ii. p. 111.

digging up the earth, what do we find? First, the ancient progenitors of the American aborigines, and then, one by one, the ancient progenitors of the American animals! It is a singular and extraordinary coincidence, that, while nearly every leading naturalist points out this remarkable feature in the faunas of America, Australia, and New Zealand, not one has the temerity to suggest their common origin, as if science should be compromised if found to add aught to the simple facts of the Bible! It is not worth while here to again go into particulars, for if I did, fifty pages of my book would be occupied with the details. Suffice it to say, that so far have all geologists carried this proposition, (of the faunas of America, Australia and numerous oceanic islands being of greater antiquity than the existing faunas of Europe, Asia, and Africa,) that some of them have even assigned a similar antiquity to the vegetation. "It is a circumstance quite extraordinary and unexpected," says Prof. Agassiz, in his work on Lake Superior, "that the fossil plants of the Tertiary beds of Œningen resemble more closely the trees and shrubs which grow at present in the eastern parts of North America, than those of any other parts of the world; thus allowing us to express correctly the difference between the opposite coasts of Europe and America, by saying that the present eastern American flora, and, I may add, the fauna also, have a more ancient character than those of Europe." The mastodon, once so numerous in America, had previously flourished in Europe and Asia, and its bones are there found in strata older than those which contain them here. The South American tapir, the hog, the puma, the lama, armadillo, (in short, all the animals now existing here), had their analogues in the Old World; and many of the species became extinct there about the very time that we can suppose them to have been introduced here. In like manner, the marsupials of Australia are no longer found in the old world, and are represented in America by the opossum and pouched rats; but extinct genera of these animals are found all over the earth, and are not confined to any particular localities, as the living species now appear to be. As the animals ran out in the old continents, they made their appearance in the new ones; and the very same fact holds good with the races of man. Many animals, especially those valuable for their fur, their hides, or their flesh, or those obnoxious to the lords of creation, have *already become extinct during the present brief settlement of America;* while the *Indians themselves are gradually disappearing with them.* The bones of the one will mingle with those of the other; as those of their *predecessors* are now found side by side with the *Mastodon, Megatherium,* the *Paleotherium,* etc. For more than three centuries, Europe, Asia, and Africa have been emptying their overflowing population on our shores; and the effects thus produced on the physical aspect of the continent need not be pointed out. Old things are eradicated—new ones introduced. The wild animals give place to the domestic ones. The aborigines fade away before the white man, like flies nipped by autumnal frosts. Nature works slowly and mysteriously; but always on the same general plan. What we see going on now, has occurred *before*, and *will* occur again. One race of animals must give way to another, as one race of *man* is superseded by another. The seat of ancient art and population in Asia and Africa, has long been little else than a barren waste. *The time will come when it will again attract a superior population*, and the semi-barbarous people now inhabiting it *will die out, as the American Indian is trampled down beneath the advancing march of civilization!* The lion and the tiger will give way to the sheep and the ox; and the school-house and the church spire will smile on plains now traversed by wandering caravans, or dotted with fragile tents and with temples of heathen gods.

All the smaller animals, including the deer, the horse, the buffalo, the tapir, the sheep, the lama, and many others, were brought here by the earliest colonies, and their hides, and wool, or fur, constituted then, as they have since, a profitable article of commercial traffic. The *increase of population in the old world* prevented their mature development there. Among nations who could go to war with a million soldiers under arms, as was repeatedly the case with Assyria, Babylonia, Egypt, Carthaginia, Persia, and Greece, (that of Xerxes, when he lay before Thermopylæ, was estimated at several millions,) it may readily be surmised that the inferior animals, if not absolutely extinct, would soon become so; and hence the policy of *transplanting them to the new world would suggest itself as a scheme worthy the sagacity and unlimited enterprise of those great nations.* But, independent of the commercial aspect of the scheme, the *religion* of the Egyptians, and other nations, was such that their people would naturally take with them animals which they were taught to regard with superstitious reverence. The dog, the cat, the oxen, and many others, were *absolutely inseparable* from them,—they were household gods. And who can doubt but that the Almighty suffered this abominable superstition to prevail, for the *purpose of effecting some ultimate and unseen good?* He seems to have *used these very idolators* for the express purpose of *transplanting his creation to other worlds;* while he suffered them to rear up their everlasting monuments of folly, merely to exhibit to *succeeding generations* the ineffaceable gloom and desolation that now brood over them. In the very height of their power and glory, and at a time when the true God had been utterly forgotten, they were yet *unconsciously carrying out his decrees!* They were allowed strength that they might the better *work;* they were allowed indomitable enterprise that the whole world might be *subdued;* they were allowed exalted learning to show how contracted was their *vision,* when, unable to fathom the works of the Creator, they bowed down to beasts and sculptured *idols.* All this, I repeat, was tolerated that ultimate good might flow from it; for every seeming evil in the plan of creation, invariably carries with it a compensating virtue. Poison has its antidote; affliction brings consolations; and the terrors of death itself vanish before the angels of hope, which bear away the undying soul.*

I now come to the direct consideration of the seventh day or Sabbath. God rested the seventh day for the purpose of showing that his work was done. Had he not *designated* such a period of rest, it would have appeared that his creative work was *continuous;* but we know that such is not the fact. *Changes* occur on the surface of the earth, both in organic life and in organic forms; but there are no *new and special creations.* All the effects now produced are the results of *law,* not of *new creative acts.* When, therefore, God *finished* his creation, its successive stages were symbolized by the days allotted to man, and

---

* We had no idea, at the outset, of lingering so long on this branch of the subject. We have many other suggestions which might properly be introduced; but the bulk of manuscript already accumulated, admonishes me of the necessity of leaving it, at least for the present. Many of the views that I have presented are new—some of them novel. I do not know what the world may think of them, if, indeed, anybody shall even think of them at all. But I indulge a hope that I, or some one else, may be able, at some future time, to set them forth in a better light. The truths of nature are like game—they must be hunted for; and even when found out, it is not every marksman that can "*shoot* folly as it flies, or *catch* the manners, living, as they rise." If my friend, the reader, should coincide with me, and be amused and entertained by the facts and positions I have presented to his consideration, I shall be well rewarded for all the paper, ink, candles, and so forth, that have been consumed!

he hallowed the *seventh*, for the purpose of saving him from the evil effects of continuous labor. No one could understand the complicated and delicate mechanism of man better than the Creator himself. He knew every artery and bone; every little treacherous nerve and vessel, and how easily they could all be unstrung, disorganized, and ruptured. He knew the delicate sensations of the brain—the inimitable fibres of the eye—the drum-like caverns of the ear—the pendulum vibrations of the heart, (like "muffled drums, beating funeral marches to the grave.") Knowing man's constitution and physical organization, the Creator saw the necessity for periods of *alternate rest from toil;* and experience has demonstrated that the seventh day is as essential to his health, and to the maintenance of vigor of body and mind, as sleep and regular periods for meals. And not only is this requisite for man and all laboring animals, but it applies with nearly equal propriety to machinery, especially locomotives and steam engines. When these are kept constantly employed for long periods, the iron loses its polish, elasticity, and strength; there is an inequality between the active working parts, and those less occupied—between those exposed to the action of heat and steam, and those more exempt. To preserve such machines in good running order, they require occasional relaxation, during which time their surfaces may be cleaned, polished, and lubricated. But if this be true of inanimate machines, it has ten times the force when applied to animals. No one need be told that a horse, driven consecutively on a long journey, without Sabbath-rests, would soon become weakened and exhausted. The animal requires *fixed times to recuperate his system.* A bow, when it remains bent for a considerable time, cannot relax its fibres so as to resume its originally straight form;—but if, after use, the string be removed, it will resume its natural form, and

NECESSITY FOR SABBATH RESTS.

thus preserve all its strength and elasticity. After a man works during the day, sleep will restore him in the night; but if he continued on, without giving his mind and body an opportunity to recuperate on the seventh day, he would soon degenerate, and the human family, in a few generations, would become a race of enervated, over-worked pigmies—a race of parched, dried-up mummies, devoid of vitality and manly energy of body and mind. God knew this, and hence hallowed the Sabbath.

But while he hallowed the *seventh* day, he appointed all the others for *labor*. "Six days shalt thou *work!*"—not six days shalt thou fritter away in idleness, dissipation, and sin! Some people, of late, have manifested a great deal of zeal in behalf of the Seventh day, (or rather, in behalf of the Christian Sabbath, which, however, is all the same,) but seem entirely to overlook the *first* part of the command. Many of them would appear to think that they are obeying the injunction by *keeping the one day*, when, in fact, they are equally bound to keep *all* of them. The Creator never contemplated *man* as an idler. He made him for *work!*—he put him on the earth to cultivate, and embellish, and subdue it.

And speaking of cultivation: The wisdom of the Almighty is demonstrated in another way, in connection with this very subject. He has not only enjoined a Sabbath upon *man*, but has applied it in a somewhat similar manner to the *very ground whereon we tread*. The twenty-fifth chapter of Leviticus opens with these remarkable words:

"And the Lord spake unto Moses, on Mount Sinai, saying, Speak unto the children of Israel, and say unto them, When ye come into the land which I give you, then shall the land keep a Sabbath unto the Lord. Six years thou shalt sow thy field, and six years thou shalt prune thy vineyard, and gather in the fruit thereof: But in the seventh year shall be a *Sabbath of rest unto the land*, a Sabbath for the Lord; thou shalt neither sow thy field, nor prune thy vineyard. That which groweth of its own accord of thy harvest thou shalt not reap, neither gather the grapes of thy vine undressed; for it is a year of rest unto the land. And the Sabbath of the land shall be meat for you; for thee and for thy servant, and for thy maid, and for thy hired servant, and for thy stranger that sojourneth with thee. And for thy cattle, and for the beasts that are in thy land, shall all the increase thereof be meat."

The recent progress made in agricultural science has established, in the most overwhelming manner, the wisdom of this law; which, like ten thousand others, tends to elevate the Bible so high in the philosophy of life and nature, that nothing short of direct inspiration can account for its authenticity. The science of chemistry was unknown at the time Moses lived—but more particularly agricultural chemistry. Even the art of compounding medicine, if it extended beyond the admixture of the juices of roots and vegetables, was not based upon the fixed principles of chemistry, and had no systematic application to any of the arts. And yet these instructions of Moses are based upon *true chemical principles*—principles which man has been more than *four thousand years in finding out*. His practical experience never would have revealed the *cause* of the exhaustion and impoverishment of soils, and the means to be applied for their restoration, had it not been for the demonstrations of *chemistry;*—chemistry, a modern science. Why does the farmer rotate his crops? Why does he divide his farm into seven or eight principal fields, and then observe a systematic change and alternation of crops from one to the other? Because the earth is made up of various mineral substances, derived from the decomposition of rocks and organic matter, which, attracting the carbon, oxygen, hydrogen, and nitrogen of the atmosphere, unite in ever varying proportions, and as they erect vegetable tissues, supply to it sugar, gum, starch, and the various acids, salts, alkalies, and oils. All plants are mainly made up of potassa, soda, lime, magnesia, and sesquioxyd of iron, which are invariably combined in greater or less proportions with carbonic acid, sulphuric acid, silicic acid, phosphoric acid, and various chlorids. The chemical combinations thus formed between the elements of the earth and of the air; the power they exert of precipitating chemical acids, and of stimulating vegetable vitality, in all its wonderful forms, is *derived mainly from the soil.* For if the *mineral ingredients* necessary to attract the *elements of the air*, are wanting in the soil, no vegetable *acids* can be evolved, and consequently crops cannot be raised. This is especially the case when a succession of crops of the *same kind* are planted in the *same soil*—the particular elements which enter into such crops being *exhausted*, there is no longer remaining sufficient strength in the mineral particles of the soil to *attract* the carbon, oxygen, and hydrogen of the air, and form vegetable acids with them.

Wheat usually contains an average of 56 parts of starch, 14 of gluten, 8 of sugar, 5 of gum, 2 of bran, and from 10 to 13 of water. The bran generally occupies about one-fourth of the entire weight of the grain, and is even more nutritious (in gluten) than the white flour. When the grain is burned, there is left behind about 2 per cent. of ash, nearly one-half of which consists of phosphoric acid, the other constituents being potash, silica, magnesia, soda, oxyd of iron, lime, etc., etc. These mineral substances are minutely diffused throughout the whole seed, but the bran contains the most. The quantity of starch in corn meal varies from 70 to 80 per cent.; in rye flour, from 50 to 60; in buckwheat, about 50; in peas and beans, 42; and in potatoes, from 13 to 15 (with the addition of nearly 70 parts of water). All the cereals are thus distinguished for the production of starch and gluten; while in others the leading element is sugar; in others, gum, resin, and mineral oils; in others, alkalies, poisonous extracts, coloring principles, etc., etc. Over two hundred distinct acid compounds, the products of vegetation, have been isolated and described by chemists. They are all composed (but in ever-varying degrees of combination) of carbon, hydrogen, and oxygen, with the latter generally in excess. They are almost invariably found in combination with the varied mineral bases of the soil, as potash, soda, lime, etc.

Now, it is plain that as *all* the cereal crops absorb a large amount of *starch and gluten*, the mineral particles in the soil which elaborate these substances would ultimately lose their vitality, and the *ground become impoverished by uninterrupted succession of such crops*. To overcome this tendency to exhaustion, all farmers find it absolutely necessary to return to the ground some of the elements thus absorbed from it, and to enable it anew to attract moisture and heat by which to maintain the fermenting and acid principle of vegetation. Animal manures, bone-dust, lime, wood-ashes, straw, and decomposed vegetation accomplish this office in a great measure, not only by their acids dissolving the coherent particles of earth and rock, and rendering the soil porous and open to the operation of the air and moisture, but to enable it to attract *from* the air the particular qualities demanded for the support of the crop. The substances which make up the great bulk of the structure of all plants—cellulose, lignine, starch, sugar, and gum—contain oxygen and hydrogen in exactly the same proportions as they exist *in water*, and they may in fact be regarded as water in combination with *carbon*. Plants absorb through their roots much more water than is applied to the enlargement of their structure, and in such cases a constant evaporation takes place from their *leaves*. By this process all the solid parts of the plant are *assimilated from the sap*, which is itself rendered *liquid* by the water. But the carbonic *acid* which plants absorb is mainly received through their *leaves*—all of which are furnished with innumerable pores to effect this object. "These pores are found mainly upon the under side of the leaf. In the white lily, where they are unusually large, and are easily seen by a simple microscope of moderate power, there are about 60,000 to the square inch on the epidermis of the lower surface, and only about 3,000 in the same space upon the upper surface. Direct sunshine being unfavorable to their operation, they are more commonly found on the lower surface of leaves. Although but two measures of carbonic acid gas are contained in 5,000 of air, its aggregate supply, by reason of the great extent of the atmosphere, is very large, and has been estimated to exceed seven tons for each acre of the earth's surface. The immensely-extended *surface* presented by the leaves of plants enables them to withdraw carbonic acid from the atmosphere in a very rapid manner. But carbonic acid, when the soil is rich in decomposing vegetable matter, is also evolved in the ground, and is fed to the roots of the plants in the form of a *solution*. The acid is decomposed, and its carbon constituents being retained by the plant, the oxygen originally combined with it is restored to the atmosphere. Thus, plants will grow in proportion to the power of the soil to supply such carbon; and this is primarily dependent on the amount of material furnished to it so as to enable it to attract, and then to assimilate, the elements when obtained."*

Now, the experience of farmers in all ages has demonstrated that, notwithstanding all the manures annually returned to the soil, and in addition to the regular alternation of crops, it is yet absolutely essential to let one of the fields lie fallow every year, and thus, by a system of

---

* Wells' Chemistry.

rotation, all of them obtain, in their turn, a *Sabbath of rest every seventh year*, agreeably to the Mosaic requirement, It will thus be seen that the institution of the Sabbath is founded upon a *direct law of Nature*, and, in the case of the ground, involves the nicest and most complex principles of chemistry—the whole phenomena of earthy constituents, air, water, chemical combinations and acids, and the functions and secretions of all plants. As these things were unknown to Science during the age that Moses lived, it follows that, like many other laws founded on similar philosophical knowledge, they must have been *revealed* to him by the Creator himself, who, as we have before remarked, was in closer intercourse with the early races of mankind than he has been at any time since, except during the holy mission of the Saviour. If civilized man kept no Sabbath of rest, it is not too much to infer that he would, in the course of a few generations, relapse into a savage or barbarous condition, respecting neither the laws of nature nor the governmental institutions of man, framed for the common decorum and protection of society. Nearly all the laws of Moses, which, at this remote age, appear so anomalous and singular, were based upon the peculiar habits and principles of mankind at that early day, and were doubtless the best that could have been devised to lead the Israelites from the superstitions and worship of idols with which, during their long sojourn among the Egyptians, they had become prepossessed. This is evinced from the fact that, on their return from that God-afflicted land, during the temporary absence of Moses on one of his missions to receive new instructions from God, they compelled Aaron to set up a golden calf, which they immediately fell down and worshiped in the true Egyptian spirit. This shows very conclusively the necessity for the adoption of forms and sacrifices, that their minds might be constantly directed to God, and that their whole study and their worldly possessions should, as it were, be devoted to the service of the true God, instead of the absurd and demoralizing worship of images and animals. And the experience of modern nations and communities has proved the fact, that wherever the Sabbath, both for man, beast, and the soil, has been observed in religious purity, the land and the people have been rendered prosperous in the highest degree; and that, wherever it has *not* been observed, the land has sunk into barrenness and the people relapsed into semi-barbarism. The brief experience of our own country has already sufficed to demonstrate this unerring law.

God has thus not only directed man in the true way to domestic happiness, health, and prosperity, individual as well as national, as based upon the *cultivation of the earth;* but he has also anticipated the future multifarious wants of the human family, by providing underneath the plowshare, seams of fossil coal, of mineral ores, and earthy substances for building and the useful arts, besides gems and jewels, and elevating them in advance, by volcanic agency, so as to *render them conveniently accessible.* While there sometimes seems to be a confusion and unnecessary complication in the stratification of the earth, a *benevolent design* may be thus observed in the plan—the wise forethought and provision of a good father for his dependent and erring children. For, it is hardly worth while to suggest, had these varied mineral substances been suffered to remain in the horizontal position in which they were originally deposited, nothing is more certain than that they could not have been available to man, and that the surface would not have presented that diversity of valley and mountain which now distinguishes the greater portion of it, and effects the most essential benefits in the diversification of the atmosphere. Wherefore, exclaims Milton, with more than poetic inspiration:

> "Wherefore hath Nature poured her bounties forth
> With such a full and unwithdrawing hand;
> Covering the earth with odors, fruits, and flocks—
> Thronging the seas with spawn innumerable,
> But all to please, and sate the curious taste,
> And give unbounded pleasure unto man?"

Every portion of the globe is thus supplied with some peculiar mineral, agricultural, commercial, or industrial resource, and he must be blind indeed that cannot see in the arrangement, not only a *geological order*, but a *geographical distribution* well calculated to enable man, agreeably to the divine command, to subdue and replenish the earth, and to exercise his humane dominion over all its creatures. The whole fraternity of man and creation is thus drawn together in a common chain of union—

> "We are but parts of one stupendous whole,
> Whose body Nature is, and God the soul.
> All Nature is but art, unknown to thee;
> All chance, direction, which thou can'st not see;
> All discord, harmony, not understood;
> All partial evil, universal good."

But there is, in the whole creation of the earth, not only a pre-arranged *order of Nature* in the distinct and successive cosmogonal eras, as we have attempted to show; but there is a continuous development, from first to last. Organic forms and structure invariably proceed according to fixed laws of proportion and number. Had there, indeed, been no *law* of progression, the work of creation would have been stationary, without power of change or of recuperation. Without progression, the work of six eras might have been confined to one—to a day—an hour—a minute. But it happens that the same principle of progression which impels forward the human family, also impels, though in a modified degree, every other species of organic life. Herein man resembles his Maker, whose works continually attract him onward and upward. Hence it is that the beautiful in the arts of design, in sculpture, architecture, painting, mechanics, and the law of mathematics, is only *attainable* according to the *preexisting models of Nature*. The *mind* of man, in other words, is directed by the *works* of God, and in following and fathoming those works, he merely imitates his Creator, and becomes God-like as he improves upon their teachings and purposes. The elaborately-carved frieze or cornice that we so much admire in the capitals of ancient art, are mere copies of the sculptured and ornate shells of the still more ancient Ammonites and Trilobites, who erected *their* marble palaces in the bottoms of primitive seas. The scars and ribs of the Sigillaria, the oval canopy and pendant tassels of the Lepidodendria, the stellar net-work and rounded dots of the Stigmaria—all these, in their innumerable varieties, form figures which the printers of calico and paper-hangings may imitate, but not surpass. Even the speaking marbles of Phidias; the eloquence of Demosthenes; the graceful figures and gorgeous coloring of Raphael; the poetic anthems of Milton; the dramatic worldlings of Shakspeare; the mathematical demonstrations of Newton; the great laws of Copernicus, Kepler, and Leverrier;—what, in fact, are all these but *reflected visions* or harmonies of Nature—reflected on minds which, like the prints of the sun upon the daguerreotype, were sufficiently transparent to retain them. What is thought but the reflected or embodied *idea which God has imprinted* on Nature!

Our senses thus detect beauty of form and structure, of curved line and coloring mixture, in all the shells of the sea, and in all the trees, flowers, and fruits of the land. There is beauty everywhere!—nothing *but* beauty and harmony. The papered walls of our houses, the figured carpets we sometimes *fear* to tread upon, the printed delaines and rustling silks so gracefully expanded around the human form divine—these are only tolerable imitations of things and pictures which the Almighty sketched millions of years ago. Even the delicately-wrought laces of the ladies—the Chantilly and Honiton fabrics which it affords husbands so much pleasure to buy at ten or fifty dollars per yard; why, really, these are little more than the gossamer webs of the spider, and it would be difficult to tell which of them catches the largest number of fluttering victims! St. Peter's has a large dome; but who can paint one like that visible over our heads on a starlight night, or during the meridian of the sun, or when he sinks down beyond the western hills to take a peep at the inhabitants of the antipodes? Our minds are God-like because we can admire God's works. Sometimes Nature teaches us what we have the effrontery to regard as *original* designs; but experience always proves that God has preceded us. There is, in fact, no such thing as originality in man. Even *sin*, in which he excels, is not *original* with him! In the arts of government, in domestic thrift, industry, and order, the "little busy bee" gives us significant lessons. What nation, or statesman, or philosopher, from Lycurgus to Buchanan, has laid down a nicer scheme of governmental order and decorum than that which they illustrate in their little pendant worlds of woven paper.

> "They have a King, and officers of sorts,
> Where some, like Magistrates, correct at home;
> Others, like Merchants, venture trade abroad;
> Others, like Soldiers, armèd in their stings,
> Make boot upon the summer's velvet buds,
> Which pillage, they, with merry march, bring home
> To the tent-royal of their Emperor,
> Who, busied in his tent, surveys
> The singing Mason, building roofs of gold;
> The Civil Citizens kneading up the honey;
> The poor, mechanic Porters crowding in
> Their heavy burdens at his narrow gate;

> The sad-eyed Justice, with his surly hum,
> Delivering o'er to executors pale,
> The lazy, yawning Drone."

St. Pierre observed, and with truth, that the "importance which we assign to our talents, proceeds not from their *utility*, but from our *pride*. We should take a material step toward its humiliation did we consider that the animals which have no skill in agriculture, and know not the use of fire, attain to the greatest part of the objects of *our arts and sciences*, and even surpass them. I will say nothing here of those which build, which spin, manufacture paper, and cloth, and practice a multitude of other trades, of which we do not so much as know. The *Torpedo* defended himself from his enemies by means of the electric shock, before Academies thought of making experiments in electricity; and the Limpet understood the power of the pressure of the air, and attached itself to the rocks by forming the vacuum with its pyramidal shell, long before the air-pump was set in motion. The Quails which annually take their departure from Europe, on their way to Africa, have such a perfect knowledge of the autumnal equinox, that the day of their arrival in Malta, where they rest for twenty-four hours, is marked on the almanacs of the Island, about the 22d of September, and varies every year as the equinox itself. The Swan and Wild Duck have an accurate knowledge of the latitude where they ought to stop, when every year they reascend in spring to the extremities of the North, and can find out, without the help of compass or octant, the spot where the year before they made their nests. The Frigate, which flies from East to West, between the tropics, over vast oceans interrupted by no land, and which regains at night, at the distance of many hundred leagues, the very rock, hardly emerging out of the water, which he left in the morning, possesses means of ascertaining his longitude hitherto unknown to our most ingenious astronomers."*

The wisdom with which Nature has settled the proportions and functions of animals is not less worthy of admiration. On a careful examination, we shall find no one deficient in its members, regard being had to its manners, and the situation in which it is destined to live. This fact is singularly illustrated in the fossil remains of animals —every formation having produced species peculiarly

---

* Studies of Nature, from the French of J. H. B. De St. Pierre; London, 1798.

adapted to the circumstances then existing for their accommodation—as the condition of the climate, the degree of warmth, the purity of the water, its vegetable, calcareous, or sedimentary qualities, and the differences in the plants and trees, and many other similar features—all of which continually varied the habits, instincts, functions, and movements of marine and terrestrial animals. Thus, the large and long bill of the Toucan, and his tongue "formed like a feather, were necessary to a bird who hunts for insects, scattered about over the humid sands of the American shores. It was needful that he should be provided at once with a long mattock, wherewith to dig, with a large spoon to collect his food, and a tongue fringed with delicate nerves, to enjoy the rich relish of it. Long legs and a long neck were necessary to the *Heron*, to the *Crane*, to the *Flamingo*, and other birds which have to walk in marshy places, and to seek their prey under the water. Nature has also infinitely varied the means of defense, as well as of subsistence. The tardiness of the Sloth is no more a paralytic affection than that of the Turtle and the Snail. The cries which he utters when you go near him, are not those of pain. But, some being destined to roam over the earth, others to remain fixed on a particular post, their means of defense are varied with their manners. Some elude their enemies by flight; others repel them by hissings, by hideous figures, by poisonous smells, or lamentable cries. There are some which deceive the eye, such as the Snail, which assumes the color of the walls, or of the bark of trees, to which he flees for refuge; others, by a magic altogether inconceivable, transform themselves at pleasure into the color of surrounding objects, as the Chameleon.

"Oh! how sterile is the imagination of man, compared to the intelligence of Nature! Genius itself,—about which such a noise is made,—this creative genius, which our wits fondly imagine they brought into the world with them, and have brought to perfection in learned circles, or by the assistance of books, is neither less nor more than *the art of observing*. Man cannot forsake the path of Nature, even when he is determined to go wrong; we are wise only with her wisdom, and we play the fool only in proportion as we attempt to derange her plans."*

Pope, inferring the instruction of man from the instinct of the lower animals, whereby he has improved the arts of industry, exclaims:

"See him from Nature rising slow to Art,
To copy Instinct then was Reason's part.
Thus then to man the voice of Nature spake—
" Go, from the creatures thy instructions take:
Learn from the birds what food the thickets yield,
Learn from the beasts the physic of the field;
Thy arts of building from the bee receive;
Learn of the mole to plow, the worm to weave;
Learn of the little Nautilus to sail,
Spread the thin oar, and catch the driving gale.
Here, too, all forms of social union find,
And hence let Reason, late, instruct mankind.
Here subterranean works and cities see;
There towns aerial on the waving tree;
Learn each small people's genius, policies,
The ant's republic and the realm of bees.
How those in common all their wealth bestow,
And anarchy, without confusion, know;
And these forever, though a monarch reign,
Their separate cells and properties maintain.
Mark what unvaried laws preserve each state;
Laws wise as Nature, and as fixed as fate.
In vain thy Reason finer webs shall draw,
Entangle justice in her net of law,
And right, too rigid, harden into wrong,
Still for the strong too weak, the weak too strong.
Yet go! and thus o'er all the creatures sway,
Thus let the wiser make the rest obey;

---

* Studies of Nature, from the French of J. H. B. De St. Pierre; London, 1798.

And for those arts *mere instinct* could afford,
Be crowned as monarchs, or as gods adored!'"

Lessons of wisdom, of invention, of patient industry, and social decorum can thus be learned from every species of created being, and even from the insects and half vegetable zoophytes and coral reefs; but more especially from the domestic animals, many of whom are doomed to drudge their brief lives away in the service of ungrateful and unappreciating man. How faithfully the horse plows; how intelligently he draws the wagon or the carriage; how noble is his amble under the rein of his rider—sharing with him "all the pleasure and the pride!" If his mission on earth is that of serving man, who will say that it is not well performed? And what is his reward? Is it to be presumed that animals so elevated in the scale of intelligence and usefulness, shall have no future rest—no Sabbath from the lashes, and neglect, and hard tasks of man? Heaven forbid! The hard tasks and cruelties which a *Christian* people inflict on these devoted animals, might at least be compensated so far as to concede, according to the spirit of our religion, that the same *benevolence* which provides for our future happiness, will not deal unkindly with the poor brutes. But what a contrast does the treatment of these creatures in civilized countries afford to that of the Arabs! *They* never beat their horses; they manage them by means of kindness and caresses, and render them so docile, that there are no animals of the kind in the world to be compared with them in beauty and in goodness. They do not fix them to a stake in the fields, but suffer them to pasture at large around their habitation, to which they come running the moment that they hear the sound of the master's voice. Those tractable animals resort at night to their tents, and lie down in the midst of the children, without ever hurting them in the slightest degree. If the rider happens to fall while coursing, his horse stands still instantly, and never stirs till he has mounted again. These animals are the first coursers of the universe. "It is related by *D'Hervieux*, in his "Journey to Mount Lebanon," that the French consul at Said offered to purchase a most beautiful mare, which comprised the whole stock of a poor Arabian of the desert, with the intention of sending her to his sovereign, Louis the Fourteenth. The poor Arab, pressed by want, hesitated a long time, but at length consented, on condition of receiving a very considerable sum, and which the king authorized him to advance. The consul sent notice to the Arab, who soon after made his appearance, mounted on his magnificent courser, and the gold which he had demanded was paid down to him. The Arab, covered with a miserable rug, dismounts, looks at the money, then, turning his eyes to the mare, he sighs, and thus accosts her: 'To whom am I going to yield thee up? To Europeans, who will tie thee close, who will beat thee, who will render thee miserable! Return with me, my beauty! my darling! my jewel! and rejoice the hearts of my children.' And as he pronounced these words, he sprung upon her back, and scampered off toward the desert."* Even the

> "Poor Indian, whose untutored mind
> Sees God in clouds, or hears him in the wind,—
> Yet thinks, admitted to that equal sky,
> His faithful dog will bear him company!"

But while lessons of profit and benevolence may often be learned from the various species of animal life, a scrutiny of the habits and instincts of others, also reveals much to deplore and condemn. For, like vain, tyrannical

---

* St. Pierre, Studies of Nature.

man, they too have their assassins, their robbers, and their imperious masters;—their Shylocks, and Peter Funks, and Mawworms, as well as their ambitious Cæsars and Napoleons! It is perhaps a sad reflection; but we know very well that there are villains lying in their native jungles, watching their inoffensive prey with glaring eyes and blood-thirsty jaws. There are villains soaring in the air, ready to pounce, with the stealthy cunning of Reynard, upon the bleating fold or the unsuspecting poultry. There are villains rioting in their sculptured and painted shells—dissipating in drunken frolics, or stalking forth with drawn swords, or keen-edged dagger, far down in the "dark unfathomed caves of ocean." The powerful everywhere persecute, pillage, and prey upon the weak. Whole communities of human fish, human fowl, and human quadrupeds, wage eternal, bloody, and exterminating wars upon each other!

Now, as the mind of man may properly be regarded as the combined and concentrated intelligences of all the different animal species, crowned with God-like Reason, his observation alone should lead him to just discriminations—teaching him what to avoid and what to follow— what to do, and what to leave undone. God has surrounded him, on every hand, with admonitions and guides; and has given him such intellectual powers as, justly exercised, not only enable him to maintain his dominion over all the animals, but to approach nearer and nearer to the character of angels.

> "These are thy glorious works, Parent of good,
> Almighty, thine this universal frame,
> Thus wondrous fair; thyself how wondrous then!
> Unspeakable, who sitt'st above these heavens,
> To us invisible, or dimly seen
> In these thy lowest works; yet these declare
> Thy goodness beyond thought, and power divine.

Speak ye who best can tell, ye sons of light,
Angels, for ye behold him, and with songs
And choral symphonies, day without night,
Circle his throne rejoicing; ye in heaven,
On earth join all ye creatures to extol
Him first, him last, him midst, and without end!"

ALPHABETICAL INDEX.

(For a Breviary of the Argument of the several days, see page 4.)

Adam and Eve, marriage of, 355.
Adam and Eve, fall and punishment of, 367.
Adam, sons and daughters of, 372.
Agassiz, remarks of, 259, 316.
African, Ethiopian or Negro race, 352.
Air, weight of, 200.
Antelope, 311—Apes, 347—Ants, 233.
Animals, law of succession, or alternation of species, 465.
Animals, number of species now living, 378.
Animals, how arranged in Noah's Ark, 383.
Animals inhabiting vegetable juices, 420.
Animals, relations of the living to extinct species considered, 428.
Animals, extinct species of South America, 434.
Animals, fallacious assumptions of geologists in reference to, 422.
Animal remains, absence of in the coal measures, 108.
Animals, classification of, Zoologists, 209.
Animal life introduced after vegetable life, 242.
Animals of the New Red Sandstone era, 243.
Animal life first appearing in the seas on the fifth day, 205.
Animals in Noah's Ark — were they young or old ? 381.
America, discovery of in the ninth century by the Northmen, 476.
America, colonies planted in Massachusetts and New York by the Icelanders, 447.
Ancient inhabitants of America, 448.
Ancient modes of writing—the cursive and the cuneiform, 452.
Animated Nature, birth of, 44.
Ancient vegetation, transmission of the seeds of, 117.
Alleghany mountains, origin of the, 67, 174.
Animalcules inherent in water, 213.
America geologically the *old* world, 65.
Anthracite coal basins described, 130.
Anthracite derived from bituminous coal, theories proposed, 161.
Ante-mundane phenomena, impossibility of man to explore, 15.

Ark of Noah, how constructed, 374, 380, 383.
Ark of Noah, compared with the Great Eastern, 375.
Ark of Noah, proofs that it was filled with young, and not adult animals, 381.
Asphalt of New Brunswick, 139.
Aqueous origin of the earth, theory of, 42.
Attraction of the sun, law of, 184.
Artesian wells, 390.
Articulate division of the animal kingdom, 228.
Art, works of the ancient Egyptians considered, 461.
Arabs, their treatment and love of horses, 485.
Astronomy, theories, speculations, discoveries, and laws of Hipparchus, Ptolemy, Copernicus, Brahe, Kepler, Galileo, Newton, Herschell, Leverrier, etc., 176.
Astronomical observations and calculations, value of, 28.
Astronomy, progress of discovery in the Seventeenth century, remarks of Baron Humboldt, 187.
Asiatic derivation of the ancient or original inhabitants of America, 451.
Athens founded by Cecrops of Egypt, 454.
Asphalt, 106—Amber, origin of, 158.
Atmosphere of the coal period, 172, 175, 193, 196.
Atmosphere, extraordinary effects produced by the, 190.
Atmosphere, wonders of the, 196.
Atmosphere and the ocean, currents and climates of the, 188.
Atmosphere, power of the, compared with the steam-engine, 190.
Avalanches, effects of, 495.

Bald Eagle, Dr. Franklin's opinion of the, 340.
Basis, the, of Christianity, 369.
Behring's Straits, former union of the two continents at, 451.
Breccia or Mosaic marble, 241.
Botanical classification, systems of Gessner, Jussieu, Tournefort, Linnæus, etc., 78.

(489)

ALPHABETICAL INDEX.

Bears, Buffaloes, Bisons, Beavers, etc., 313.
Bone-cells of different animals, 272.
Bible, truths of the, manifest with increased knowledge of natural law, 22, 202, 273.
Bible, gratuitous apologies for the, rebuked, 305.
Bible versus Geology, testimony of the witnesses in this case, 251.
Boulders, erratic, how distributed, 394.
Bituminous coal, how converted into anthracite, 160.
Bituminous shale in Wisconsin, 139.
Brown coal or lignite, nature and origin of, 158.
Bird tracks in New Red Sandstone, 268.
Bird tracks classified by Prof. Hitchcock, 270.
Brandy from coal oil, 144.
Byron, Lord, epitaph on his dog Boatswain, 334.

Coal basins, origin of, 64.
Coal basins, extent and geographical distributions of, 254.
Coal, fossil vegetation of the, described, 80.
Coal oil of Kentucky, 142.
Coal oil whiskey, 144.
Coal oil for fuel in Ocean steamers, 145.
Coal oil, speculations in, 145.
Coal oil, geographical extent of, 146, 147.
Coal oil strata, description of, 147.
Coal oil, how formed, 147–50.
Coal oil vegetation, 149, 150, 151.
Coal oil, production of, 152.
Coal oil, commercial value of, 152.
Coal oil of the Alleghanies, 153.
Coal, process of distillation from the ancient forests, 124, 280.
Coal measures, alternation of strata, 126.
Coal, identity of origin of various mineral combustibles, 132.
Coal, chemical changes of, 154.
Coal period, climate of the, 172, 175, 196, 206, 251.
Coal basins of Texas, Illinois, Missouri, Virginia, Pennsylvania, Rhode Island, Nova Scotia, etc., 68.
Coral animals, mechanical operations of, 212.
Calorific sublimation of the ancient rocks, 42.
Caves, the Mammoth of Kentucky, and others, 69.
Coniferous trees, extinct and living, 110.
Creation, original seat of, 414.
Creative plan or design manifest in Nature, 35.
Creative tableaux revealed to Moses as described in his Cosmogony, 38.
Cambrian system of rocks, character and distribution of, 60.
Calendar of solar time, origin of, 18.
Conglomerate and sandstone rocks, origin of, 63.

Copernicus not intimidated by priestly persecution, 181.
Church, false views of Cosmogony formerly held by the, 43, 167, 202.
Cohesion of atoms, 36.
Cubit measure of Scripture, 376.
Cuvier, Baron, on the animals of South America, 427.
Crops of the farmer, necessity for their alternation, 472.
Crystalline rocks the basis of the globe, 46.
Currents of the atmosphere, 188.
Cosmogonal eras, days, or circles of time, 17, 24, 194, 206.
Church, its errors external, not internal, 22.
Cretaceous and chalk strata, their character and distribution, 284.
Civet, 342—Camel, 313—Cattle, domestic and wild, 312.
Christ, his divinity, doctrines, acts, triumph over sin, etc., 369.
Christian doubters rebuked, 373.
Continents, submergence of, explained, 408.
Chiriqui, golden anaglyphs of, 441.
Cypress trees, 114.

Day, meaning of the word as used by Moses, 17, 20, 194, 206.
Devonian system of rocks, 53.
Devonian lakes and rivers, 70, 122.
Distribution of coal beds, 72, 167.
Dry land, emergence of from the primitive seas, 73.
Deposition of coal seams, theories of geologists, and a new one proposed, 93, 105, 125, 128.
Distillation of tar, pitch, turpentine, oils, etc., 119, 121.
Drop of water, animals inhabiting a, 214.
Discovery of a Saurian *What-is-it?* 256.
Dinotherium, 309—Dromedary, 293—Deer, 311.

Earth, origin of the, 13, 40.
Earth, antiquity of the, proved, 16, 21, 25.
Earth, original nebulosity and fluidity of the, 36, 41.
Earth, rotundity of the, 43.
Earth, supposed to be an animal by Kepler, 43.
Earth, surface of the, a vast tar pit during the coal period, 124.
Elks, 319—Elephants, 318.
Evening or night, symbolically used by Moses, 18.
European, American, and all other pine trees, 111.
Ehrenberg, Prof., microscopic investigations of, 213.
Echinodermatian animals described, 216, 306.
Error unlocking the door for Truth, 180.
Egyptians, their probable knowledge and colonization of America, 462.

ALPHABETICAL INDEX.   491

Egyptians, their religious veneration for animals, 455.
Egyptian superstitious, mythology and enterprise furnish a key by which to account for the distribution of Asiatic animals in America, 444.
Ethiopian or Negro race, 352.
Eden, garden of, 355, 363.
Eve, seduced by sin, 364.

First day of Creation described by Moses, 13.
Firmament, origin of the, 43.
Fœtus of life in the metamorphic era, 53, 63.
Fossil vegetation, character of, 71.
Fossil plants, systems of classification, 78.
Fossils, what do they teach? 91.
Fossil trees in coal veins, 96, 98, 104, 107, 123.
Fossils, how accumulated over the coal veins, 126.
Fossil remains of coral, molluscan, and other marine animals, 212.
Fossil Radiata and Mollusca, table of species of the, in each geological formation, 214.
Fossils of the chalk and cretaceous strata, 250.
Fruits and fruit trees absent in the ancient strata, 108.
Forests of the coal period, density and prolificacy of growth, 123.
Fire-damp explosions in mines, 166.
Fourth day of Creation described by Moses, 161—Fifth day, 209.
Fogs and vapors of the ancient climate, 171.
Foramenifera, works of the, 215.
Footprints of extinct animals, 243.
Fowls of the air introduced on the fifth day, 209.
Fishes of the Tertiary, and their classification, 296—the Ganoid, Cycloid, Placoid, and Ctenoid orders, 299.
Flood of Noah, how produced, and its universality proved, 372.
Fountains of the great deep, how broken up, 389.

God's spirit moving the nascent seas, 39.
Granite rocks, group of the, 42.
Granite, geographical distribution over the earth, 47.
Geological classification of rocks, (and nomenclature,) 51.
Geological survey of Pennsylvania, scientific trash, 54, 259.
Geological agencies of the coal period, 68.
Geological formations or eras have distinct orders of vegetation — a new theory, 116.
Geological writers, errors of, and how spread abroad, 163.
Geographical knowledge and commercial enterprise of the ancient races of man, 459.

Galileo, astronomical laws, and discoveries of, 182, 184.
Gravitation, law of, discovered by Newton, 184.
Gulf stream, remarkable features of, described, 189.
Grecian cosmogony contrasted with that of Moses, 43.
Gulf of the (Devonian) Atlantic, formerly located in Nebraska, Kansas, Missouri, etc., 70.
Gasteropoda, animals of the family of, 219.
Glyptodon, the, described, 321.
Garden of Eden, description of the, 355, 363.
Giants of the antediluvian period, 376.
Glaciers, mountain and polar, 391, 396.
Gospel, ministers of the, 358.
Geological pretenders, 412.
Generation of animalcules and insects, 421.

History of the earth written in its fossils, 93.
Hipparchus, astronomical theories of, 177.
Humboldt, Baron von, remarks of, on various topics, 179, 188, 448.
Humboldt on the unity of the human species, 464.
Humboldt glacier, described by Kane, 398.
Horse, the family of, 319.
Halley's comet, 28.
Hitchcock, Professor, on bird tracks, 255.
Herschell, Sir John, telescopic observations of, 29.

Infidelity, efforts to ally itself with the natural sciences, 23.
Important errors corrected as to the origin and phenomena of coal, 104.
Infusorial animals described, 213.
Insects, living and fossil, 229.
Intermediate planetary distance, law of, 33.
Ichthyosaurus, 264—Iguanodon, 281—Ichneumon, 243.
Insects, Cuvier's twelve orders of, 240.
Indian, the North American, race of, described, 374.
Immaculate Conception, theory and object of, 369.
Icebergs, geological influences of, 393.
Insects, how provided for in Noah's Ark, 419, 434.
Insect fecundation, extraordinary peculiarities of, 421.
Indians of Central America, whence derived, 441.

Jefferson, Thomas, opinions of, 436.
Juniper, and other allied trees, 115.

Known and the unknown, 35.
Kane, Dr., explorations of the Arctic regions, 395.

Key to unlock the mysteries of American antiquities, 435.
Kepler, astronomical laws and discoveries of, 179.
Knowledge of the universe expanding, 182.
Kangaroo, the, 307.
King, Dr., discovery of supposed footprints in the coal measures of Pennsylvania, 245.

Light, diffusion of, in space, 27, 34.
Light, inherent in nebulous bodies or aggregations, 39.
Land, dry, emergence of the, from the nascent waters, 49.
Land or air-breathing animals, their existence previous to the fifth day denied, 244.
Lake Superior, iron and copper of, 57.
Land surface of the Devonian era, 65.
Lepidodendria, fossils of, 83.
Lesquereux, Mr., remarks of, 89.
Larch, the, and other similar trees, 103.
Lignite or brown coal, how formed, 158.
Light of the sun, effects of on the earth, air, and sea, 188.
Limestone, Magnesian, 241.
Leverrier, discovery of the planet Neptune, 29.
Lakes—descriptions of Superior, Huron, Michigan, St. Clair, Erie, Ontario, etc., 69, 122, 132.
Law of planetary movement, 180.
Law of universal gravitation described, 185.
Lyell, Sir Charles, remarks on footprints, rain-drops, sun-cracks, ripples, etc., 245, 250, 271, 277.
Lyell, Sir Charles, on the submergence of continents, 408.
Lyell, Sir Charles, on the extinct animals of South America, etc., 434.
Lee, Isaac, description of supposed animal foot-prints in the sandstone of Mount Carbon, 246.
Lee, Isaac, answer to the foregoing, 252.
Limestone, origin of beds of, 283.
Lubricating oil, 144.

Man, the last effort of the Creator, 354.
Man receiving lessons from the inferior animals, 480.
Man, duty of, to investigate Nature, 16.
Man, first appearance on the stage, 344.
Man, different groups of, 352.
Man, his relation to the Creator, 394.
Man, ancient and modern races compared, 459.
Man, animals and machinery, necessity of alternate Sabbaths of rest, 470.
Man's mind, wherein it becomes Godlike, 479.
Man and animals, tribes and nations, alternating with each other, 465.
Man, the servant of God, 16, 354.
Mongolian race, the, 353.
Moraines, how formed, 394.

Mountains, elevations of, explained, 410.
Mastodon, Megatherium, Glyptodon, etc., antiquity of, 436, 467.
Mounds of the West described, 438.
Mœris, lake of, in Ancient Egypt, 462.
Monkeys, family of, 349.
Milky Way, grandeur and extent of the, 28.
Morphology, or law of forms, 32.
Milton's First Day, 40—Second day, 49—Third Day, 158—Fourth Day, 207—Fifth Day, 254—Sixth day, 356.
Moses in advance of scientific discovery, 45, 199.
Metamorphic rocks, 49, 56, 71, 239.
Miller, Hugh, writings and erroneous theories of, 63, 301, 406, 413, 417, 424, 427.
Metamorphic coal of Rhode Island and Scandinavia, 76.
Microscopic examinations of coal, 164.
Mountains, upheaval of on fourth day, effects produced on the climate, 173, 175.
Mary, the mother of Christ, 369.
Moses and the old system of astronomy, 177.
Maury, Lieut., on the geography of the sea, 189.
Microscope, revelations and wonders of the, 213.
Molluscan animals described, 217.
Molluscan fossils, number of species in each geological formation, 223.
Mollusca, number of surviving species, 224.
Mosaic days or eras, correspondence with geological formations, 24.
Mosaic ideal tableaux consistent with physical law, 38.
Miracles of Nature, 418.
Mississippi river, floods and peculiarities of the, 129.
Marine animals described, fossil and living, 209.
Mutual Admiration Society, how it works, 270.
Mammalian division of animals described, 306.
Microlestes, fossil remains of, 244.

Noah's Ark, was it occupied with young or old animals? 381.
Nature, teachings of, 23.
Nebular hypothesis of the origin of worlds, 30.
Nebular theory indicated by Moses, 38.
Niagara Falls, geological description of, 60.
Niagara Falls, discovery, retrogression, and antiquity of, 61.
Norfolk Island, and other pines, 115.
Newton, Sir Isaac, discoveries of, 184.
New red sandstone described, 241.
New red sandstone, animals of the, 243.
Neptune, discovery of by Leverrier, 29.
Nature, mysteries of, 418.
Noah's Ark, description of, 381.

ALPHABETICAL INDEX. 493

Noah's Ark as compared with the Great Eastern, 380.
Northmen, their discovery of America before Columbus, 447.
Noah, dispersion of his family after the flood, 454.
Nature, study of, by St. Pierre, 481.
Nature, harmony of, 476.
Nations, origin of, 458.

Oracles and Seers, their influence in Greece, 38.
Organic life, beginning of, on the earth, 51.
Ocean, wonders of the, 196.
Ocean, saltness of the, 416, 192.
Ocean navigation, value of astronomy to, 29.
Old red sandstone described, 55.
Ocean, rivers, currents, and climates of the, 188.
Ocean currents, how produced, 191.
Oil springs of Trinidad, 133.
Oil springs of Cuba, 134.
Oil springs in Venezuela, 136.
Oil springs in New Grenada, 136.
Oil springs in Mexico, 137.
Oil lake in Texas, 137.
Oil springs in Kansas, 139.
Oil of Pennsylvania, 140.
Oil springs of Virginia, 141.
Oolitic rocks, description of the, 278.
Ox or cattle, 312—Ounce, 329—Ouran-outang, 324—Otter, 345.
Ouran-outang, supposed resemblance to man disproved, 346.
Origin of worlds, 30.

Progressive creation, but not development, 25.
Planets, distance from the sun, 26.
Planets, revolutions of the, 32.
Planets, density of the, 33.
Plutonic or igneous rocks, origin of the, 42.
Plutonic rocks, family of, described, 46.
Paleozoic formation, 53, 56.
Pilot knob iron, origin of, 58.
Primitive rocks of the Eastern States, 77.
Peat-bog theory of the deposition of coal, 93.
Pine trees, 109—Pitch pine, 112, and various other pines.
Pitch lake of Trinidad, 133.
Paleozoic formation, vegetable character of, 166.
Ptolemy, astronomical theories of, 177.
Paul explaining revelation, 200, 201.
Poetry of the Bible, 203, 204, 205.
Polyparia, description of, 212.
Promulgators of the Divine Word, harmony among the, 20.
Planets move in one direction only, 31.
Pitch, rosin, oil, and tar, intimate relation to coal, 133.
Pleiades, mysteries of the, 203.
Professional jealousy, a case in point, 259.

Pickwick Controversy, the, 261.
Plesiosaurus, 267—Pterodactyle, 267.
Progressive development disproved, 273, 298, 301.
Pelasgian inhabitants of the Mediterranean, character of, 453.
Petroleum in Oregon, 139.
Pyramids of Egypt and of Central America, 460.
Pelasgic-Egyptians, did they plant colonies in America? 462.

Religion and Science, want of harmony between, 22, 24, 25.
Rain-drops, sun-cracks, etc., in ancient rocks, disproved, 71, 245, 248, 250.
Radiata division of animals described, 209.
Radiata, fossil, exhibiting number of species in each formation, 223.
Reptiles, remains of, 243, 244.
Rocks, how polished and grooved by icebergs, 394.
Rocky Mountains, the, 65.
Rhode Island coal basin, extraordinary geological features of, 74.
Radiated heat of the earth during the coal period, 124.
Rubbish, detritus, and fragments of the coal forests, how disposed of, 127.
Railway cross-ties, how preserved, 160.
Rain, absence of, during the coal period, 195.
Ripple-marks, sun-cracks, etc., explained, 253.
Reindeer, 311—Rhinoceros, 317—Rats, 323—Raccoon, 340—Ratel, 341.
Raleigh, Sir Walter, on Noah's Ark, 377.
River freshets, effects of, 404.
Rome, geological character of the " seven hills of," 433.

Space, immensity of, 56.
Sun, the centre of the planetary system, 31, 180.
Sun, the source and parent of worlds, 32.
Second day described by Moses, 40.
Silurian rocks, geographical distribution of, 59, 75.
St. Clair Flats, origin of the, 62.
Sigillaria, fossils of the, in coal, 88.
Sigillaria and Stigmaria, their identity disproved, 91.
Spruce, silver firs, and other trees, 113.
Sun, stars, and moon introduced on fourth day, 59, 177.
Sun introduced on the fourth day, effect on climate, 194.
Salt of the sea, 192.
Science of the Bible vindicated, 199, 203.
Serpent, the, symbolical of sin and immortality, 366.
Spongiaria, animals of the, 211.
Spiders, family of, 228.
Salt springs, 241.
Solar days, impossibility of their existence before the fourth day of Moses, 18.

Stratification of the earth's crust, want of parallel order proof of its antiquity, 24.
Spontaneous fructification of the ancient grasses, 73.
Sun and moon made to *rule* over the earth, 176, 195, 206.
Seventeenth century, discoveries of, 182.
Salt of the sea, solid quantity held in solution, 192.
Salt springs of the new red sandstone, 242.
Sauropus primævis, description of footprints of the, 237.
Silk-worms and silk manufactures, 231.
Secondary formation, or fifth day, geographical distribution over the earth, 285.
Sixth day or Tertiary, description of the, 290.
Sheep, 312—Squirrels, 323—Sloths, 320—Sable, 343.
Satan, rebellion of, in heaven, 361.
Seventh day, sanctified, 358, 459.
Spontaneous life, theory of the, disproved, 421.
Spontaneous life, experiment of Weeks and Crossi explained, 420.
Stephens, travels in Central America, 446.
Sabbath, objects contemplated in the, 469.
Submergence of continents conceded by geologists, 408.
Satan in the garden of Eden, 362.
Sensational preachers, 359.
Squiers, E. J., remarks of, on the Central American antiquities, 443.

Thought and vision, limited range of, 14.
Third day of Moses, 49.
Trees, resinous gums and oily secretions of, 114.
Tar, how manufactured, 109.
Third day, concluding review of, 167.
Telescope, discovery of the, 181.
Telescope of Lord Rosse, 183.
Tropical vegetation and forests, description of, 123.
Tentaculifera, described, 222.
Tumbling runensineus, 256.
Tertiary formation, description of, 290.
Tertiary, animals of, described, 306.
Trinity, the, Father, Son, and Holy Ghost, 360.

Universal gravitation, law of, 185.
Universe, man peering into the mysteries of the, 182.
Unity of the human race, remarks of Humboldt, 464.
Universality of the Noachian flood considered, 373.

Vision and thought, limited range of, 14.
Vice flourishing in the domains of Christianity, 23.
Volcanic rocks described, 46.
Vegetation, diversity of accounted for, 117.
Vegetable origin of coal, 154.
Vegetable structure of coal explained, 164.
Volcanic action on fourth day, effects of on climate, 172.
Volcanic action at the close of the sixth day of Tertiary period, 400.
Vegetable growth, and chemical phenomena of, 473.
Vegetation, animals in the juices of, 237.
Vegetable preceding animal life, 239.
Volcanic Islands, emergence of, 46.
Vegetable life, beginning of, 71.
Vegetation of the Tertiary, 294.
Volcanic action in the polar regions, effects of, 401.
Vegetation of America, antiquity of, 466

Worlds within worlds, 37.
Wood not converted into coal, 155, 156.
Water of Alleghanies, distribution of, 175.
Worlds, origin of, 30.
Worlds, original unity of, 36, 361.
Worlds compound aggregates, united by affinity, 37.
World, the, too much controlled by scientific flummery, 157.
Weight or pressure of the atmosphere, 199.
Waters, the, commanded to bring forth life on the fifth day, 209.
World, the, how humbugged by its so-called science, 266, 274.
Weald Rocks, description of, 280.
Whales, 210—absence of their fossils in the cretaceous rocks explained, 277, 303.
Whiskey from coal oil, 144.
Water, subterraneous rivers and lakes, 410.

Zoology, science and classification of, 209.

THE END.

www.ingramcontent.com/pod-product-compliance
Lightning Source LLC
Chambersburg PA
CBHW021416300426
44114CB00010B/518